MECHANISMS OF INORGANIC REACTIONS

MECHANISMS OF INORGANIC REACTIONS

DIMITRIS KATAKIS

Department of Chemistry
University of Athens
Athens, Greece

GILBERT GORDON

Department of Chemistry
Miami University
Oxford, Ohio

A WILEY-INTERSCIENCE PUBLICATION

JOHN WILEY & SONS

New York / Chichester / Brisbane / Toronto / Singapore

Library of Congress Cataloging in Publication Data:

Katakis, Dimitris, 1931–
  Mechanisms of inorganic reactions.

  "A Wiley-Interscience publication."
  Includes bibliographies and index.
  1. Chemical reaction, Conditions and laws of.
2. Chemistry, Inorganic. I. Gordon, Gilbert. II. Title.

QD501.K316 1987   541.3'9   86-22407
ISBN 0-471-84258-3

To our teachers . . .

PREFACE

This text is directed to helping students understand the details of the mechanisms of various types of common inorganic reactions. It is not intended as a comprehensive review of the current literature; instead, it stresses the underlying principles and gives examples (along with additional "homework" problems) in each category.

Most of the examples given are from homogeneous solution chemistry. For special mechanistic features of gas-phase chemistry, solid-state chemistry, or heterogeneous reactions, more specialized texts should be consulted. Our purpose is to use solution chemistry to illustrate the fundamental principles of various reaction mechanisms.

We avoided burdening the book with many references, but a brief list of relevant books and review articles for further study is given at the end of each chapter.

This book was written to serve as a basis for a one-semester course. Knowledge of inorganic chemistry, thermodynamics, kinetics, atomic and molecular structure, analytical chemistry, and calculus is important as general background. The student should, in particular, have some knowledge related to mechanisms, such as what a simple rate law is, what the kinetic order of a reaction is, how temperature affects reactions, and so forth.

At the end of the semester the student should know

1. The concepts that have been developed in mechanistic studies, how to correlate them, and how to distinguish them from similar but different concepts. For example: What is an elementary reaction, and how is an elementary reaction distinguished from a nonelementary process?
2. How mechanisms are classified.
3. The methods, both theoretical and experimental, used in the study of mechanisms, the possibilities each technique offers, the kind of information that can be obtained, and the limitations.

4. How the information for the clarification of a mechanism is organized and treated.
5. How to solve theoretical and practical problems and, even better, how to conceive such problems and express them clearly.

In other words, the reader should be able to apply this knowledge.

Many excellent books are available that describe the basic principles of chemical kinetics both in the gas phase and in solution. More and more often, chapters on inorganic mechanisms are included in advanced texts in inorganic chemistry, and chapters (or portions of chapters) are even included in texts at the beginning level of chemistry. Also, more and more papers involving mechanistic detail are appearing in current inorganic chemical literature and at related conferences. Synthetic chemists (inorganic, organometallic, polymer, bioinorganic) are turning to mechanistic detail to gain a better understanding of the role (and properties) of the intermediates. It is our hope that this systematic text will help its readers (especially advanced level undergraduates and beginning graduate students) to gain a better understanding of these mechanisms.

DIMITRIS KATAKIS
GILBERT GORDON

Athens, Greece
Oxford, Ohio
March 1987

ACKNOWLEDGMENT

The authors would like to pay tribute to the really important people—all those who did the original work and made this secondary source of information possible.

D.K.
G.G.

CONTENTS

1. INTRODUCTION   1

    HISTORICAL PERSPECTIVE   1
    A Chemical Reaction ~2500 Years Ago, 1
    An "Alchemical" Reaction ~1000 Years Ago, 2
    A Chemical Reaction ~50 Years Ago, 2
    A Chemical Reaction Today, 2

2. MECHANISMS, KINETICS, AND EQUILIBRIA   5

    2.1. A BRIEF REVIEW OF BASIC CONCEPTS   5
        Factors Determining the Energy Barriers, 5
        Elementary Reactions of a Mechanism and Molecular
          Processes, 6
        Classification of Elementary Reactions, 7
        General Characteristics of Elementary Reactions, 9
        Relations among Elementary Reactions in a
          Mechanism, 9

    2.2. THE RELATION BETWEEN THE COMPLETE RATE LAW
        AND THE MECHANISM   10
        Mathematical Difficulties, 11
        Steady-State Approximation, 11
        Numerical Methods, 14
        Experimental Difficulties, 16
        Incomplete Data, 19
        Concentrations Used to Test the Rate Law, 22

2.3. STOICHIOMETRY AND THE RATE LAW   24
   Mechanistic Implications of Variable Stoichiometry, 26

2.4. CHAIN REACTIONS   27
   Initiation, Propagation, and Termination, 27
   Explosions, 29
   An Industrial Incident, 30

2.5. SOLVENT EFFECTS   30
   Classification of Solvents, 30
   Empirical Multicorrelation between Rate Constants and
      the Properties of Solvents, 31
   A Comparison of Rates in Different Solvents, 31
   Transfer Functions, 32
   Ionic Strength Effects, 33

2.6. KINETICS AND EQUILIBRIUM   34

2.7. MICROSCOPIC REVERSIBILITY   35

2.8. STEADY STATE AND EQUILIBRIUM   35

2.9. LINEAR FREE ENERGY RELATIONSHIPS   36

2.10. THERMODYNAMIC AND KINETIC STABILITY   37

   GENERAL REFERENCES   39

   PROBLEMS   39

3. EVENTS AT THE MOLECULAR LEVEL—THE
   ACTIVATED COMPLEX   44

3.1. A QUALITATIVE DESCRIPTION OF MOLECULAR
   COLLISIONS   44
   Illustrations: The Methane + Tritium System, 45
   Reactions of Potassium Atoms with Methyl Iodide,
      Chloroform, and Trifluoromethyl Iodide, 46

3.2. STOCHASTIC APPROACH TO REACTIONS   47
   Monomolecular Reactions, 48
   Bimolecular Reactions, 49
   Nonelementary Reactions—The Transition Matrix, 50

3.3. TEMPERATURE DEPENDENCE: THE ARRHENIUS
   EQUATION   51
   Elementary Reactions, 51
   Composite Rate Constants, 52

3.4. COLLISION THEORY   53
     Outline of the Theory, 53
     A Metaphor, 54
     Some Hints of Practical Value, 55

3.5. BASIC CONCEPTS IN TRANSITION STATE THEORY   55
     A "Topographical" Description, 55
     A More Accurate Picture, 56
     Some Dialectics, 56
     Comments, 57

3.6. THE EYRING EQUATION   58
     The Equation, 58
     Units, 59

3.7. APPLICATION OF TRANSITION STATE THEORY TO
     NONELEMENTARY REACTIONS   60
     The Activated Complex in Nonelementary Reactions, 60
     The Composition of the Activated Complex, 60
     Ambiguities Regarding the Activated Complex, 61

3.8. A QUALITATIVE COMPARISON OF DIFFERENT
     ACTIVATION ENERGIES   63
     Molecular and Statistical Activation Energies, 63
     Practical Considerations, 64

3.9. PRESSURE EFFECTS AND THE VOLUME OF ACTIVATION   65
     Definitions and Concepts, 65
     Interpretation of the Volume of Activation, 65

3.10. THE "STRUCTURE" OF THE ACTIVATED COMPLEX   67

3.11. METHODS FOR STUDYING THE STRUCTURE OF
      ACTIVATED COMPLEXES   68

      GENERAL REFERENCES   75

      PROBLEMS   76

4. EXPERIMENTAL METHODS AND HANDLING OF
   THE DATA   78

   4.1. TIME SCALE FOR CHEMICAL REACTIONS AND THE
        CHOICE OF EXPERIMENTAL TECHNIQUE   78
        Time Scale, 78
        Choice of Experimental Technique, 79

   4.2. EXPERIMENTAL METHODS FOR SLOW REACTIONS   80

4.3. APPLICATION OF A SIMPLE TITRATION TECHNIQUE   80

4.4. SPECTROPHOTOMETRIC METHODS   80

4.5. ISOTOPIC TRACERS   81
Exchange Reactions, 81
McKay Plots, 82
Nonlinear McKay Plots, 82

4.6. GENERAL METHODS FOR HANDLING EXPERIMENTAL
DATA FOR SLOW REACTIONS   83
The Method of Finite Differences, 83
The Method of Integration, 84
The Guggenheim Method, 86

4.7. HANDLING COMPLICATED RATE LAWS   87
Ostwald's Method of Isolation, 87
Inhibition, 88
Two-Term Rate Laws, 90
Obtaining Rate Constants by Using Concentration–Time
    Integrals, 93
Simplification of the System of Differential Rate
    Equations, 94
Laplace Transforms for Solving Differential Rate
    Equations, 97

4.8. EXPERIMENTAL METHODS FOR FAST REACTIONS   99

4.9. FLOW METHODS   100
General Description, 100
Treatment of the Data, 100
Stopped Flow, 102
Observation Techniques, 102

4.10. FLASH PHOTOLYSIS—PULSE RADIOLYSIS   103

4.11. PICOSECOND AND FEMTOSECOND SPECTROSCOPY   104

4.12. TEMPERATURE JUMP AND ACOUSTICAL METHODS   104
Treatment of Relaxation Data: Simple Relaxation, 105
Multiple Relaxations, 108
Generalization, 110
The Use of Indicators in Relaxation Studies, 111
Fitting the Relaxation Times to a Mechanism, 112

4.13. EMISSION SPECTROSCOPY   113

4.14. ELECTROCHEMICAL METHODS   114
      Parameters and Processes, 114
      Time Scale of Electrochemical Methods, 115
      Special Characteristics, 115

4.15. NUCLEAR MAGNETIC RESONANCE (NMR)   116
      Resonance and Relaxation, 116
      NMR Parameters, 118
      Mechanistic Information from NMR, 119
      An NMR Application: Exchange Processes by
         ^{17}O NMR, 120

4.16. ELECTRON PARAMAGNETIC RESONANCE (EPR)   121
      A Brief Comparison of NMR and EPR, 121
      Parameters of EPR, 122
      EPR of Transition Metal Ions, 123
      Mechanistic Information from EPR, 123

      GENERAL REFERENCES   124

      PROBLEMS   125

5. MECHANISM AND STRUCTURE   129

5.1. GEOMETRICAL STRUCTURE AND MECHANISM   129
     The Steric Factor, 129
     The Size, 129
     The Shape, 130
     Stereospecific Effects, 131

5.2. HOMO AND LUMO   131

5.3. MINIMUM ENERGY PATHWAYS AND THE MAXIMUM
     OVERLAP CRITERION   132
     The Overlap, 132
     Sigma and Pi Donors and Acceptors, 133
     Asymmetric Donor–Acceptor Bonds, 133

5.4. WHICH IS THE DONOR AND WHICH THE ACCEPTOR?   134
     Chemical Tendency for Electron Displacement, 134
     The Polarization of the Orbitals, 136

5.5. HOMO AND LUMO IN REACTIONS OF TRANSITION
     METAL COMPLEXES   137

5.6. HOMO AND LUMO IN REACTIONS OF COMPOUNDS
     OF REPRESENTATIVE ELEMENTS   138

5.7. WHAT HAPPENS AFTER THE OVERLAP   138

5.8. ATTRACTIVE AND REPULSIVE INTERACTIONS   141
      Attractive Interactions, 141
      Repulsive Interactions, 141
      Zero Overlap, 143

5.9. AN ALTERNATIVE APPROACH BASED ON THE
      CONCEPT OF FORCE   143
      Stable Configurations, 143
      The Hellmann–Feynman Theorem, 144
      Forces and Potential Energy, 145
      Application to Dynamics, 145
      Forces in HOMO, LUMO Interactions, 146

5.10. THE CONSERVATION OF ORBITAL SYMMETRY   147
      Wigner–Witmer and Woodward–Hoffmann Rules, 147
      Application of the Orbital Symmetry Conservation
        Rules, 148

5.11. WHAT IS THE CONNECTION BETWEEN HOMO
      AND LUMO, AND THE CONSERVATION OF ORBITAL
      SYMMETRY?   151

5.12. MECHANISTIC CONSEQUENCES OF THE PAULI
      PRINCIPLE   152

5.13. IS THE SPIN CONSERVED?   154
      Spin Conservation Rules, 154
      Violation of the Spin Selection Rules, 156

5.14. A PARADIGM: THE REACTIONS OF DIOXYGEN   158
      Question 1: Why Is Dioxygen Inert?, 158
      Question 2: How Is Dioxygen Activated?, 161
      Question 3: How Is Dioxygen Transported?, 163

      GENERAL REFERENCES   164

      PROBLEMS   165

6. GROUP-TRANSFER AND ATOM-TRANSFER REACTIONS   167

6.1. TYPES OF SUBSTITUTION REACTIONS   168

6.2. SUBSTITUTION MECHANISMS   169
      Mechanisms, 170

6.3. MOLECULAR MECHANISM   171
      More on the A or I and D or I Dichotomies, 171

6.4. CONTOUR DIAGRAMS   172

6.5. EMPIRICAL CRITERIA FOR DECIDING THE MECHANISM
     OF SUBSTITUTION   173
     The Observed Rate Law, 173
     Dependence on the Nature of the Entering Ligand, 174
     Detection of Intermediates, 174
     The Entropy of Activation, 175
     Volume of Activation, 175
     The Effect of the Nonleaving Ligands, 176

6.6. THE MECHANISM OF A GIVEN SUBSTITUTION
     SOMETIMES CHANGES   177

6.7. CHELATE RING FORMATION   178

6.8. COORDINATION SPHERE EXPANSION, ADDITION, AND
     CONDENSATION   181

6.9. TETRAHEDRAL SUBSTITUTION   183
     Boron, 183
     Silicon, 184
     Phosphorus, 185
     Lithium, 186
     Beryllium, 187

6.10. TETRAHEDRAL TRANSITION METAL COMPLEXES   187

6.11. THE STRUCTURE CORRELATION METHOD IN
      TETRAHEDRAL SUBSTITUTION   189

6.12. SUBSTITUTION IN SQUARE PLANAR COMPLEXES   191
      The Rate Laws and the Mechanism, 191
      Dependence on the Nature of the Entering Ligand, 192
      Dependence on the Nature of the Leaving Ligand, 193
      Dependence on the Nature of the Metal Center, 193

6.13. TRANS EFFECT IN SQUARE PLANAR SUBSTITUTION   194
      Trends, 194
      A Simple Electrostatic Model, 194
      The cis Effect, 195
      Applications to Synthesis, 195
      Antitumor Activity of Pt^{II} Complexes, 196

6.14. PATHWAYS IN SQUARE PLANAR SUBSTITUTION   196
      Qualifications, 198

6.15. SUBSTITUTION IN OCTAHEDRAL Co^{III} COMPLEXES   199
      Effect of the Leaving Ligand, 199
      Stereochemical Hindrance, 199

Nonleaving Ligand Effects, 199
Solvent Effects, 200
Rate Laws, 200
Stereochemical Changes during Substitution, 204

6.16. ACID AND BASE HYDROLYSIS IN OCTAHEDRAL Co^{III}
COMPLEXES   204

6.17. ASSOCIATIVE AND DISSOCIATIVE MECHANISMS IN
OCTAHEDRAL Cr^{III} COMPLEXES   205

6.18. LABILITY OF AQUA IONS   206

6.19. AQUATION OF OCTAHEDRAL ORGANOCHROMIUM(III)
COMPLEXES   207
Acid Catalysis, 209

6.20. INTRAMOLECULAR METATHESIS   211
Nucleophilic Attack on Transition Metal π-Complexed
    Olefins, 211

6.21. INSERTION REACTIONS   213
The Olefin Insertion Reaction, 213
Other Insertion Reactions, 214
α-Hydride Ion Transfer, 215

6.22. TOPOLOGICAL MECHANISMS   215
The Berry Mechanism, 216
Inversion of Configuration of Pyramidal Molecules, 217
Trigonal Twist, 218
Other Topological Mechanisms, 220

6.23. INTER- AND INTRAMOLECULAR PROTON TRANSFER   222
General Description, 222
HOMO and LUMO in Proton Transfer, 222
Covalent Hydrate and Pseudobase Formation, 223
Lewis Acids and Bases, 223

GENERAL REFERENCES   224

PROBLEMS   225

7. ELECTRON-TRANSFER REACTIONS   229

7.1. BACKGROUND   229
Historical Note, 229
Definition, 230
Relation Between Oxidation Number, Molecular
    Geometry, and the Composition of the First
    Coordination Sphere, 231

Useful Generalizations, 232
Tendency toward Electroneutrality: Formal Oxidation
    Numbers and Real Charges, 234

7.2. MECHANISMS AND RATE LAWS   235

7.3. CLASSIFICATIONS   236
Direct and Indirect Electron Transfer, 237

7.4. SOLVENT MEDIATION AND SOLVATED ELECTRONS   239

7.5. OXIDATION–REDUCTION REACTIONS OF OXO AND
HYDROXO COMPOUNDS   240
The Role of Lability, 240
Acid and Base Catalysis, 241
Coordination Sphere Expansion, 243
Electron Transfer without Significant Structural
    Change, 244
Qualitative Consideration of Electronic Factors, 244

7.6. OXIDATIVE ADDITION   245
Definition, 245
Clarification, 246
The Multiple Bond Addition Controversy, 247
The Intimate Mechanism, 248
Alkyl Halides, 250
Cyclometalation, 251

7.7. REDUCTIVE ELIMINATION   252
Classification and Clarification, 252
Mechanisms, 253
Reductive Disproportionation, 255

7.8. REDUCTION OF MULTIPLE BONDS BY METAL IONS OF
LOW VALENCE   256

7.9. INTRAMOLECULAR ELECTRON TRANSFER: MIXED
VALENCE COMPOUNDS   259
Some Mechanistic Implications, 261

7.10. "TWO-ELECTRON" TRANSFER   261
Complementary and Noncomplementary Reactions, 261
Two-Electron, Atom, and Positive Ion Transfer, 262

7.11. MODELING AND SIMULATION   263

7.12. MAPPING THE COURSE OF A BIMOLECULAR REDOX
REACTION   265
The Encounter, 266
The Franck–Condon Principle, 266

Optical Electron Transfer, 267
Thermal Electron Transfer, 269
Solvent Reorganization, 270
Tunneling, 270

7.13. A DONOR–ACCEPTOR MODEL   270
Differences in Bond Lengths, 272
The Strength of Interaction, 272
Differences in Electronegativity, 273
"Nonsymmetric" Reactions, 273
Orbital Combinations and Effective Charge, 274
Inverted Donor–Acceptor Sites, 275
Summary, 276
Outer-Sphere, Inner-Sphere, and Intramolecular
    Electron Transfer, 276

7.14. SOME USEFUL FORMULAS AND CORRELATIONS   276
Formulas for the Precursor Formation Constant, 276
Electron Transfer within the Precursor, 277
Linear Free Energy Relations, 278
The Marcus Cross-Correlation, 279
Examples of Calculations, 279
Correlation between Thermal and Optical Electron
    Transfer, 280

GENERAL REFERENCES   283

PROBLEMS   284

8. CATALYSIS   289

8.1. GENERAL CONSIDERATIONS   289
The Catalyst, 289
Homogeneous and Heterogeneous Catalysis, 290
Promoters, Inhibitors, Poisons, and Precursors, 291
Catalysis and Excitation, 291
Ways of Achieving Activation, 291
The 16- or 18-Electron Rule, 293
Reactions of the Energized Substrate, 293
Focusing on the Catalyst, 293
Simple Cycles, Catalytic Cycles, 294
Mystical Connotations, 295

8.2. BIG CYCLES, SMALL CYCLES   295
The Kinetics of a Small Cycle, 296
The Kinetics of a Big Cycle, 300
Alternative Methods for the Study of Big Cycles, 301

8.3. SELECTIVITY   301
Definitions, 302
Selectivity of Big Cycles, 304
Stereoselectivity, 304

8.4. MONO- AND POLYNUCLEAR CATALYSTS   305
Some Sources of Nonselectivity, 305
Collective Modes of Binding and Activation, 306
From Polynuclear Species to Metal Surfaces, 306

8.5. HETEROGENIZED HOMOGENEOUS CATALYSTS   308

8.6. COMMON FEATURES IN THE HYDROGENATION OF
OLEFINS AND RELATED COMPOUNDS   310
Activation of Dihydrogen, 310
Activation of the Olefin, 311
What Happens after Insertion? 314
Extensions, 316

8.7. THE HYDROGENATION OF CYCLOHEXENE USING
THE WILKINSON CATALYST   318

8.8. REACTIONS INVOLVING CARBON MONOXIDE   318
Activation of Carbon Monoxide, 319
The Fischer–Tropsch Method, 320
Some Energetics, 321
Hydroformylation, 322
Carbonylation of Methanol, 323

8.9. THE WACKER PROCESS   323

8.10. ISOMERIZATION   324

8.11. OLIGOMERIZATION AND ZIEGLER–NATTA
POLYMERIZATION OF ALKENES   326
The Mechanism, 326
Carbene Intermediates, 327

8.12. NITROGEN FIXATION   328

8.13. AN OVERVIEW   329
Comparisons, 329
Classification Based on Interactions, 330

GENERAL REFERENCES   332

PROBLEMS   333

9. INORGANIC PHOTOCHEMISTRY   337

9.1. GENERAL CONSIDERATIONS   337

9.2. EXCITATION MODES IN TRANSITION ELEMENT COMPLEXES   338
The d–d Transitions, 339
Bond Length and Bond Energy Changes Associated with
d–d Transitions, 340
Charge-Transfer Transitions, 341
Intraligand Transitions, 341
Selection Rules for Electronic Transitions, 341

9.3. A SIMPLIFIED ELECTROSTATIC DESCRIPTION OF
PROMOTIONAL EXCITATION   343
Configuration d^1, 343
Configuration d^2, 344
Configuration d^3, 345
A Useful Generalization, 346
Configuration d^4 (High Spin), 346
Configuration d^5 (High Spin), 347
Configuration d^6, d^7, d^8, d^9 (High Spin), 347
Low-Spin Configurations, 347

9.4. WHAT HAPPENS AFTER THE ELECTRONIC EXCITATION   348
Physical Modes of Deactivation, 348
Photon Emission, Fluorescence, Phosphorescence, 348
Intermolecular Energy Transfer, 348
Vibrational Relaxation, Energy Degradation, 349
Internal Conversion, Intersystem Crossing, 350
Photochemical Processes, 352

9.5. THREE CASES: PHOTOSUBSTITUTION,
PHOTOSTEREOCHEMISTRY, AND PHOTOCHEMICAL
ELECTRON TRANSFER   353
Photosubstitution in $Rh(NH_3)_5X^{2+}$ (X = I, Br, Cl), 353
Stereochemistry of Photosubstitution: Cr^{III} Complexes, 354
Redox Reactions of Excited $Ru(bipy)_3^{2+}$, 355

9.6. COMPETITION   356
Continuous Illumination, 356
Pulsed Illumination, 357
What to Measure, 358
Counting Losses, 358

9.7. SOLAR ENERGY CONVERSION AND STORAGE   359
Alternatives, 359
Solar Light, 361

9.8. PHOTOCHEMICAL SPLITTING OF WATER   361
Basic Chemistry and Thermodynamics, 361
Number of Photons, 362
Direct Routes, 363

Photosensitizers, Reversible Donors and Acceptors,
   Relays, Catalysts, 365
A Return to Catalytic Cycles, 367

GENERAL REFERENCES   367

PROBLEMS   368

INDEX   **371**

MECHANISMS OF INORGANIC REACTIONS

1

INTRODUCTION

HISTORICAL PERSPECTIVE

A Chemical Reaction ~2500 Years Ago

About 2500 years ago, Plato described a "gasification" (!) reaction as a transformation of fire into air in an aqueous solution:

$$2 \text{ Fire } \xrightarrow[\text{solution}]{\text{Aqueous}} 1 \text{ Air}$$

Figure 1.1 expresses this reaction with Plato's "stereochemical" formulas.

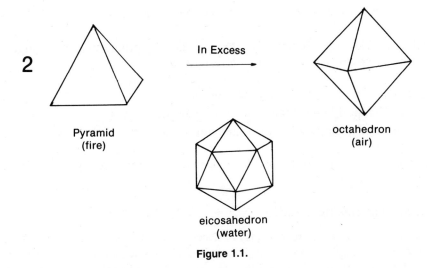

2

In Excess

Pyramid
(fire)

octahedron
(air)

eicosahedron
(water)

Figure 1.1.

1

Fire is not, of course, tetrahedral, water is not icosahedral, and air is not octahedral, but the great adventure—the attempt to correlate properties (chemical and physical behavior) and structure—had begun.

An Alchemical Reaction ~1000 Years Ago

The alchemists had a better grasp of the diversity of elements, but they were still far from reaching an understanding of natural processes. Notice that in Figure 1.2, which shows a typical alchemical explanation for the existence of elemental gold, there is less clarity than in Plato's equations, more mystery, and no attempt to relate properties to structure.

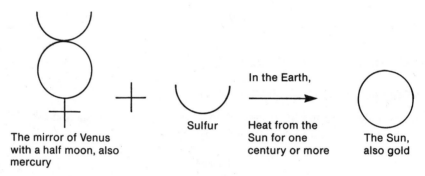

The mirror of Venus with a half moon, also mercury

Sulfur

In the Earth,

Heat from the Sun for one century or more

The Sun, also gold

Figure 1.2.

A Chemical Reaction ~50 Years Ago

In old textbooks, a reaction would be written

$$Cr_2O_7'' + 14H^{\cdot} + 6Fe^{\cdot\cdot} \longrightarrow 2Cr^{\cdot\cdot\cdot} + 6Fe^{\cdot\cdot\cdot} + 7H_2O$$

Where the dots represent positive charges and the primes negative charges. There was no apparent interest with respect to

How the charges (electrons) are transferred
How chemical bonds break and new bonds are formed
What intermediates are formed and *how*

In short, no questions were raised as to *how* the reaction takes place; its course remained a mystery.

A Chemical Reaction Today

Careful!! *One* chemical reaction is not necessarily only *one* reaction anymore!! For example, Figure 1.3 presents Henry Taube's classic inner-sphere electron-transfer reaction (see Chapter 7 for more on such reactions). The effort today

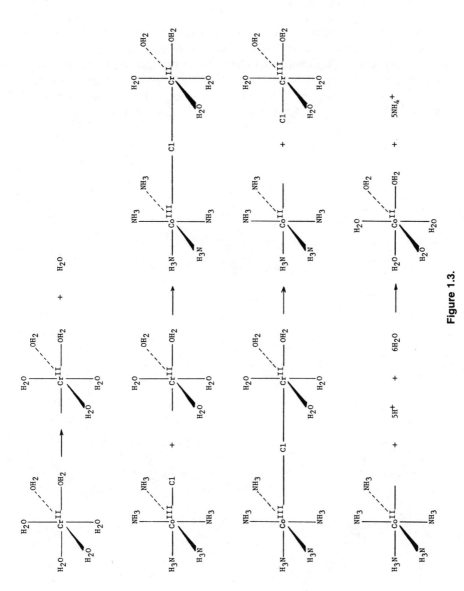

Figure 1.3.

3

is to follow the course of the reaction as closely as possible and to correlate the observations to structure. Historic evolution in this respect was slow. Only a few decades ago many influential chemists considered "mechanisms" to be some sort of chemical mythology. And then the explosion came. Now even synthesis is often accompanied by mechanistic considerations, and mechanisms are gradually becoming integrated into general chemistry textbooks.

2

MECHANISMS, KINETICS, AND EQUILIBRIA

2.1. A BRIEF REVIEW OF BASIC CONCEPTS

The formation of water vapor from dihydrogen and dioxygen is exothermic by 227 kJ mol^{-1} corresponding to an equilibrium constant of 10^{40}. Yet under ordinary conditions the reaction does not take place; it is too slow. The reason in the last analysis is that before the new H—O bonds are formed, the old bonds (H—H and O=O) must be broken.

$$2HH + OO \longrightarrow 2HOH$$

The final state is energetically lower than the initial state, but in order for the system to pass from a higher energy level to a lower level it must pass over an *energy barrier*.

Factors Determining the Energy Barriers

In the example of the reaction between dihydrogen and dioxygen, the barrier is determined by the energy required in the process of breaking some of the existing bonds, which somehow *precedes* the formation of the final product. Other factors that may contribute to the energy barrier are:

An increase or decrease in the lengths and/or angles of some bonds.

Electrostatic repulsion between ions or dipoles and weaker "closed-shell" repulsion. Stereochemical hindrance is also conveniently included here.

In the gas phase, if the collision between molecules A and B is not head-on, there may be a "rotational" barrier; B does not approach A close enough because of centrifugal forces.

In reactions in the condensed phase, energy is also required for the reactant molecules to diffuse toward each other.

Elementary Reactions of a Mechanism and Molecular Processes

Fenton's reagent is obtained by mixing Fe^{2+}(aq) and H_2O_2, and it is a stronger oxidizing agent than H_2O_2 alone. It can oxidize dihydrogen, initiate polymerization, and oxidize organic substances that are difficult to oxidize. In the absence of a substrate, the overall reaction

$$2Fe^{2+}(aq) + H_2O_2 + 2H^+ \longrightarrow 2Fe^{3+}(aq) + 2H_2O$$

is accompanied by catalytic decomposition of H_2O_2. This, however, does not explain the enhanced oxidizing power. The explanation is clear only after the overall reaction has been analyzed into simpler reactions:

$$Fe^{2+} + H_2O_2 \longrightarrow Fe^{3+} + OH^- + OH$$

$$OH + H_2O_2 \longrightarrow HOH + HO_2$$

$$HO_2 + HOOH \longrightarrow HOH + O_2 + OH$$

$$Fe^{3+} + H_2O_2 \longrightarrow Fe^{2+} + HO_2 + H^+$$

$$Fe^{3+} + HO_2 \longrightarrow Fe^{2+} + O_2 + H^+$$

$$Fe^{2+} + OH \longrightarrow FeOH^{2+}$$

$$OH + S \longrightarrow P$$

where S denotes the substrate to be oxidized (e.g., H_2) and P denotes the products. These reactions may be generally analyzed further into *elementary reactions,* but even this limited analysis indicates that the enhanced oxidizing power of Fenton's reagent is due mainly to the intermediate OH radicals.

The set of all elementary reactions constitutes the *mechanism* of the overall reaction.

Each elementary reaction represents a process that takes place at the molecular level, that is, it represents changes in molecules or molecule-like species, but it gives no explicit information about the *intrinsic (molecular) mechanism*—about how the change actually takes place or what happens when the molecular species come close to each other.

For a complex reaction, the mechanism has two aspects: the elementary reactions that constitute the overall reaction *and* the molecular processes for each of these elementary steps.

It must be stressed that before trying to understand the molecular processes we must first be sure that we know what elementary reactions are taking place. If the set of elementary reactions is not the right set, whatever is said about the molecular processes will be irrelevant. An example can illustrate this point. For many years the gas-phase formation of hydrogen iodide from dihydrogen and diiodine was quoted as a typical example of a thermal reaction involving a simple collision of two molecules. The existing experimental data were interpreted on the basis of a mechanism involving a single elementary reaction:

$$H_2 + I_2 \longrightarrow 2HI \qquad\qquad 2.1$$

The corresponding molecular mechanism could be schematically represented as

$$
\begin{array}{ccc}
\text{I——I} & \text{I} & \text{I} \\
\vdots\;\;\vdots & \backslash & / \\
\text{H—H} & \text{H} & \text{H}
\end{array}
\qquad 2.2
$$

More careful measurements, however, proved that this simple mechanism does not correspond to reality. Rather, on the basis of the new data, it was proposed that the reaction

$$H_2 + I + I \longrightarrow 2HI$$

takes place, preceded by the decomposition of a diiodine molecule into atoms. The earlier molecular mechanism is irrelevant.

Classification of Elementary Reactions

The usual elementary reactions can be classified in one of the following categories:

(a) *Electron transfer, without breaking or forming of bonds.* Consider the reaction

$$Fe(dipy)_3^{2+} + Ru(dipy)_3^{3+} \longrightarrow Fe(dipy)_3^{3+} + Ru(dipy)_3^{2+}$$

where dipy = dipyridyl. There are changes in the strength and length of the bonds, but there is neither breaking of the original bonds nor formation of new ones.

The molecules that participate in an elementary reaction are not necessarily simple; they can be complex. The process of the chemical change is simple— as simple as possible. This process must be conceived to occur during the course of a simple collision or encounter or unimolecularly. The transfer of one electron is such a "simple" process.

(*b*) *Breaking or forming of one chemical bond.* In the formation of dibromine from bromine atoms,

$$\text{Br}^\cdot + {}^\cdot\text{Br} + \text{M} \longrightarrow \text{Br}:\text{Br} + \text{M}^*$$

M is a third body carrying away the excess energy, so that the bromine molecule does not break again. It is easy to conceive of two bromine atoms coming to a triple collision with an M (or on a reaction vessel wall) and sticking together. Each of the bromine atoms contributes one electron to the new bond. In another example,

$$\text{H}^+ + :\text{H}^- \longrightarrow \text{H}:\text{H}$$

both electrons for the new bond come from only one of the reactants. Correspondingly, the breaking of an existing bond can take place *homolytically*, as in the reaction

$$\text{I}:\text{I} \longrightarrow \text{I}^\cdot + {}^\cdot\text{I}$$

or *heterolytically*,

$$(\text{CH}_3)_3\text{N}:\text{BF}_3 \longrightarrow (\text{CH}_3)_3\text{N}: + \text{BF}_3$$

In the category of forming or breaking of a bond, we may conveniently include many cases of adsorption or desorption on the surface of a solid and cases of breaking or forming bonds of adsorbed species, such as

$$\text{H}_3\text{N} + \text{surface} \longrightarrow \text{H}_3\text{N (adsorbed)}$$

$$\text{H}_2 \text{ (adsorbed)} \longrightarrow 2\text{H (adsorbed)}$$

(*c*) *Breaking of a bond and simultaneous formation of a new bond.* Hydrogen atom abstraction by hydroxyl radicals,

$$\text{RH} + \text{OH} \longrightarrow \text{R} + \text{HOH}$$

belongs in this category. The old bond breaks and the new bond forms in a single step.

(*d*) *Simultaneous breaking of two bonds and formation of two new bonds.* Entries in this category are rare. The reaction

$$\overset{\displaystyle \text{O}}{\underset{\displaystyle \text{O}}{\diagdown\!\diagup}}\!\text{NO} + \text{ONO} \longrightarrow \overset{\displaystyle \text{O}}{\underset{\displaystyle \text{O}}{\diagdown\!\diagup}}\!\text{N} + \text{OO} + \text{NO}$$

can be considered as an example, provided that the double bond in dioxygen is counted as two bonds. Better documentation may prove that some reactions in this category are not elementary after all and that they can be analyzed further.

General Characteristics of Elementary Reactions

From the examples of elementary reactions given in the preceding pages, two facts become obvious:

(*a*) *The molecularity seldom exceeds 3.* The probability of a three-body collision is small, the probability of a collision involving four or more bodies is negligible. So, for reactions such as

$$4I_2 + S_2O_3^{2-} + 10OH^- \longrightarrow 8I^- + 2SO_4^{2-} + 5H_2O$$

it can be stated from the outset that they are complex and that they cannot be elementary. Any reaction with large stoichiometric coefficients is necessarily complex, since it is practically impossible for these molecules or ions all to meet at the same time at the same point in space. The stoichiometry of a reaction gives information about its complexity.

(*b*) *During an elementary reaction, the change in structure is the simplest possible.* It is improbable that a reaction such as

$$\underset{\underset{H}{|}}{\overset{\overset{H}{|}}{H-C}} + O{=}O \longrightarrow \underset{\underset{O}{\|}}{\overset{\overset{H}{|}}{H-C}} + H{-}O$$

in which one double bond (O=O) and one single bond (C—H) break and a new double bond (C=O) and a single bond (O—H) form will occur in only one step.

Relations among Elementary Reactions in a Mechanism

In formulating a mechanism it is useful to recognize that elementary reactions may be related to each other, that is, that the mechanism may contain opposing reactions, parallel reactions, and/or successive reactions. In the mechanism

$$Cl_2 \underset{(2)}{\overset{(1)}{\rightleftharpoons}} 2Cl$$

$$CO + Cl \underset{(4)}{\overset{(3)}{\rightleftharpoons}} COCl$$

$$COCl + Cl_2 \overset{(5)}{\longrightarrow} COCl_2 + Cl$$

reactions (1) and (2) are opposing reactions, (3) and (5) are successive, (4) and (5) or (2) and (3) are parallel, and so on.

2.2. THE RELATION BETWEEN THE COMPLETE RATE LAW AND THE MECHANISM

The most important method for studying mechanisms is undoubtedly the study of the kinetics of the reaction. Powerful modern techniques, which we will be dealing with later on, can give considerable mechanistic information. Here it should only be emphasized that old-fashioned kinetic investigations still remain the most important tool in the study of mechanisms, and kinetics should not be neglected—even if it is sometimes onerous and time-consuming to collect the data.

Let us examine more closely the relation between kinetics and the mechanism, and first of all the relationship between the rate law and the mechanism.

Consider a simple mechanism consisting of two successive monomolecular reactions

$$X \xrightarrow{k_1} Y \tag{2.3}$$

$$Y \xrightarrow{k_2} Z \tag{2.4}$$

If x, y, z are the concentrations of X, Y, Z at time t, and the volume does not change during the reaction, *the system* of differential equations describing these reactions is

$$\frac{dx}{dt} = -k_1 x \tag{2.5}$$

$$\frac{dy}{dt} = k_1 x - k_2 y \tag{2.6}$$

$$\frac{dz}{dt} = k_2 y \tag{2.7}$$

X disappears by reaction 2.3, and since this elementary reaction is monomolecular, the rate is proportional to the first power of x. Z is formed in reaction 2.4 with a rate as noted in equation 2.7. However, Y is formed in equation 2.3 but decomposes in equation 2.4; for this reason there are two terms on the right-hand side of equation 2.6.

The above system of differential equations describes the time dependence for all the species at all concentrations, from zero time ($t = 0$) until the reaction is complete ($t = \infty$), and it can be called the complete set of rate equations.

If the mechanism is known, it is possible to derive this set in a direct, one-to-one way. Only one set of rate equations can be derived from a given mechanism. In practice, however, rates and their dependence on the concentrations are determined experimentally, and we attempt to deduce the mechanism from them.

If it is possible to determine experimentally the complete time dependence for all species in the reaction from $t = 0$ to $t = \infty$, then the mechanism can be specified without any doubt. If in the example given it were possible to determine empirically all three equations of the system [i.e., equations 2.5–2.7], the mechanism would be determined uniquely.

In order not to have any doubts about the mechanism, we need an exact measurement of the concentration of all species, including the intermediates, at all times. Unfortunately, this is often impossible in practice.

To avoid confusion we will reserve use of the term *rate law* for equation(s) describing the experimentally determined time dependence. For equations derived from a proposed mechanism the term *rate equation(s)* is preferable.

Mathematical Difficulties

In practice, infinitesimal changes in concentration (dc) and time (dt) cannot be measured. Only finite differences Δc and Δt are measured, which is already an approximation, a compromise. Alternatively, the differential equations can be integrated and verified experimentally. However, analytical (exact) integration of a system of nonlinear differential equations is possible only in special cases.

Two alternatives remain: (1) simplifying approximations and (2) numerical methods. In either case, the difficulties in handling the mathematics introduce some ambiguity regarding the mechanism.

Steady-State Approximation

The system of differential equations 2.5, 2.6, and 2.7 can be simplified by using mass balance:

$$x + y + z = a = \text{constant}$$

A system of two differential equations and two unknowns is thus obtained:

$$\frac{dy}{dt} = k_1 a - (k_1 + k_2)y - k_1 z$$

$$\frac{dz}{dt} = k_2 y$$

According to the steady-state approximation,

$$\frac{dy}{dt} = 0 \qquad\qquad 2.8$$

and y is estimated from the expression

$$k_1 a - (k_1 + k_2)y - k_1 z = 0 \qquad\qquad 2.9$$

The condition for the application of the steady-state approximation is that $k_2 \gg k_1$. It should be noted that equations 2.8 and 2.9 are not equivalent. Equation 2.8 is a simple differential equation, which when integrated gives constant y, whereas equation 2.9 is an algebraic equation and the value of y depends on z—it is not constant. In the steady-state approximation, y is estimated from equation 2.9, not from 2.8. It is therefore essentially assumed that y changes with time, but slowly, not that it remains constant throughout the reaction. In this respect the term "semisteady state" is more accurate, but the term "steady state" is in common use.

Finally, the system of differential equations is reduced to a single equation:

$$\frac{dz}{dt} = \frac{k_1 k_2}{k_1 + k_2}(a - z) \simeq k_1(a - z) \qquad\qquad 2.10$$

since $k_1 k_2/(k_1 + k_2)$ can be put approximately equal to k_1 because $k_2 \gg k_1$. For sufficiently long times, the solution of equation 2.10 coincides almost completely with the solutions of the complete system. For short times there are differences.

Generalization

Analysis similar to the above can also be applied to more complicated systems. In fact, for two first-order consecutive reactions the approximation is not really necessary, because the system of differential equations can be solved analytically. In more complicated systems, an analytical solution is not possible and a comparison of the approximate solution to the exact one cannot be made. Whether the approximation is justified or not is finally determined not by comparing the approximate solution to the exact one, but by checking experimentally the conclusions reached after having used the approximation.

Application of the steady-state approximation for a component is generally justified if the variation with time of the concentration of this component is small compared to the rate of the overall reaction. If s is the concentration of the component S, ds/dt is put equal to zero. If, in addition, S is an intermediate having zero initial concentration, there is a small time interval at the beginning of the reaction during which s increases until it reaches a small but finite value. During this initial time interval the rate $(ds/dt)_{t\to0}$ is obviously not zero. The criterion, therefore, for applying the steady-state approximation can be satisfied only after this initial time has passed. If r is the concentration of a component R whose rate of formation or disappearance can be taken as a characteristic measure of the overall rate, the general criterion for applying the steady-state approximation can be expressed with the relation

$$\left|\frac{ds/dt}{dr/dt}\right| = \epsilon \qquad\qquad 2.11$$

where ϵ is a very small number $(0 < \epsilon \ll 1)$. From equation 2.11, after integration it follows that

$$s = \epsilon r$$

which implies that

$$s \ll r \qquad\qquad 2.12$$

In general we can make the following statement: *If the kinetics of a chemical system are such that the contribution of one of the components to the total mass is always negligible, then for this component we can use the steady-state approximation.* For further clarification it must be emphasized that at time $t \to 0$ there is a discontinuity; concentration s is assumed to change suddenly from a value equal to zero to a small but finite nonzero value. The approximation is therefore physically justified only if the change in the amount of matter involved is not large. A sudden, discontinuous change in the amount of matter is physically unacceptable, unless it is very small indeed.

EXAMPLE

For reactions of the type:

$$(H_2O)_5CrR^{2+} + SCN^- \rightleftharpoons (H_2O)_4CrR(SCN)^+ + H_2O$$

where $(H_2O)_5CrR^{2+}$ is an organochromium compound, the following mechanism has been proposed:

$$(H_2O)_5CrR^{2+} \underset{k_{-1}}{\overset{k_1}{\rightleftharpoons}} (H_2O)_4CrR^{2+} + H_2O$$

$$(H_2O)_4CrR^{2+} + SCN^- \underset{k_{-2}}{\overset{k_2}{\rightleftharpoons}} (H_2O)_4CrR(SCN)^+$$

The complete system of differential equations for this mechanism is

$$-\frac{d[\text{H}_2\text{OCrR}]}{dt} = k_1[\text{H}_2\text{OCrR}] - k_{-1}[\text{CrR}] \qquad 2.13$$

$$\frac{d[\text{CrR}]}{dt} = k_1[\text{H}_2\text{OCrR}] - k_{-1}[\text{CrR}] - k_2[\text{CrR}][\text{SCN}]$$

$$+ k_{-2}[\text{CrR(SCN)}] \qquad 2.14$$

$$\frac{d[\text{CrR(SCN)}]}{dt} = -\frac{d[\text{SCN}]}{dt} = k_2[\text{CrR}][\text{SCN}] - k_{-2}[\text{CrR(SCN)}] \quad 2.15$$

In these equations the charges and the water molecules that are not directly involved have been omitted to simplify the presentation. The concentration of water (solvent) is also omitted, because it remains essentially constant and is incorporated into the rate constants.

Assuming now that $[(\text{H}_2\text{O})_4\text{CrR}^{2+}]$ reaches steady state, we put $d[\text{CrR}]/dt$ in equation 2.14 equal to zero and estimate the concentration of CrR from the expression

$$k_1[\text{H}_2\text{OCrR}] - k_{-1}[\text{CrR}] - k_2[\text{CrR}] [\text{SCN}] + k_{-2}[\text{CrR(SCN)}] = 0 \quad 2.16$$

obtaining

$$[\text{CrR}] = \frac{k_1[\text{H}_2\text{OCrR}] + k_{-2}[\text{CrR(SCN)}]}{k_{-1} + k_2[\text{SCN}]} \qquad 2.17$$

Substituting equation 2.17 into 2.13, we have

$$-\frac{d[\text{H}_2\text{OCrR}]}{dt} = \frac{k_1 k_2[\text{H}_2\text{OCrR}] [\text{SCN}] - k_{-1}k_{-2}[\text{CrR(SCN)}]}{k_{-1} + k_2[\text{SCN}]} \qquad 2.18$$

In other words, we obtain an expression that contains only the concentrations of the initial reactants and final products. The concentration of the intermediate has been eliminated. This is a simplification, but it should be kept in mind that it also introduces some uncertainty regarding the mechanism. If the concentration of the intermediate is eliminated, we cannot claim that we have any direct information about it.   □

Numerical Methods

Numerical methods are particularly useful for complicated mechanisms. Here the actual calculations will not be described; such descriptions can be found in specialized texts. Only a generalized example will be given to illustrate the possibilities and the pitfalls.

Consider the following complicated mechanism:

$$A + B \underset{k_{-1}}{\overset{k_1}{\rightleftharpoons}} C$$

$$C \underset{k_{-2}}{\overset{k_2}{\rightleftharpoons}} D$$

$$A + B \overset{k_3}{\longrightarrow} E + F$$

$$E + C \overset{k_4}{\longrightarrow} G + B$$

$$G \underset{k_{-5}}{\overset{k_5}{\rightleftharpoons}} A + E$$

This mechanism involves seven species, eight rate constants, parallel and successive reactions, reformation of reactants, and intermediates that cannot be detected. In addition, the rates depend on factors that do not appear explicitly in the mechanism, such as pH and the solvent. The complete system of differential equations is nonlinear, and it cannot be solved analytically. Approximations, such as the steady-state approximation, cannot reduce the system to a manageable form. In general, we cannot neglect any of the reactions or consider only one reaction as being the slow determining step. The only option left is to use numerical methods.

In the design of the experiments, we can in principle choose conditions that will simplify the situation and allow separate study of the various parameters. In "real life" problems, for example, when studying biological systems, this step-by-step study may not be possible.

Experimentally one or more of the physical properties of the system are usually measured as a function of time under different initial conditions. Then the values of the rate constants are guessed, the concentrations at various times are estimated, and the calculated concentrations are fitted to those observed.

Returning to the mechanism under consideration, the problem is somewhat simplified if the experiments are performed under conditions of a large excess of B. Then the concentration of this species can be considered to remain constant. Species F is also kinetically unimportant, because it appears only as a product and does not participate in any reaction as a reactant.

In Figure 2.1, the circles represent experimental points and the solid lines passing through the points are calculated. The curves for the concentrations of A, C, D, and G are also calculated. The fit achieved appears to be consistent with the mechanism, but it does not prove it. *There is no way to prove, beyond any doubt, that the chosen rate constants and elementary reactions are correct.*

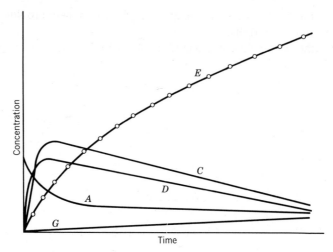

Figure 2.1. Fit of numerically calculated curves (solid lines) with the experimental data (circles). Experimentally, the concentration of only one component has been measured (curve *E*). The concentrations of the other species are calculated (curves *A, C, D, G*).

However, more confidence would be gained if A, C, D, and G could be determined independently and if the numerical fit of the data is equally good.

Experimental Difficulties

The most serious difficulties for the deduction of the mechanism from kinetic measurements are experimental. Three kinds of experimental difficulties can be distinguished:

Limitations in studying the system during very short times.

Limitations in detecting very small concentrations.

Limitations in varying the concentrations of some of the participating species.

Each experimental technique has a characteristic lowest time limit. In principle, any process that is over within this short time can be included in the mechanism but without having direct kinetic information about that process. With picosecond spectroscopy (see Chapter 4), reactions can be studied that are completed within 10^{-12} s (the time scale of molecular vibrations). This technique, however, is a specialized one. In most of the techniques used for ordinary reactions, the times are considerably longer. The characteristic time for a stopped-flow experiment, for example, is of the order of 1 ms. What happens during this millisecond is not tested.

Similar remarks can be made for the sensitivity of each of the experimental techniques used in kinetic measurements. We cannot be certain about species having concentrations below the detection limit of the technique.

Finally, it is obvious that in order to determine whether or not the rate depends on the concentration of a component, one should be able to vary this concentration, which is impossible for the solvent because it is present in excess. The mechanism is therefore ambiguous as to the participation of the solvent or of any other component whose concentration remains constant during the reaction.

EXAMPLE I

The reaction of ethyl acetoacetate with bromine is catalyzed by copper(II) ions. The rate is first-order in the concentration of the ester and first-order in the concentration of copper(II). The following mechanism has been proposed:

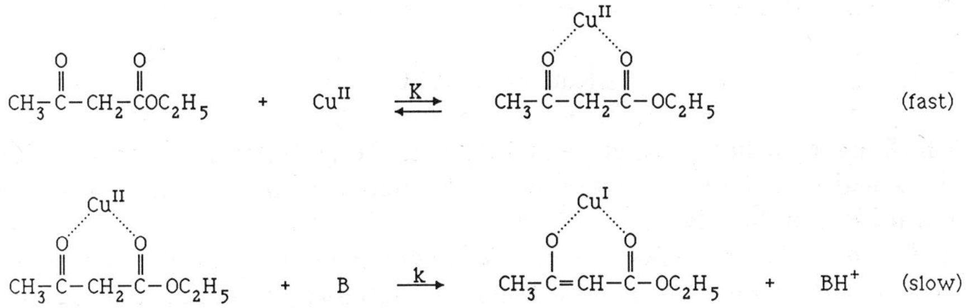

where B = base. The second (slow) reaction determines the overall rate. The product of this reaction (the Cu^I complex) reacts rapidly with bromine to give the final brominated products.   □

* * * * *

In general, *if in a series of consecutive elementary reactions one of the steps is considerably slower than the others, this step determines the rate of the overall reaction.* The overall reaction cannot proceed faster than the slowest of its consecutive elementary steps: The rate-determining (or "rate-controlling") process is the step with the smaller intrinsic rate, that is, the smaller rate constant, not the smaller overall rate. In a steady-state situation established by two successive first-order reactions, the rate-determining reaction has the smaller rate constant. The overall rates are equal.

For parallel reactions, the fast process (rather than the slow one) dominates and determines the overall rate.

* * * * *

On the basis of the above mechanism, the rate of product formation is

$$Rate = k[C][B]$$

where [C] is the concentration of the Cu^{II} chelated complex and [B] is the concentration of the base. This rate expression, however, contains the concentration of the intermediate, and it would be preferable to have an expression in terms of only the concentrations of the initial reactants. From the fast equilibrium preceding the rate-determining step we have

$$[C] = K[A][Cu^{II}]$$

where K is the equilibrium constant and [A] is the concentration of the ester. The rate then becomes

$$\text{Rate} = kK[A][B][Cu^{II}]$$

If the base is in excess, [B] is constant, and then

$$\text{Rate} = k'[A][Cu^{II}]$$

which agrees with the experimental data. It should be noted, however, that the kinetics of the fast reaction of the formation of the Cu^{II} complex were not measured directly.

In general, the rate law gives no information about fast equilibria preceding the rate-determining step. The mechanism is therefore uncertain in this respect.

EXAMPLE II

The oxidation of uranium(IV) by iron(III),

$$2Fe^{III} + U^{IV} \longrightarrow 2Fe^{II} + U^{VI}$$

proceeds with the rate

$$\frac{d[U^{VI}]}{dt} = k[Fe^{III}][U^{IV}]$$

This rate law is consistent with the mechanism

$$Fe^{III} + U^{IV} \longrightarrow Fe^{II} + U^{V} \quad \text{(slow)}$$

$$Fe^{III} + U^{V} \longrightarrow Fe^{II} + U^{VI} \quad \text{(fast)}$$

but it is also consistent with the mechanism

$$Fe^{III} + U^{IV} \longrightarrow Fe^{I} + U^{VI} \quad \text{(slow)}$$

$$Fe^{III} + Fe^{I} \longrightarrow 2Fe^{II} \quad \text{(fast)}$$

On the basis of the rate law alone it is impossible to choose between these two mechanisms. The choice might be possible only if there were an analytical method to detect the small amounts of U^V or Fe^I that may be formed.

It can be argued that U^V is known and has been characterized in other studies, under different conditions, and that this is indeed the intermediate, and not Fe^I. This argument, however, is not based on the rate measurements.   □

EXAMPLE III

The reaction

$$2V^{3+} + Cl_2 + 2H_2O \longrightarrow 2VO^{2+} + 2Cl^- + 4H^+$$

illustrates the third category of ambiguity originating in experimental limitations. The rate law is

$$\text{Rate} = \frac{k_{obs}[V^{3+}][Cl_2]}{[H^+]}$$

but since in the kinetic experiments the concentration of the solvent remained constant, it is possible that the observed rate constant is the product of the real rate constant and some power of the concentration of water, for example,

$$k_{obs} = k[H_2O]^a$$

This means that *there is also uncertainty as to whether or not the solvent participates in the reaction as a reactant.*   □

Incomplete Data

Sometimes the experimental "difficulties" are of our own making: We just could not or would not collect enough experimental data or did not design the experiments correctly.

EXAMPLE I

Consider the reaction

$$A + 2B \rightleftharpoons C + 2D$$

which is supposed to proceed by the mechanism

$$A + B \underset{k_{-1}}{\overset{k_1}{\rightleftharpoons}} X + D$$

$$X + B \underset{k_{-2}}{\overset{k_2}{\rightleftharpoons}} C + D$$

The rate for reactant A is

$$-\frac{d[A]}{dt} = k_1[A][B] - k_{-1}[X][D]$$

Assuming steady state for the concentration of X,

$$[X]_{ss} = \frac{k_1[A][B] + k_{-2}[C][D]}{k_{-1}[D] + k_2[B]}$$

and

$$-\frac{d[A]}{dt} = \frac{k_1k_2[A][B]^2 - k_{-1}k_{-2}[C][D]^2}{k_{-1}[D] + k_2[B]}$$

As has been stated, experimental verification of this last expression can give no assurance that X is indeed formed. In fact, many other mechanisms will give a similar expression, as, for example, the mechanism

$$A + B \rightleftharpoons M + N + D$$
$$M + N + B \rightleftharpoons C + D$$

However, if the data are incomplete, the difficulties are even more serious. In the first place, if the equilibrium lies far to the right ($k_1k_2 \gg k_{-1}k_{-2}$) at some concentration range, the second term in the numerator becomes negligible compared to the first. Under these conditions,

$$-\frac{d[A]}{dt} \simeq \frac{k_1k_2[A][B]^2}{k_{-1}[D] + k_2[B]}$$

This essentially means that the reverse reaction (k_{-2}) is not tested.

If, on the other hand, the experiments are performed under conditions for which

$$k_2[B] \gg k_{-1}[D]$$

the rate law becomes

$$-\frac{d[A]}{dt} \simeq k_1[A][B]$$

Then more mechanisms can be added to the list of those consistent with the experimental data.

If, instead,

$$k_{-1}[\text{D}] \gg k_2[\text{B}]$$

the rate becomes

$$-\frac{d[\text{A}]}{dt} = \frac{k_1 k_2 [\text{A}][\text{B}]^2}{k_{-1}[\text{D}]}$$

and again the number of mechanisms consistent with the data increases. □

EXAMPLE II

Another example involving ambiguities of our own making is provided by the reaction

$$2\text{Ag}^{\text{II}} + \text{Tl}^{\text{I}} \longrightarrow 2\text{Ag}^{\text{I}} + \text{Tl}^{\text{III}}$$

If excess Ag^{I} is added to the reaction mixture, the rate law takes the following form:

$$-\frac{d[\text{Ag}^{\text{II}}]}{dt} = \frac{k_{\text{obs}}[\text{Tl}^{\text{I}}][\text{Ag}^{\text{II}}]^2}{[\text{Ag}^{\text{I}}]} \qquad 2.19$$

Then at least two mechanisms consistent with this rate law can be proposed.

Mechanism 1

$$2\text{Ag}^{\text{II}} \underset{k_{-1}}{\overset{k_1}{\rightleftarrows}} \text{Ag}^{\text{I}} + \text{Ag}^{\text{III}}$$

$$\text{Ag}^{\text{III}} + \text{Tl}^{\text{I}} \overset{k_2}{\longrightarrow} \text{Ag}^{\text{I}} + \text{Tl}^{\text{III}}$$

Mechanism 2

$$\text{Ag}^{\text{II}} + \text{Tl}^{\text{I}} \underset{k_{-3}}{\overset{k_3}{\rightleftarrows}} \text{Ag}^{\text{I}} + \text{Tl}^{\text{II}}$$

$$\text{Ag}^{\text{II}} + \text{Tl}^{\text{II}} \overset{k_4}{\longrightarrow} \text{Ag}^{\text{I}} + \text{Tl}^{\text{III}}$$

By assuming steady state for Ag^{III} and Tl^{II}, respectively, the rate equations 2.20 and 2.21 are derived:

For mechanism 1:

$$-\frac{d[\text{Ag}^{\text{II}}]}{2dt} = \frac{k_1 k_2 [\text{Tl}^{\text{I}}][\text{Ag}^{\text{II}}]^2}{k_{-1}[\text{Ag}^{\text{I}}] + k_2[\text{Tl}^{\text{I}}]} \qquad 2.20$$

For mechanism 2:

$$-\frac{d[Ag^{II}]}{2dt} = \frac{k_3 k_4 [Tl^I][Ag^{II}]^2}{k_{-3}[Ag^I] + k_4[Ag^{II}]}$$

2.21

With a large excess of $[Ag^I]$, $k_{-1}[Ag^I] \gg k_2[Tl^I]$ and $k_{-3}[AgI] \gg k_4[Ag^{II}]$, respectively, and both mechanisms yield the same empirical rate law (Eq. 2.19). This choice of conditions precludes distinction. Distinction could have been achieved if additional experiments had been carried out, for example, at very high $[Tl^I]$, in which case mechanism 1 would yield a rate simply proportional to $[Ag^{II}]^2$, whereas the rate law corresponding to mechanism 2 would still be given by equation 2.19.

The two mechanisms are also differentiated at $t \rightarrow 0$, without Ag^I added ($[Ag^I]_0 = 0$). Then for mechanism 1 we obtain the limiting law

$$-\left[\frac{d[Ag^{II}]}{2dt}\right]_{t \to 0} \simeq k_1[Ag^{II}]^2$$

and for mechanism 2,

$$-\left[\frac{d[Ag^{II}]}{2dt}\right]_{t \to 0} \simeq k_3[Tl^I][Ag^{II}]$$

Finally, if experiments with excess Ag^{II} were possible, mechanism 1 would still give the same rate law, but mechanism 2 would give

$$-\frac{d[Ag^{II}]}{2dt} \simeq k_3[Tl^I][Ag^{II}]$$

The general conclusion to be drawn from all this is: Never do experiments under only one set of limited conditions. Vary the conditions as widely as possible. Cutting corners may lead to wrong conclusions.   □

Concentrations Used to Test the Rate Law

What concentrations should be used to test the rate law: the initial, the equilibrium, or the . . . right ones? The question sounds strange, and the answer seems obvious, yet it is sometimes overlooked. For example, if the initial rate is measured (at $t \rightarrow 0$), the concentrations are the initial concentrations. If the rate is measured at any other time t, the concentrations must be measured simultaneously.

The reacting substances often exist in solution in more than one form, and some of these forms may not be reactive. Then the rate law must include only the concentration(s) of the form(s) actually participating in the reaction. If, for example, each of the reacting substances exists in two conjugate acid-base forms with different reactivities, the difference must be recognized and the total concentration should not be used indiscriminately.

EXAMPLE I

Consider a rate law of the form

$$\text{Rate} = k[\text{AH}^+][\text{B}] \qquad 2.22$$

The conjugate of AH^+ is A,

$$\text{AH}^+ \xrightleftharpoons{K_1} \text{A} + \text{H}^+$$

and the conjugate of B is BH^+,

$$\text{BH}^+ \xrightleftharpoons{K_2} \text{B} + \text{H}^+$$

If only AH^+ and B are reactive, the concentrations in the rate law 2.22 are the concentrations of these species, not the total concentrations $[\text{A}]_{\text{tot}}$ and $[\text{B}]_{\text{tot}}$. In order to transform equation 2.22 into a form containing total concentrations and not the concentration of the reactive species, we make use of the following relations:

$$[\text{A}]_{\text{tot}} = [\text{A}] + [\text{AH}^+] = [\text{AH}^+]\left(1 + \frac{K_1}{[\text{H}^+]}\right) \qquad 2.23$$

$$[\text{B}]_{\text{tot}} = [\text{B}] + [\text{BH}^+] = [\text{B}]\left(1 + \frac{[\text{H}^+]}{K_2}\right) \qquad 2.24$$

Solving these equations for $[\text{AH}^+]$ and $[\text{B}]$ and substituting into 2.22, we obtain

$$\text{Rate} = k\left[\left(1 + \frac{K_1}{[\text{H}^+]}\right)\left(1 + \frac{[\text{H}^+]}{K_2}\right)\right]^{-1}[\text{A}]_{\text{tot}}[\text{B}]_{\text{tot}}$$

If the experiments are performed at constant hydrogen ion concentration, there is no way to get kinetic information as to which acid-base forms really participate. Information about any component can be obtained only by varying the concentration of that component.   □

* * * * *

In conclusion, there are mathematical difficulties, there are experimental difficulties, and there are uncertainties. That is why we do not say, "This is the mechanism," but rather, "This is the proposed mechanism" or "This is a mechanism consistent with the results."

"The God whose oracle is in Delphi neither speaks out nor conceals: he indicates" (Heraclitus, 544–483 B.C.).

* * * * *

2.3. STOICHIOMETRY AND THE RATE LAW

In any attempt to suggest a mechanism, it must be remembered that the sum of the individual steps must be consistent with the overall reaction(s), and the postulated rate-determining step and associated preequilibria must agree with the observed rate law.

EXAMPLE I

In the oxidation of iodide ion by hypochlorite ion,

$$I^- + OCl^- \longrightarrow OI^- + Cl^- \qquad\qquad 2.25$$

in 1 M potassium hydroxide, the rate expression is

$$\frac{d[OI^-]}{dt} = \frac{k_{obs}[I^-][OCl^-]}{[OH^-]} \qquad\qquad 2.26$$

The following mechanism is suggested:

$$OCl^- + H_2O \overset{K}{\rightleftharpoons} HOCl + OH^- \qquad \text{(fast)} \qquad 2.27$$

$$I^- + HOCl \overset{k}{\longrightarrow} HOI + Cl^- \qquad \text{(slow)} \qquad 2.28$$

$$\underline{OH^- + HOI \longrightarrow H_2O + OI^- \qquad \text{(fast)} \qquad 2.29}$$

$$I^- + OCl^- \longrightarrow OI^- + Cl^- \qquad \text{(overall)} \qquad 2.25$$

Addition of the elementary reactions gives the overall reaction. The rate is determined by the second reaction, 2.28. Therefore,

$$-\frac{d[OCl^-]}{dt} = \frac{d[OI^-]}{dt} = k[I^-][HOCl]$$

The equilibrium constant (K) for equation 2.27 is equal to the dissociation constant for water ($K_w = 1.0 \times 10^{-14}$) divided by the acid dissociation constant for hypochlorous acid (K_a)

$$K = \frac{K_w}{K_a} = \frac{1.0 \times 10^{-14}}{3.4 \times 10^{-8}} = 2.9 \times 10^{-7}$$

Therefore,

$$\frac{d[OI^-]}{dt} = \frac{kK[I^-][OCl^-]}{[OH^-]}$$

and $k_{obs} = kK$. ☐

EXAMPLE II

Another example is provided by the oxidation of iodide ion by hydrogen peroxide in acidic solution according to the net equation

$$3I^- + H_2O_2 + 2H^+ \longrightarrow I_3^- + 2H_2O \qquad 2.30$$

which proceeds according to the rate law

$$\frac{d[I_3^-]}{dt} = k_a[H_2O_2][I^-] + k_b[H_2O_2][H^+][I^-] \qquad 2.31$$

Clearly, the reaction occurs by two independent, parallel pathways. The path corresponding to the first term on the right-hand side may be described by the mechanism

$$H_2O_2 + I^- \longrightarrow HOI + OH^- \qquad \text{(slow)} \qquad 2.32$$

$$H^+ + OH^- \rightleftharpoons H_2O \qquad \text{(very fast)} \qquad 2.33$$

$$HOI + H^+ + 2I^- \longrightarrow I_3^- + H_2O \qquad \text{(fast)} \qquad 2.34$$

The sum of the reactions gives the net equation 2.30, and the slow step results in the proper rate law. The rapid steps following the rate-determining step are more speculative. Reaction 2.34 is by itself complex, and it can be analyzed further.

The path corresponding to the second term on the right-hand side of the rate law can be described by

$$H^+ + I^- \rightleftharpoons HI \qquad \text{(fast)} \qquad 2.35$$

$$HI + H_2O_2 \longrightarrow H_2O + HOI \qquad \text{(slow)} \qquad 2.36$$

followed by reaction 2.34. On the other hand, the first step could just as well be

$$H_2O_2 + H^+ \rightleftharpoons [H_2O_2H]^+ \qquad \text{(fast)}$$

followed by

$$[H_2O_2H]^+ + I^- \longrightarrow H_2O + HOI \qquad \text{(slow)}$$

or we could even imagine simultaneous attack by H^+ and I^- on H_2O_2.   □

Mechanistic Implications of Variable Stoichiometry

In some reactions, the stoichiometry changes markedly as the concentration of the initial reactants is varied. The order of the reaction may also vary. Both of these variations are a result of competing reactions contributing differently to the overall stoichiometry, depending on the initial concentrations and the concentration ratio.

EXAMPLE

An example of this type is found in the formation of ClO_2 as a result of the oxidation of $HClO_2$ by Cl_2 and/or $HOCl$. An unbalanced equation that simply indicates the products formed is

$$Cl_2 + HClO_2 \longrightarrow ClO_2 + ClO_3^- + Cl^- + H^+$$

The rate of disappearance of reactants is first-order in the concentrations of Cl_2 (or $HOCl$) and $HClO_2$, but the products (ClO_2 and ClO_3^-) appear at different rates.

Some stoichiometric data, related to kinetic behavior, are given in Table 2.1. A mechanism consistent with these data is

$$Cl_2 + HClO_2 \longrightarrow Cl_2O_2 + H^+ + Cl^- \qquad \text{2.37}$$

$$Cl_2O_2 + H_2O \longrightarrow Cl^- + ClO_3^- + 2H^+ \qquad \text{2.38}$$

$$2Cl_2O_2 \longrightarrow 2ClO_2 + Cl_2 \qquad \text{2.39}$$

Reaction 2.37 accounts for the observed kinetics of disappearance of the reactants. The products ClO_3^- and ClO_2 are formed from the intermediate Cl_2O_2 by the competitive reactions 2.38 and 2.39. At higher reactant concentrations, Cl_2O_2 is produced faster and its concentration is higher. Under these conditions, the second-order reaction 2.39 is favored over the first-order reaction 2.38 and the $[ClO_2]/[ClO_3^-]$ ratio is higher. In fact, if the concentra-

TABLE 2.1 Product Ratios in the Reaction between Cl_2 and $HClO_2$

Molarity of Initial Reactants[a] (in 0.20 M $HClO_4$)	$[ClO_2]/[ClO_3^-]$ Ratio
10^{-2}	9
3×10^{-3}	4.5
3×10^{-4}	1.2
10^{-4}	0.4
10^{-5}	<0.1

[a]$[Cl_2] = [HClO_2]$.

tions of the reactants are high enough, for all practical purposes only ClO_2 is produced.

Process 2.39 is the basis of most commercial production of the ClO_2 used to purify drinking water.   □

2.4. CHAIN REACTIONS

The characteristics listed below are shown by chain reactions. In rare cases, they may also be observed in reactions that do not proceed by a chain mechanism.

The rate laws are generally complicated and have characteristic forms.

The rate is generally affected by the surface area of the vessel, because the termination reactions are often surface reactions.

There is often a detectable induction period, during which the concentration of the carriers needed to sustain the chain reaction build up and reach "steady" state.

Certain substances can inhibit the reaction.

Explosion limits appear; that is, limiting values of pressure and/or temperature beyond which the reaction proceeds explosively.

The reactions are generally very sensitive to impurities, even at the trace level. The trace impurities may accelerate or decelerate the reaction—by increasing or decreasing, respectively, the number of carriers.

Initiation, Propagation, and Termination

It is generally accepted that the oxidation of uranium(IV) to uranium(VI) by dioxygen,

$$U^{IV} \xrightarrow{\ O_2\ } U^{VI}$$

proceeds by a chain mechanism. A simplified version of this mechanism is the following:

Initiation:

$$U^{IV} + O_2 \xrightarrow{k_1} U^V + HO_2 \qquad\qquad 2.40$$

Propagation:

$$U^V + O_2 \xrightarrow{k_2} U^{VI} + HO_2 \qquad\qquad 2.41$$

$$U^{IV} + HO_2 \xrightarrow{k_3} U^V + H_2O_2 \qquad\qquad 2.42$$

Termination:

$$U^V + HO_2 \xrightarrow{k_4} U^{VI} + H_2O_2 \qquad\qquad 2.43$$

In this mechanism, U^{IV} denotes U^{4+}, U^V denotes UO_2^+, and U^{VI} denotes UO_2^{2+}. Water, protons, and the rapid proton-transfer reactions are omitted.

The chain reaction is initiated with reaction 2.40, where the *chain carriers* U^V and HO_2 are produced. Subsequently, U^V reacts with O_2 to form HO_2 and U^{VI}, and HO_2 reacts with U^{IV} to regenerate U^V. These two reactions are the *chain-propagation* steps. The chain is *terminated* by removal of the carriers as shown by equation 2.43. The hydrogen peroxide formed in reactions 2.42 and 2.43 is not accumulated. It reacts rapidly with U^{IV}:

$$U^{IV} \xrightarrow{H_2O_2} U^{VI} \qquad\qquad 2.44$$

without intermediate formation of chain carriers.

In general, recycling is possible only if there are *at least two carriers.*

The initiation step adds new carriers to the system. Without the termination step, which removes carriers, there would be a continuous increase in the number of carriers, acceleration of the overall reaction, and eventually explosion. The termination step is necessary for the carriers to reach a steady state concentration, which can be written as follows:

$$\frac{d[HO_2]}{dt} = O = k_1[U^{IV}][O_2] + k_2[U^V][O_2] - k_3[U^{IV}][HO_2] - k_4[U^V][HO_2]$$

$$\frac{d[U^V]}{dt} = O = k_1[U^{IV}][O_2] - k_2[U^V][O_2] + k_3[U^{IV}][HO_2] - k_4[U^V][HO_2]$$

adding and subtracting these equations, we obtain the equivalent system

$$k_4[U^V][HO_2] = k_1[U^{IV}][O_2] \qquad\qquad 2.45$$

$$k_2[U^V][O_2] = k_3[U^{IV}][HO_2] \qquad\qquad 2.46$$

Dividing equation 2.45 now by equation 2.46 yields

$$[HO_2] = \left[\frac{k_1 k_2}{k_3 k_4}\right]^{1/2} [O_2] \qquad\qquad 2.47$$

On the other hand, from the mechanism equations 2.40 and 2.42,

$$-\frac{d[U^{IV}]}{dt} = 2(k_1[U^{IV}][O_2] + k_3[U^{IV}][HO_2]) \qquad\qquad 2.48$$

The factor 2 is included to account for the additional U^{IV} oxidized by H_2O_2 according to equation 2.44. By substituting equation 2.47 into 2.48, the following equation is obtained.

$$-\frac{d[U^{IV}]}{dt} = 2\left[k_1 + \left(\frac{k_1 k_2 k_3}{k_4}\right)^{1/2}\right][U^{IV}][O_2]$$

This is indeed the rate law observed experimentally at constant hydrogen ion concentration.

In this relatively simple example, in each of the propagation steps one carrier is destroyed and one is formed. There are other cases where more than one carrier is formed for each destroyed; the chain is then said to be *branched*.

Explosions

Chain reactions are fast but not necessarily explosive. They become explosive if the termination step(s) are not effective enough and the new carriers produced in the initiation step are not removed fast enough. In this case, the number of carriers builds up quickly, and the result may be an explosion. On the other hand, not all explosions involve chain reactions. If the heat produced in a fast exothermic reaction (not necessarily a chain reaction) is not dissipated efficiently to the environment, the temperature increases, and as a result the reaction is accelerated, heat is produced faster, and the temperature increases further until eventually an explosion may occur.

An Industrial Incident

If termination takes place on the walls of the vessel, the rate of the chain reaction depends on the surface-to-volume ratio. Dissipation of heat also depends on this ratio. Consequently, there is a greater probability of explosion if the vessel is large.

In the laboratory, product Y was formed from reactant X in solvent Z. The reaction was known to proceed smoothly at a given temperature. Scaling up to pilot plant presented no problems, but when the industrial plant was constructed the reactor was blown up on the first try—before it even reached the "optimum" temperature. The reaction was exothermic, and the vessel was too large to efficiently dissipate the heat produced.

The rate of dissipation of heat to the environment is approximately inversely proportional to the ratio of the volume of the vessel to its surface area (V/A). For a spherical vessel with radius r, V/A is proportional to r, and hence the rate is inverse in r (the vessel is supposed to be nearly full).[1]

There are exothermic reactions that can take place smoothly on the laboratory scale by air cooling. On an industrial scale, same reactions frequently require more vigorous water cooling.

2.5. SOLVENT EFFECTS

The rate and the mechanism, among other things, depend also on the solvent. Sometimes the effect of the solvent can be dramatic; for example, it can change the rate by many orders of magnitude. In competitive reactions the choice of solvent may be crucial in determining which reaction will dominate. In industry and in the research laboratory, the right choice may also save time and money.

Classification of Solvents

Many solvent classification schemes have been proposed. According to one scheme, solvents are classified into four categories on the basis of their polarity and the extent to which they ionize.

1. Protic solvents are ionized by proton transfer.

$$H_2O + H_2O \rightleftharpoons OH^- + H_3O^+$$

$$NH_3 + NH_3 \rightleftharpoons NH_2^- + NH_4^+$$

$$H_2SO_4 + H_2SO_4 \rightleftharpoons HSO_4^- + H_3SO_4^+$$

[1]Whales have relatively small heat losses and they manage to swim in ice-cold water because their V/A ratio is large.

The inherent acidity or basicity differs in each case, and so does the ability of these solvents to protonate or deprotonate dissolved molecules. The solvents in this category, in addition to being strong proton donors or acceptors, are also strongly complexing.

2. Benzene, dioxane, tetrahydrofuran, and carbon tetrachloride belong in the category of nonprotic, nonpolar, noncomplexing, nonionized solvents.

3. Acetonitrile, dimethylsulfoxide, dimethylformamide, and pyridine belong in the category of polar solvents, which are not appreciably self-ionized, and are not proton donors or acceptors. Because of their polarity, these solvents form complexes with dissolved metal ions.

4. Finally, solvents like bromine trifluoride (BrF_3) and phosphorus oxochloride ($OPCl_3$) are very polar and nonprotic, and they self-ionize. These solvents are usually very reactive and are difficult to keep pure.

Empirical Multicorrelation between Rate Constants and the Properties of Solvents

Correlation of a rate constant with the properties of a series of solvents can be made by using an empirical equation of the form:

$$\log k = a + bX + cY + dZ + \cdots$$

where X, Y, Z are the values of the properties and a, b, c, d correspond to empirical constants. Properties affecting the rate include the dielectric constant, density, acidity, basicity, donor–acceptor properties, hydrogen bonding, size of the solvent molecules, stereochemical characteristics, and viscosity.

A Comparison of Rates in Different Solvents

A comparison of reaction rates in different solvents is indeed a difficult matter. Even the choice of units used to express the rate constant is important. Consider one particular second-order reaction that was studied in three solvents: water, methanol, and isopropanol. The data were expressed in two different fashions where the concentrations were noted in units of mole fraction X and in units of molarity M. The conversion factor is

$$\frac{k_M}{k_X} = \left(\frac{V_1^\circ}{10^3}\right)^{n-1}$$

where k_M is the rate constant in units of $M^{-1}\,s^{-1}$, k_X is the rate constant in units of (mole fraction)$^{-1}\,s^{-1}$, and V_1° is the molar volume of the solvent. The observed results are given in Table 2.2.

TABLE 2.2 Rate Constants in Solvents Expressed in Different Units

Solvent	V_1°, mL mol^{-1}	k_M, M^{-1} s^{-1}	k_X, (mole frac)$^{-1}$ s^{-1}
H_2O	18.0	5.7	316
CH_3OH	40.5	10.0	247
iPrOH	79.5	18.7	146

In which solvent does the reaction occur most rapidly? This question evidently cannot be answered from the trend in rate constants, which depends on the choice of units. In fact, the product of the rate constant and the appropriate concentrations, which gives the reaction rate, still does not settle the question, because the relative rate of change of molarity from solvent to solvent may still be different from the relative rate of change in mole fractions.

Transfer Functions

Figure 2.2 gives the free energy levels for a reaction taking place in solvents A and B, one of which (e.g., solvent A) can be taken as the reference solvent. It is seen from this figure that the differences $\delta G^{(r)}$ and $\delta G^{(tr)}$ for the reactants

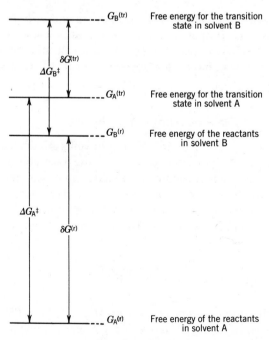

Figure 2.2. Free energies of the reactants and the transition states in two solvents A and B.

and the transition state, respectively, are given by the relations:

$$\delta G^{(r)} = G_B^{(r)} - G_A^{(r)}$$

$$\delta G^{(tr)} = G_B^{(tr)} - G_A^{(tr)}$$

It is also seen that the difference $\delta\Delta G^{\ddagger}$ between the activation energies in the two solvents is given by

$$\delta\Delta G^{\ddagger} = \Delta G_B^{\ddagger} - \Delta G_A^{\ddagger} = (G_B^{(tr)} - G_B^{(r)}) - (G_A^{(tr)} - GA^{(r)})$$
$$= \delta G^{(tr)} - \delta G^{(r)}$$

The difference in the activation energies depends both on the difference in the energy of the reactants and on the corresponding difference in the transition states.

Ionic Strength Effects

Rate constants often depend on ionic strength. The presence of a salt that does not participate directly in the reaction decreases the rate of a simple bimolecular reaction between oppositely charged ions and increases the rate if the ions have charges of the same sign.

For a limited range of low ionic strengths, the dependence of the rate constant on ionic strength is given by the Brönsted–Bjerrum equation

$$\log\left(\frac{k_0}{k}\right) = 2Az_a z_b \mu^{1/2}$$

where k_0 is the rate constant of a bimolecular reaction under ideal conditions and k is the value of the same rate constant in a solution having ionic strength μ. The reacting species have charges z_a and z_b. Parameter A depends on the dielectric constant and the temperature.

Usually the ionic strength is too high for this relation to be valid. Reactant concentrations in mechanistic studies generally range from 10^{-1}–10^{-5} M depending on the rate and the detection method used. Under these conditions it is advisable that ionic strength be kept fairly high (0.5–3.0) in order to minimize solution effects resulting from changes in concentration of reactants and products during the reaction.

A better expression for higher ionic strengths is the Davies equation:

$$\log\left(\frac{k_0}{k}\right) = \frac{2Az_a z_b \mu^{1/2}}{1 + \mu^{1/2}}$$

More sophisticated expressions are also available, but their discussion is beyond the scope of this book.

2.6. KINETICS AND EQUILIBRIUM

In many general chemistry books, the concept of equilibrium is introduced by using the *law of mass action*. For a reaction of the general form:

$$A + B \underset{k_{-1}}{\overset{k_1}{\rightleftharpoons}} C + D$$

the rate from left to right is taken as equal to $k_1[A][B]$, and for the reverse (opposite) reaction, as equal to $k_{-1}[C][D]$. At equilibrium there is no overall change, and these two rates are equal:

$$k_1[A][B] = k_{-1}[C][D]$$

and hence

$$\frac{[C][D]}{[A][B]} = \frac{k_1}{k_{-1}} = K$$

Thus, the equilibrium constant is expressed in terms of the rate constants. Obviously, for $K > 1$, $k_1 > k_{-1}$, and for $K < 1$, $k_1 < k_{-1}$.

This derivation of the expression for the equilibrium constant was first made by the Norwegian chemists Guldberg and Waage, and its great importance is that it explicitly recognized the dynamic nature of equilibrium. However, the *rate law can be derived from the chemical equation only for elementary reactions*. Only for an elementary reaction can we be sure that the rate is proportional to the concentration of the reactants. The *rate law of a complex reaction must be determined experimentally*. For a reaction of the general form

$$aA + bB \underset{k_{-1}}{\overset{k_1}{\rightleftharpoons}} cC + dD$$

the equilibrium constant is always given by the expression

$$K = \frac{[C]^c[D]^d}{[A]^a[B]^b}$$

but it is not at all certain, or necessarily true, that the rates of the two opposing reactions are equal to $k_1[A]^a[B]^b$ and $k_{-1}[C]^c[D]^d$, respectively.

2.7. MICROSCOPIC REVERSIBILITY

For nonelementary reactions, the relationship between the equilibrium constant and the rate constants is derived by making use of the principle of *microscopic reversibility,* which dictates that if an overall reaction has reached equilibrium, each one of the elementary reactions contributing to the overall reaction will have also reached equilibrium. In the example given earlier, if equilibrium has been established for the overall reaction, then

$$\frac{[CrR(SCN)]_{eq}}{[H_2OCrR]_{eq}[SCN]_{eq}} = K_{eq}$$

Because of microscopic reversibility, the elementary reactions of the mechanism also must have reached equilibrium:

$$\frac{[CrR]_{eq}}{[H_2OCrR]_{eq}} = K_1$$

$$\frac{[CrR(SCN)]_{eq}}{[CrR]_{eq}[SCN]_{eq}} = K_2$$

Multiplying these two expressions, we obtain

$$\frac{[CrR(SCN)]_{eq}}{[H_2OCrR]_{eq}[SCN]_{eq}} = K_1K_2 = K$$

but K_1 and K_2 refer to elementary reactions and hence $K_1 = k_1/k_{-1}$ and $K_2 = k_2/k_{-2}$. Therefore,

$$K_{eq} = \frac{k_1k_2}{k_{-1}k_{-2}}$$

The overall equilibrium constant is expressed in terms of rate constants, but the rate constants of the elementary reactions of the mechanism and not of the overall reaction. A prerequisite for this derivation is a detailed knowledge of the mechanism.

2.8. STEADY STATE AND EQUILIBRIUM

At the molecular level, chemical equilibrium is dynamic. The overall composition of the system does not change, but reactions still go on. In the steady state the concentration of one (or some) reactant for all practical purposes does not change, but the concentrations of other components continue to change. A system in which some reactant has reached steady state is a non-equilibrium system.

At equilibrium some thermodynamic function has reached an extremum; for example, the potential energy has reached a minimum, or the entropy has reached a maximum. This means that the *rate* of entropy production, for example, has become equal to zero. Before reaching equilibrium, however, the reaction is accompanied by entropy production. In a steady-state system not far from equilibrium (i.e., with concentrations close to the equilibrium concentrations), the rate of entropy production is a minimum but not zero.

Consider the two cases

$$A \underset{k_2}{\overset{k_1}{\rightleftharpoons}} B \qquad\qquad 2.49$$

and

$$A \xrightarrow{k_1'} B \xrightarrow{k_2'} C \qquad\qquad 2.50$$

Assuming that all these reactions are elementary, at equilibrium for equation 2.49 we have

$$\frac{[B]_{eq}}{[A]_{eq}} = \frac{k_1}{k_2} = K$$

If, on the other hand, B is in a steady state, from 2.50 we obtain

$$k_1' [A]_{ss} - k_2' [B]_{ss} = 0$$

or

$$\frac{[B]_{ss}}{[A]_{ss}} = \frac{k_1'}{k_2'} = K'$$

which is again a constant. In fact, K' is a constant resembling the equilibrium constant, since both are given as ratios of rate constants. Yet in spite of this formal resemblance there are important differences, the basic one being chemical, namely, that in 2.49 B gives back A, whereas in 2.50 B gives another species C. K and K' also differ mathematically. The ratio $[B]_{ss}/[A]_{ss}$ is constant, but the concentrations themselves vary—slowly, but they do vary. The concentrations $[B]_{eq}$ and $[A]_{eq}$, under given conditions, are always constant.

2.9. LINEAR FREE ENERGY RELATIONSHIPS

There is no general correlation between thermodynamics and kinetics. What this means is simply that if the equilibrium constant of a reaction or its standard free energy change is given, there are no equations allowing the calculation of the energy of activation or the rate constant. However, in a series of reactions (not just one reaction) between closely related compounds, certain *differences* in thermodynamic and kinetic quantities are sometimes related.

A relatively simple kind of correlation is a *linear free energy relation.* In some cases, a plot of $\Delta G°$, the differences between the *standard free energies of the products and those of the reactants* versus $\Delta G^{\ddagger}$, the differences between *the free energies of the transition state and those of the reactants* for a series of similar reactions, results in a straight line. In terms of rate and equilibrium constants, for two reactions of the series;

$$\log \left(\frac{K_1}{K_2}\right) \quad \propto \quad \log \left(\frac{k_1}{k_2}\right)$$

For dissimilar compounds and dissimilar reactions, even such differences cannot be correlated.

2.10. THERMODYNAMIC AND KINETIC STABILITY

A classic example of a kinetically stable but thermodynamically unstable system is the dihydrogen–dioxygen mixture. The reaction between these two gases is thermodynamically very favorable, but it does not take place under ordinary conditions unless it is accelerated by a catalyst or by some other means.

The thermodynamic stability of a complex ion is often expressed in terms of the "*instability constant*," the equilibrium constant for its decomposition. The complex $Cu(NH_3)_4^{2+}$ is thermodynamically stable, because the constant for the equilibrium

$$Cu(NH_3)_4^{2+}(aq) \rightleftharpoons Cu^{2+}(aq) + 4NH_3(g)$$

is small, but this equilibrium is established rapidly (the complex is labile). In contrast, the complex $Co(NH_3)_6^{3+}$ is thermodynamically unstable but kinetically stable (inert).

In a solution containing a 6-coordinated metal ion M and a ligand L, the following equilibria are eventually established:

$$M(H_2O)_6 + L \xrightleftharpoons{K_1} M(H_2O)_5L + H_2O$$

$$M(H_2O)_5L + L \xrightleftharpoons{K_2} M(H_2O)_4L_2 + H_2O$$

$$\vdots$$

$$M(H_2O)L_5 + L \xrightleftharpoons{K_6} ML_6 + H_2O$$

K_1, K_2, . . . are called *successive stability constants*. *The overall stability constants* or *stability products* are defined by the equations:

$$\beta_1 = K_1 = \frac{[M(H_2O)_5L]}{[M(H_2O)_6][L]}$$

$$\beta_2 = K_1K_2 = \frac{[M(H_2O)_4L_2]}{[M(H_2O)_6][L]^2}$$

$$\vdots$$

$$\beta_6 = K_1K_2K_3K_4K_5K_6 = \frac{[ML_6]}{[M(H_2O)_6][L]^6}$$

In an analogous manner, the *successive rate constants* k^f for the formation reactions or k^d for the decomposition reactions can be defined:

$$M(H_2O)_6 + L \underset{k_1^d}{\overset{k_1^f}{\rightleftharpoons}} M(H_2O)_5L + H_2O$$

$$M(H_2O)_5L + L \underset{k_2^d}{\overset{k_2^f}{\rightleftharpoons}} M(H_2O)_4L_2 + H_2O$$

$$\vdots$$

$$M(H_2O)L_5 + L \underset{k_6^d}{\overset{k_6^f}{\rightleftharpoons}} ML_6 + H_2O$$

For the overall reaction, we can write

$$M(H_2O)_6 + 6L \rightleftharpoons ML_6 + 6H_2O$$

Since this reaction is not elementary, a single rate constant cannot be given unless one of the steps is considerably slower and determines the rate of the overall reaction.

For the complex $M(H_2O)L_5$, for example, in the presence of excess L there are two possibilities: decomposition back to $M(H_2O)_2L_4$ or formation of ML_6; the corresponding tendencies are measured by the rate constants k_5^d and k_6^f, respectively. By analogy, k_5^d can be taken as the measure of the kinetic stability of $M(H_2O)L_5$ toward removal of one L and k_6^f toward addition of another L.

The complexes are usually characterized as *stable* if the formation constant has a value of $K > 1$, *unstable* if $K < 1$, and *labile* if the equilibrium is established during the time required to mix the reactants. If the establishment of equilibrium is slow, the complex is characterized as *inert*. For a second-order formation reaction, labile complexes have rate constants of formation of the order of $0.1\ M^{-1}\ s^{-1}$ or larger. For first-order formation reactions, the rate constants for a labile complex are larger than $\sim 0.01\ s^{-1}$.

GENERAL REFERENCES

S. W. Benson, *The Foundations of Chemical Kinetics,* McGraw-Hill, New York, 1960.

E. Buncel and H. Wilson, Solvent Effects on Rates and Equilibria, *J. Chem. Educ.* **57**, 629, (1980).

M. R. J. Dack, The Influence of Solvent on Chemical Reactivity, *J. Chem. Educ.* **51**, 231, (1974).

A. A. Frost and R. G. Pearson, *Kinetics and Mechanisms,* 2nd ed., Wiley, New York, 1961.

A. Haim, The Ambiguity of Mechanistic Interpretations of Rate Laws: Alternate Mechanisms for the Vanadium(III)–Chromium(II) Reaction, *Inorg. Chem.* **5**, 2081, (1966).

E. L. King, *How Chemical Reactions Occur,* Benjamin, New York, 1963.

K. J. Laidler, *Chemical Kinetics,* 2nd ed., McGraw-Hill, New York, 1965.

R. Schaal, *La Cinetique Chimique Homogene,* Presse Université de France, Paris, 1971.

N. N. Semenov, *Some Problems of Chemical Kinetics and Reactivity* (Engl. transl. by J. E. S. Bradley), Pergamon, New York, 1959.

PROBLEMS

1. Write the complete set of rate equations for the mechanism:

$$X + Y \longrightarrow Z$$

$$Z \longrightarrow W$$

$$X + Z \longrightarrow Q$$

Simplify by using the steady-state approximation for Z.

2. What is the mechanism corresponding to the following complete set of rate equations?

$$\frac{dx}{dt} = -k_1 xy + k_2 z$$

$$\frac{dy}{dt} = -k_1 xy + k_2 z$$

$$\frac{dz}{dt} = kxy - k_2 z - k_3 z$$

$$\frac{dw}{dt} = 2k_3 z$$

3. The reaction $A \rightarrow C$ proceeds through an unstable intermediate B. During the reaction, B remains in a steady state. What happens to B? After the reaction is over and the equilibrium between A and C has been established, will B's concentration go to zero or not? Justify your answer.

4. What is the mechanistic significance of having a sum of terms on the right-hand side of a rate law? Give an example.

5. The desired product of a reaction reacts further with the initial reactant to give another, undesirable product. What form does the rate law for the formation of the desirable product take? In practice, how would the problem of maximizing the yield be solved?

6. Derive the rate for the following chain reaction (assuming steady state for C, D, and F):

   Initiation:

   $$A + B \xrightarrow{k_1} C + D$$

   Propagation:

   $$C + A \xrightarrow{k_2} D + E$$

   $$D + A \xrightarrow{k_3} F + E$$

   $$F + A \xrightarrow{k_4} D + 2E$$

   Termination:

   $$F + F \xrightarrow{k_5} G$$

7. Consider the Earth as a huge reaction vessel where complicated chemical reactions take place. As a rule the final stable products are harmless. Some thermodynamically unstable intermediates are harmful and responsible for the pollution. Although, for example, a mixture of carbon monoxide and oxygen under the conditions of the atmosphere is unstable toward formation of nonpoisonous carbon dioxide, the reaction is slow and there is accumulation of carbon monoxide. Hydrogen sulfide and nitrogen oxides are also toxic and unstable, but again their removal takes time. Thus, the removal of pollutants from the environment poses essentially a kinetic and mechanistic problem. Taking this into account, comment on the practice of disposing of poisonous and other harmful materials by dispensing them into the atmosphere or into the hydrosphere to "dilute" them and therefore make them less harmful. Is their destruction (e.g., by air oxidation) facilitated by this process? Suggest solutions to this problem.

8. In the late 1970s the "steady" amount of CO in the atmosphere of the Earth was 5×10^{11} kg. Every year 2×10^{11} kg of CO is added and an equal quantity is removed mainly by surface reactions. Estimate:

   (a) The "steady state" concentration of CO in parts per million. Take the total mass of the atmosphere to be 2.2×10^{19} kg.

   (b) The average amount of CO destroyed per square meter of the Earth's surface per day. The diameter of the Earth is $\sim$12,000 km, and land is one third of its surface.

9. At usual temperatures the equilibrium

$$N_2 + O_2 \rightleftharpoons 2NO$$

lies to the left, but at high temperatures it is shifted to the right. This is why the preparation of NO in the nitric acid industry is carried out at high temperatures, and then the mixture is cooled quickly. Why should the cooling be done quickly?

10. If fast equilibria precede the rate-determining step, the rate law often contains concentrations with negative exponents. Show that this is the case for the reaction of H_3AsO_3 with I_3^-,

$$H_3AsO_3 + H_2O + I_3^- \rightleftharpoons H_3AsO_4 + 3I^- + 2H^+$$

for which the following mechanism has been proposed:

$$H_3AsO_3 \underset{k_{-1}}{\overset{k_1}{\rightleftharpoons}} H_2AsO_3^- + H^+ \qquad \text{(fast)}$$

$$I_3^- + H_2O \underset{k_{-2}}{\overset{k_2}{\rightleftharpoons}} 2I^- + H_2OI^+ \qquad \text{(fast)}$$

$$H_2OI^+ + H_2AsO_3^- \underset{k_{-3}}{\overset{k_3}{\rightleftharpoons}} H_2AsO_3I + H_2O \qquad \text{(slow)}$$

$$H_2AsO_3I \underset{k_{-4}}{\overset{k_4}{\rightleftharpoons}} I^- + H_2AsO_3^+ \qquad \text{(fast)}$$

$$H_2O + H_2AsO_3^+ \underset{k_{-5}}{\overset{k_5}{\rightleftharpoons}} H_3AsO_4 + H^+ \qquad \text{(fast)}$$

Express the equilibrium constant in terms of the rate constants of the elementary reactions of the mechanism. If the value of this equilibrium constant and the value of the rate constant of the overall reaction are given, can the value of the rate constant of the reverse reaction be calculated? Justify your answer.

11. The absorption of a drug by an organism often follows simple kinetics, in spite of the fact that the mechanism is very complicated. Consider a drug introduced into the organism orally in a common capsule at zero time. Call [A] its concentration in the stomach, and consider that the rate of absorption (e.g., introduction into the blood stream) is first order in this concentration. Furthermore, assume that the rate by which the drug is used or discarded by the body is proportional to its concentration in the blood. Give plots of the concentration in the stomach and in the blood as a function of time.

12. The reaction

$$R(NH_2)_2 + H_2O_2 \longrightarrow R(NH)_2 + 2H_2O$$

is catalyzed by I_2. The mechanism suggested is

$$R(NH_2)_2 + I_2 \underset{k_{-1}}{\overset{k_1}{\rightleftharpoons}} RNH_2NHI + I^- + H^+ \qquad \text{(fast)} \quad (1)$$

$$RNH_2NHI \overset{k_2}{\longrightarrow} I^- + H^+ + R(NH)_2 \qquad \text{(slow)} \quad (2)$$

$$I^- + H_2O_2 + H^+ \overset{k_3}{\longrightarrow} HOI + H_2O \qquad \text{(slow)} \quad (3)$$

$$HOI + I^- + H^+ \underset{k_{-4}}{\overset{k_4}{\rightleftharpoons}} H_2O + I_2 \qquad \text{(fast)} \quad (4)$$

Assuming that the *rates* of the two slow steps are equal, derive the rate law in terms of $[R(NH_2)_2]$, $[I_2]$, and $[H_2O_2]$.

13. The rate law of the oxidation of oxalic acid by chlorine contains the following term:

$$\frac{k[Cl_2][H_2C_2O_4]}{[H^+]^2[Cl^-]}$$

Propose four other forms kinetically equivalent to this term.

14. For reactions of the type

$$A + B \longrightarrow C$$

a fairly common rate law is

$$\frac{d[C]}{dt} = \frac{k[A][B]}{1 + K[B]}$$

Show that the following three mechanisms are completely consistent with the above rate law (specify under what conditions?).

Mechanism 1:

$$A + B \underset{k_{-1}}{\overset{k_1}{\rightleftharpoons}} AB \qquad \text{(fast)}$$

$$AB \xrightarrow{k_2} C \qquad \text{(slow)}$$

Mechanism 2:

$$A \underset{k_{-3}}{\overset{k_3}{\rightleftharpoons}} A^* \qquad \text{(fast)}$$

$$A^* + B \xrightarrow{k_4} C \qquad \text{(slow)}$$

Mechanism 3:

$$A + B \underset{k_{-5}}{\overset{k_5}{\rightleftharpoons}} AB \qquad \text{(fast)}$$

$$A + B \xrightarrow{k_6} C \qquad \text{(slow)}$$

3

EVENTS AT THE
MOLECULAR LEVEL—THE
ACTIVATED COMPLEX

We have seen so far that, in general, a chemical reaction can be analyzed into one or more elementary steps. This set of steps is called the *stoichiometric mechanism*. In this chapter we will see how each elementary reaction takes place and examine the factors influencing its course; in other words, we will deal with the *molecular* or *intimate mechanism*.

After a short diversion into gas-phase reactions, necessary for gaining some insight into the physics of molecular collisions, we will return to the discussion of more real-life solution chemistry.

3.1. A QUALITATIVE DESCRIPTION OF MOLECULAR COLLISIONS

Even a simple elementary reaction, when examined on a molecular scale, is found to be complicated. In a bimolecular reaction, it is not only necessary for a collision to take place, but, in order to change chemically, the colliding species must also have the correct orientation and energy. Under ordinary chemical conditions only the average behavior of very many events is observed, and many of the details of what really happens on a molecular scale are lost. In order to study the physics of the collisions, very special conditions must be created whereby the molecules are quite isolated from each other except when colliding. Moreover, their energy and orientation must be known. Such conditions can be achieved, albeit not without a great deal of experimental sophistication. It is beyond the scope of this book to describe the techniques in detail, but relevant information can be found in other books or in review articles. Here it suffices to say that in *crossed molecular beam* experiments, rarefied beams of monoenergetic molecules, atoms, or free radicals cross each other, and the products or unreacted initial species scattered at various angles are detected as a function of the scattering angle by mass spectrometry or other means.

Illustrations: The Methane + Tritium System

(*a*) *Nonreactive collisions.* If the translational energy of tritium, *T*, is small and the impact parameter, *b*, large, there is no reaction, and the collision is nonreactive. This is what happens in most cases (see Fig. 3.1).

The interaction in a nonreactive collision may lead to the transfer of energy from the tritium atom to methane (vibrations). Energy transfer is larger for smaller impact parameters.

If the interaction is weak, there is little or no change in the trajectory of the tritium atom. For stronger interactions, tritium is scattered.

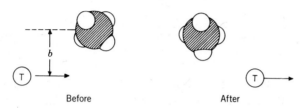

Before   After

Figure 3.1.

(*b*) *Hydrogen abstraction* can take place for smaller impact parameters. The HT molecule formed is scattered over large angles, even backwards (see Fig. 3.2). A large part of the available energy is partitioned into the HT and CH_3 vibrational modes. The duration of the interaction is of the order of 10^{-14} s.

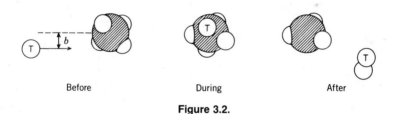

Before   During   After

Figure 3.2.

(*c*) *Displacement* is also possible for small impact parameters (see Fig. 3.3). In displacement, the configuration may or may not be retained.

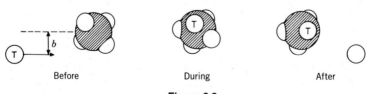

Before   During   After

Figure 3.3.

(*d*) *Fragmentation* can occur if the energies are high, as noted in Figure 3.4.

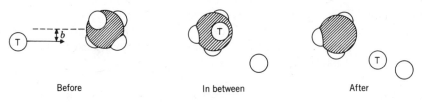

Before   In between   After

Figure 3.4.

Reactions of Potassium Atoms with Methyl Iodide, Chloroform, and Trifluoromethyl Iodide

(*a*) *CH$_3$I + K.* The experimental results indicate that the iodine end of the alkyl iodide is the more reactive. A long-lived intermediate complex between CH$_3$I and K has not been detected. Apparently K hits the I end of the molecule, and KI is formed and bounces back (see Fig. 3.5). If K attacks from the alkyl side of CH$_3$I, the reaction may involve formation of a complex, which exists, however, for only half of a rotation.

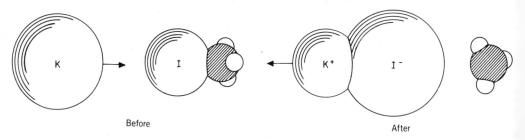

Before   After

Figure 3.5.

(*b*) *CHCl$_3$ + K.* For the reaction of potassium atoms with chloroform,

$$CHCl_3 + K \longrightarrow KCl + CHCl_2$$

the potassium atom can attack the chloroform molecule from either side of the plane defined by the chlorine atoms. The relatively small hydrogen atom on one of these sides does not hinder the approach of the incoming potassium atom or the departure of the outgoing potassium chloride. The reactivity does not depend on orientation (see Fig. 3.6).

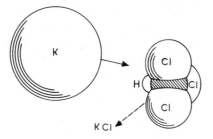

Figure 3.6.

(c) *CF₃I + K*. In the reaction CH_3I + K, the iodine end is more reactive. In the reaction $CHCl_3$ + K, there is no preference. In the reaction,

$$CF_3I + K \longrightarrow CF_3 + KI$$

it is the trifluoromethyl end rather than the iodine end that is more reactive. The trajectories shown in Figure 3.7 have been proposed for the two orientations.

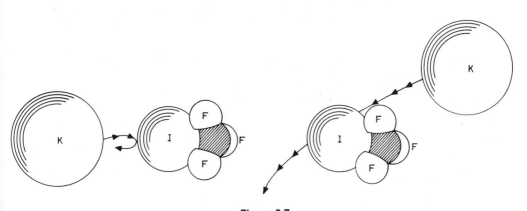

Figure 3.7.

3.2. STOCHASTIC APPROACH TO REACTIONS

Molecular beam experiments can supply detailed information about reaction dynamics in rarefied gases—how the outcome of the collision depends on energy, on the impact parameter, and on orientation. Under ordinary conditions such detailed information is impossible to obtain. Only an overall macroscopic behavior can be measured and related to the probabilities that certain events (e.g., reaction or no reaction) will take place, without necessarily stating explicitly what physical factors influence these probabilities.

Phenomenological (i.e., macroscopic) kinetics were examined in Chapter 2. It is instructive now to try to relate the phenomenological aspects to some stochastic details at the molecular level.

Monomolecular Reactions

Consider the elementary reaction:

$$A \longrightarrow B$$

The probability that in the short time interval between t and $t + \Delta t$ an individual A molecule will be transformed into a B molecule depends on the magnitude of Δt. If a very short Δt is chosen, only single decays need be considered, and "simultaneous" decays of more than one A molecule can be ignored. In this case, the probability p_{AB} per A molecule can be regarded as "instantaneous" and is simply proportional to Δt:

$$p_{AB} = \alpha \, \Delta t \qquad\qquad 3.1$$

If, in addition, the only other alternative is no change at all, the normalization condition requires

$$p_{AB} + p_{AA} = 1 \qquad\qquad 3.2$$

where p_{AA} is the probability that in the time period Δt, A will not decompose. From probabilities 3.1 and 3.2 it follows that

$$p_{AA} = 1 - \alpha \, \Delta t \qquad\qquad 3.3$$

Probabilities 3.1 and 3.3 refer to each A molecule separately. For X_A such molecules, it follows that

$$P_{AB} = \alpha X_A \, \Delta t \qquad\qquad 3.4$$

$$P_{AA} = 1 - \alpha X_A \, \Delta t \qquad\qquad 3.5$$

If the short time interval Δt is taken as the unit of time (e.g., one picosecond), the probabilities p_{AB} and p_{AA} simply become

$$p_{AB} = \alpha \qquad\qquad 3.6$$

$$p_{AA} = 1 - \alpha \qquad\qquad 3.7$$

The probability $P_x(t + \Delta t)$ that in the mixture of A and B molecules the number of A molecules in time $t + \Delta t$ is exactly X_A, equals the probability that the number was X_A in time t and there was no decay during Δt, plus the probability that the number was $X_A + 1$ in t and one decay occurred. Hence,

$$P_x(t + \Delta t) = (1 - X_A \alpha \, \Delta t) \, P_x(t) + (X_A + 1)\alpha \, \Delta t \, P_{x+1}(t) \qquad 3.8$$

From equation 3.8 we obtain

$$P_x(t + \Delta t) - P_x(t) = (X_A + 1)\alpha \, \Delta t \, P_{x+1}(t) - X_A \alpha \, \Delta t \, P_x(t)$$

and if we divide by Δt and go to the limit ($\Delta t \to 0$), we finally get

$$\lim_{\Delta t \to 0} \frac{P_x(t + \Delta t) - P_x(t)}{\Delta t} \equiv \frac{dP_x}{dt} = (X_A + 1)\alpha P_{x+1}(t) - X_A \alpha P_x(t) \qquad 3.9$$

The solution[1] to equation 3.9 is

$$P_x(t) = \exp(-X_0 \alpha t) \qquad 3.10$$

where X_0 is the number of A molecules at $t = 0$. The *expected*[1] number of molecules A at time t is

$$[X_A(t)]_{\text{expected}} = X_0 \exp(-\alpha t) \qquad 3.11$$

But equation 3.11 has the same form as the equation obtained by integrating the first-order rate law of the monomolecular reaction

$$X_A(t) = X_0 \exp(-k^{(1)}t) \qquad 3.12$$

Comparison of equations 3.11 and 3.12 leads to the conclusion that *the proportionality factor α in the probability for the decay of an A molecule concides with the rate constant; in other words, the probability p_{AB} is analogous to $k^{(1)}$.*

Thus, a connection has been established between random events on the molecular level and macroscopic kinetics (i.e., phenomenological events).

Bimolecular Reactions

Consider the bimolecular reaction

$$A + B \xrightarrow{k^{(2)}} C$$

---

[1]The mathematics for solving equation 3.9 and for obtaining the expected value are beyond the scope of this book.

and once again a very short time interval Δt is taken as the time unit. The probability p_{AC} that a given A molecule will be transformed into C will be proportional to $k^{(2)}$ but also proportional to the number of B molecules:

$$p_{AC} = k^{(2)} X_B$$

Similarly,

$$p_{BC} = k^{(2)} X_A$$

Nonelementary Reactions—The Transition Matrix

The transition probabilities can be assembled in the form of a stochastic matrix known as the *transition matrix*. For a first-order reaction,

$$
\begin{array}{c}
 \\
A \\
B
\end{array}
\begin{array}{cc}
A & B \\
\left[\begin{array}{cc}
1 - k^{(1)} & k^{(1)} \\
0 & 1
\end{array}\right]
\end{array}
$$

The first entry is the probability of no change, and the second entry in the first row is the probability of the A → B transition. The probability for the reverse reaction (B → A) is taken to be equal to zero. If the probability for the reverse reaction is different from zero, the transition matrix would become

$$
\begin{array}{c}
 \\
A \\
B
\end{array}
\begin{array}{cc}
A & B \\
\left[\begin{array}{cc}
1 - k_1^{(1)} & k_1^{(1)} \\
k_{-1}^{(1)} & 1 - k_{-1}^{(1)}
\end{array}\right]
\end{array}
$$

where the index in parentheses denotes the order.

Similarly, for the second-order process

$$A + B \xrightarrow{k^{(2)}} C$$

without considering the back-reaction, the transition matrix is

$$
\begin{array}{c}
 \\
A \\
B \\
C
\end{array}
\begin{array}{ccc}
A & B & C \\
\left[\begin{array}{ccc}
1 - k^{(2)}X_B & 0 & k^{(2)}X_B \\
0 & 1 - k^{(2)}X_A & k^{(2)}X_A \\
0 & 0 & 1
\end{array}\right]
\end{array}
$$

With a first-order back-reaction, the matrix is

$$
\begin{array}{c}
 \\
A \\
B \\
C
\end{array}
\begin{array}{ccc}
A & B & C \\
\left[\begin{array}{ccc}
1 - k^{(2)}X_B & 0 & k^{(2)}X_B \\
0 & 1 - k^{(2)}X_A & k^{(2)}X_A \\
k_{-1}^{(1)} & k_{-1}^{(1)} & 1 - k_{-1}^{(1)}
\end{array}\right]
\end{array}
$$

Stochastic matrices can also be written for more complex cases. As an example, consider the mechanism

$$A \xrightarrow{k^{(1)}} D$$

$$A + B \xrightarrow{k^{(2)}} C$$

for which the transition matrix is

$$
\begin{array}{c}
\\
A \\
B \\
C \\
D
\end{array}
\begin{bmatrix}
\begin{array}{cccc}
A & B & C & D \\
1 - k^{(1)} - k^{(2)}X_B & 0 & k^{(2)}X_B & k^{(1)} \\
0 & 1 - k^{(2)}X_A & k^{(2)}X_A & 0 \\
0 & 0 & 1 & 0 \\
0 & 0 & 0 & 1
\end{array}
\end{bmatrix}
$$

This matrix is an alternative way of presenting the complete system of differential equations describing the time evolution of the reacting mixture (Chapter 2). It contains all of the information on where each species in the reacting system comes from and on the molecularity of the corresponding elementary reactions.

3.3. TEMPERATURE DEPENDENCE: THE ARRHENIUS EQUATION

Elementary Reactions

For an elementary reaction of any order, the rate constant generally obeys the empirical Arrhenius equation

$$k_{obs} = A \exp\left(\frac{-E_a}{RT}\right) \qquad 3.13$$

where k_{obs} is the rate constant at absolute temperature T, E_a is the *activation energy* (or more specifically, the Arrhenius activation energy), and A is the *preexponential factor*. Sometimes A is also called the *frequency factor*, although it has units of frequency, s^{-1}, only if k_{obs} refers to a first-order reaction. The activation parameters E_a and A can be evaluated by appropriately fitting equation 3.13, for example, by plotting the experimentally determined ln k_{obs} versus $1/T$.

It must be emphasized that k_{obs} in the Arrhenius equation strictly refers to an elementary reaction, not to an overall complex or to a pseudo-first-order rate constant.

Equation 3.13 essentially amounts to a definition of the Arrhenius activation energy:

$$E_a = \frac{-R\partial(\ln k_{obs})}{\partial(1/T)} \qquad 3.14$$

Composite Rate Constants

Frequently k_{obs} is a composite of some type containing more than one rate constant and equilibrium constant. Whether or not a plot of $\ln k_{obs}$ versus $1/T$ will give a straight line will depend on the actual type of composite involved and perhaps on how wide a range of temperatures is covered. The departure from linearity may provide some clues regarding the mechanism, although in practice such departures are often small and difficult to diagnose. Linearity is not necessarily an indication of a single-step process. However, if there is real curvature in the $\ln k_{obs}$ versus $1/T$ plot, one should look for a complex mechanism.

A few examples will help to illustrate these points. The discussion will be based on the stochastic considerations introduced in the previous section.

EXAMPLE I

Consider the decomposition of A by two parallel first-order elementary pathways:

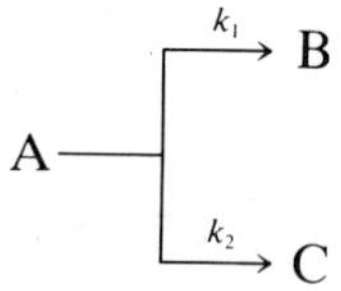

The rate constants represent the "instantaneous" probabilities for the corresponding processes to occur. The weighted average of the activation energies at each temperature is

$$E_{av} = \frac{E_1 k_1 + E_2 k_2}{k_1 + k_2}$$

Since E_1 and E_2 are assumed to be constant, the dependence of E_{av} on temperature is incorporated into k_1 and k_2:

1. If $k_1 \simeq k_2$ throughout the temperature range investigated, then

$$E_{av} \simeq \frac{E_1 + E_2}{2}$$

In this case there is no curvature in the Arrhenius plot.
2. If $k_1 \gg k_2$, then

$$E_{av} \simeq E_1$$

Again, there is no curvature.

3. If $k_1 \neq k_2$ but their values do not differ appreciably and $E_1 > E_2$, $A_1 \simeq A_2$, then

$$E_{av} = \frac{E_1 \exp(-E_1/RT) + E_2 \exp(-E_2/RT)}{\exp(-E_1/RT) + \exp(-E_2/RT)}$$

Since $E_1 > E_2$, with increasing temperature $\exp(-E_1/RT)$ will increase faster than $\exp(-E_2/RT)$. This means that with increasing temperature the larger of the two activations (E_1) will weigh more in the average, and the *observed* activation energy will increase.   □

EXAMPLE II

Consider the mechanism

$$A \xrightarrow{k_1} C$$

$$A + B \xrightarrow{k_2} D$$

The observed activation energy is given by

$$E_{av} = \frac{k_1 E_1 + k_2 E_2 X_B}{k_1 + k_2 X_B}$$

Again, if one of the terms dominates, there is no change of E_{av} with temperature. If the rates of the two paths are comparable and $E_1 > E_2$, the first may become dominant at higher temperatures. However, empirical conditions with excess B (large X_B) may smooth out the nonlinearity.   □

3.4. COLLISION THEORY

Outline of the Theory

According to collision theory, the rate constant for a bimolecular reaction in the gas phase can be expressed as

$$k = \sigma(v)v \qquad\qquad 3.15$$

where v is the relative velocity of the colliding molecules and $\sigma(v)$ the *collision cross section*.

If the molecules A and B are assumed to be like hard spheres interacting only when they are in contact, $\sigma(v)$ is given simply by

$$\sigma(v) = \pi(r_A + r_B)^2$$

where r_A and r_B are the radii of the spheres. By substituting this relationship into equation 3.15 and averaging over all velocities (assuming a Boltzmann distribution), the rate constant k_{hs} for the hard sphere model is obtained:

$$k_{hs} = \left(\frac{8\pi \mathbf{k}T}{\mu}\right)^{1/2} (r_A + r_B)^2 \qquad 3.16$$

where $\mathbf{k}$ is the Boltzmann constant, T is the absolute temperature, and μ is the reduced mass $= m_A m_B/(m_A + m_B)$.

This hard-sphere model implies that each time two molecules come into contact, they react, which is, of course, chemically unrealistic. A more realistic version of the hard-sphere model would again assume that there is no interaction at distances larger than $r_A + r_B$ but that for the cross section at a distance equal to $r_A + r_B$ there is a potential barrier E_0. If the energy E of the colliding molecules is lower than this barrier, the cross section is zero. If it is higher than the barrier, the cross section is proportional to the difference $E - E_0$:

$$\sigma(E) = 0 \qquad\qquad \text{for } E < E_0$$

$$\sigma(E) = \frac{(\text{constant}) \ (E - E_0)}{E_0} \qquad \text{for } E \geqslant E_0$$

The bimolecular rate constant then becomes

$$k'_{hs} = \left(\frac{8\pi \mathbf{k}T}{\mu}\right)^{1/2} (r_A + r_B)^2 \exp\left(\frac{-E_0}{RT}\right) \qquad 3.17$$

which differs from equation 3.16 by the factor $\exp(-E_0/RT)$, implying that only sufficiently energetic collisions are reactive.

In addition, the colliding molecules may be hard but not necessarily spherical. The rate will then depend also on orientation. A more appropriate expression in this case is

$$k''_{hs} = p\left(\frac{8\pi \mathbf{k}T}{\mu}\right)^{1/2} (r_A + r_B)^2 \exp\left(\frac{-E_0}{RT}\right) \qquad 3.18$$

where the steric factor $p \ll 1$.

A Metaphor

First we said that molecules collide like hard steel balls, then that they rather look like plastic balls, and finally that, after all, they are not balls. Collision theory can be improved further by:

Using more complicated potentials to describe the interactions.
Including vibrational and rotational levels into the energy averaging (not only the translational levels).

Some Hints of Practical Value

Under normal conditions only one out of 10^{10}–10^{11} collisions leads to chemical change. Activation of molecules in the gas phase leading to a monomolecular decomposition also is usually the result of collisions.

Triple collisions are more difficult to deal with theoretically. In chemical kinetics the difficulty in solving the so-called three-body problem is circumvented by assuming that the ratio of the probability for a triple collision to the probability for a double collision is equal to the ratio of the molecular diameter to the mean free path. On the basis of this assumption it is estimated that at most about one triple collision occurs for every 1000 double collisions. The probability for quadruple- or higher-order collisions is virtually negligible.

3.5. BASIC CONCEPTS IN TRANSITION STATE THEORY

A "Topographical" Description

The basic assumption in transition state theory is that the reactants are in equilibrium with the *activated complex*. The curve in Figure 3.8 represents the energetically least expensive path for the reaction. The configuration at the top of the curve, $[AB]^{\ddagger}$, represents an activated complex.

The reacting molecules are generally described by N independent variables,

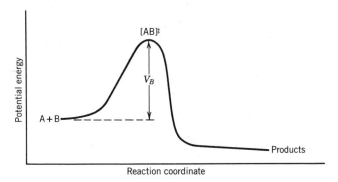

Figure 3.8. Transition state theory postulates that the reaction proceeds through formation of an activated complex $[AB]^{\ddagger}$, which is in equilibrium with the reactants. V_B is the potential energy barrier.

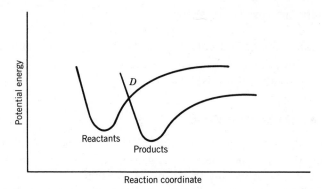

Figure 3.9. Intersection of the (two-dimensional) energy "surfaces" of reactants and products.

that is, the variables required for describing all degrees of freedom. The energy of the system is a function of all these variables, and it can be "represented" by an N-dimensional "surface" in an $(N + 1)$-dimensional space.

The energy "surfaces" corresponding to the reactant and product molecules intersect each other. At the intersection, which is $(N - 1)$-dimensional, reactants and products have the same energy. Consequently, the transformation of reactant molecules into product molecules becomes possible.

A two-dimensional reaction "surface" is represented in Figure 3.9. The intersection D is one-dimensional. Actually, the "space" is multidimensional, so Figure 3.9 is in fact a "profile."

The $(N - 1)$-dimensional intersection is not flat; it has "valleys" and "hills," and there are an infinite number of paths for crossing this "land." A path corresponding to the least energy (easiest ascent and descent) is preferred.

The maximum height in this path (the saddle point) is the *activation energy*. The *activated complex* is the configuration at the top of the saddle.

A More Accurate Picture

Figure 3.10 is a picture of the potential energy surface of a simple system: three hydrogen atoms lying on the same line, at varying distances from each other.

The reaction is represented by the equation

$$H_A + H_B\text{—}H_C \longrightarrow H_A\text{—}H_B + H_C$$

$$\text{Reactants} \qquad\qquad \text{Products}$$

Some Dialectics

The topographical description of the activated complex just given is not enough. Dialectically the concept of the activated complex must be distinguished from

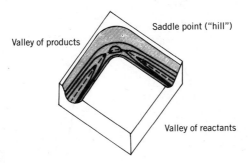

Figure 3.10. Perspective drawing of the potential energy surface of the H + H$_2$ exchange reaction. Dashed line: the preferable path. Transition state: the configuration at the saddle point. The energy corresponding to this point is called the potential energy barrier (V_B).

other concepts resembling it. Two such concepts naturally come to mind:

An ordinary complex (i.e., a coordination compound)
An intermediate

The activated complex resembles an ordinary complex in that both can decompose or form from ions or molecules that are stable under ordinary conditions. On the other hand, a coordination compound usually contains an ion or atom of metallic character; this is irrelevant for an activated complex. The basic difference, however, is that an ordinary complex is a more or less stable species, corresponding to a minimum in the potential energy surface. The activated complex corresponds to a local maximum.

This basic difference also clearly distinguishes the activated complex from an intermediate. The intermediate may be short-lived, but it corresponds to a minimum in energy, no matter how shallow. The activated complex is always in transition toward product formation or a return to the original molecules; in fact, these are the only reactions in which it participates. In contrast, an intermediate needs activation to react further (Fig. 3.11), and at least in principle it can participate in a variety of reactions with other reactants, not just the reaction under consideration. In principle too, the intermediate can be isolated, while the activated complex cannot. Isolation of an activated complex has no meaning. An intermediate may have the same composition as one or more of the activated complexes, but it is clear that there are distinct differences in energy and geometry. In Figure 3.11 the intermediate [AB] has an energy and a geometry close to those of the activated complexes $[AB]_1^{\ddagger}$ and $[AB]_2^{\ddagger}$, but not exactly the same.

Comments

The basic postulate of transition state theory is that equilibrium is established between the activated complex and the reactants—not the products, unless

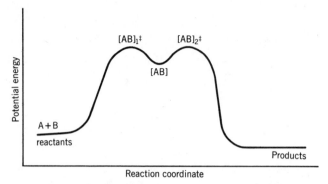

Figure 3.11. The intermediate [AB] returns to the original reactants, A + B, or is transformed into products through the activated complexes $[AB]_1^{\ddagger}$ and $[AB]_2^{\ddagger}$, respectively.

the reverse reaction is also considered. The energy of activation is assumed to be large compared to the thermal energy ($\mathbf{k}T$) of the molecules. Consequently, the population of the activated complex is small compared to the number of reactant molecules.

The assumption of a reactant–activated complex equilibrium also forces us to abandon any attempts at a *detailed* description of the molecular happenings (e.g., trajectories of pairs of molecules). There are still "microscopic" features that can be outlined, but the picture is necessarily blurred.

The rate law gives no information about the assumed equilibrium for the activated complex.

3.6. THE EYRING EQUATION

The Equation

For a second-order process, transition state theory can be summarized as follows:

$$A + B \underset{}{\overset{K^{\ddagger}}{\rightleftharpoons}} [AB]^{\ddagger}$$

$$[AB]^{\ddagger} \overset{\nu}{\longrightarrow} products$$

where $K^{\ddagger}$ is the constant for the postulated equilibrium between reactants and activated complex, and ν is the frequency of passage over the energy barrier. The rate constant is given by

$$k = \nu K^{\ddagger}$$

In an actual reaction mixture in thermal equilibrium, there is a statistical distribution of configurations, and some (very few) correspond to the activated complex. Their number is estimated by statistical mechanical methods. This is how the *Eyring equation*

$$k = \frac{\kappa RT}{Nh} \exp\left(\frac{\Delta S^{\ddagger}}{R}\right) \exp\left(\frac{-\Delta H^{\ddagger}}{RT}\right) \qquad 3.19$$

is obtained, where R is the gas constant; T, the absolute temperature; N, Avogadro's number; $R/N = \mathbf{k}$, the Boltzmann constant; h, Planck's constant; $\Delta S^{\ddagger}$, the entropy of activation; $\Delta H^{\ddagger}$, the enthalpy of activation; and κ, the transmission coefficient, which is often taken as unity.

The quantities $\Delta H^{\ddagger}$ and $\Delta S^{\ddagger}$ are the differences in enthalpy and entropy, respectively, between the transition state and the reactants, all in a standard thermodynamic state. In other words, $\Delta H^{\ddagger}$ and $\Delta S^{\ddagger}$ are the standard enthalpy and entropy changes for the bimolecular reaction

$$A + B \longrightarrow [AB]^{\ddagger}$$

Equation 3.19 can also be written in the form

$$k = \frac{\kappa RT}{Nh} \exp\left(\frac{-\Delta G^{\ddagger}}{RT}\right) \qquad 3.20$$

where $\Delta G^{\ddagger} = \Delta H^{\ddagger} - T\Delta S^{\ddagger}$ is the standard Gibbs free energy of activation. This thermodynamic potential is clearly quite different from the potential energy of a pair of interacting molecules such as that shown in Figure 3.8. It would therefore be misleading to directly associate a profile like that of Figure 3.8 with $\Delta G^{\ddagger}$ or to imply that $\Delta G^{\ddagger}$ changes along the reaction coordinate.

Units

It can be readily shown that equations 3.19 and 3.20 are also valid for mono-molecular reactions, in which case they are also dimensionally correct. For a bimolecular reaction, the dimensions of k are $M^{-1}s^{-1}$, whereas the dimensions of the right-hand sides of equations 3.19 and 3.20 are always s^{-1}. A more complete version of the equation is obtained by introducing the equilibrium constant between the activated complex and the reactants in their standard states:

$$K_{ss}^{\ddagger} = \frac{[AB]_{ss}^{\ddagger}}{[A]_{ss}[B]_{ss}}$$

Equation 3.19 then becomes

$$k = \left(\frac{\kappa RT}{Nh}\right) K_{ss}^{\ddagger} \exp\left(\frac{\Delta S^{\ddagger}}{R}\right) \exp\left(\frac{-\Delta H^{\ddagger}}{RT}\right) \qquad 3.21$$

The numerical value of k does not change, because the numerical value of $K_{ss}^{\ddagger}$ is unity, but the dimensions are now correct, since $K_{ss}^{\ddagger}$ has dimensions of $(\text{concentration})^{1-n}$, with n being the molecularity of the reaction.

3.7. APPLICATION OF TRANSITION STATE THEORY TO NONELEMENTARY REACTIONS

The Activated Complex in Nonelementary Reactions

The activated complex in transition state theory refers to an elementary reaction. For a complex reaction there are as many activated complexes as there are elementary reactions in the mechanism. However, if one of the steps determines the overall rate, it is the activated complex of this step that is of interest. If two or more steps have comparable rates, it is necessary to consider them all.

It follows that clarification of the mechanism is a prerequisite for applying transition state theory to nonelementary reactions.

In order to illustrate this point, consider the following oxidation of $Co(CN)_5^{3-}$ by hydrogen peroxide:

$$2Co(CN)_5^{3-} + H_2O_2 \longrightarrow 2Co(CN)_5OH^{3-}$$

The mechanism proposed is

$$Co(CN)_5^{3-} + H_2O_2 \longrightarrow Co(CN)_5OH^{3-} + OH \qquad \text{(slow)}$$

$$Co(CN)_5^{3-} + OH \longrightarrow Co(CN)_5OH^{3-} \qquad \text{(fast)}$$

Here it is the activated complex of the first step that is of interest. In contrast, for the mechanism

$$Co(NH_3)_5NO_3^{2+} + H_2O \longrightarrow Co(NH_3)_5OH_2^{3+} + NO_3^-$$

$$Co(NH_3)_5OH_2^{3+} + NCS^- \longrightarrow Co(NH_3)_5NCS^{2+} + H_2O$$

the two steps have, in the appropriate concentration range, comparable rates; both activated complexes must be considered.

The Composition of the Activated Complex

If the rate law is of the form

$$\text{Rate} = k_1[A]^m[B]^n[C]^l + k_2[A]^{m'}[B]^{n'}[C]^{l'} + \cdots$$

where m, n, l, etc. are positive or negative integers, each term in the summation represents the composition of an activated complex. Reagents in excess (e.g., the solvent) may or may not be included in the composition.

For example, the rate law for the oxidation of iodide ion by hydrogen peroxide contains an acid-independent term and an acid-catalyzed term. In the presence of other catalysts, additional terms would be included

$$\text{Rate} = k_1[\text{H}_2\text{O}_2][\text{I}^-] + k_2[\text{H}_2\text{O}_2][\text{I}^-][\text{H}^+] + \cdots$$

The two terms shown correspond to two activated complexes having the composition $[\text{H}_2\text{O}_2\text{I}^-]$ and $[\text{H}_3\text{O}_2\text{I}]$, respectively. The corresponding processes can be represented as follows:

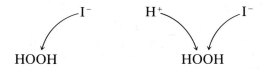

Ambiguities Regarding the Activated Complex

The rate law may specify the composition of various activated complexes, but the same composition is often consistent with more than one elementary reaction or combination of such reactions. The question of the mechanistic ambiguities has already been examined (Chapter 2), but the problem of arriving at a mechanism from kinetic data is of fundamental importance, and it is worth reviewing again.

Consider a specific example: the reduction of vanadium(III) by chromium(II). The empirical rate law is

$$\text{Rate} = \frac{a + b/[\text{H}^+]}{c + [\text{H}^+]}[\text{V(III)}][\text{Cr(II)}]$$

The limiting form of this law at high acid ($[\text{H}^+] \gg c$, $b/[\text{H}^+] \ll a$) is

$$(\text{Rate})_{\text{lim}} = a\frac{[\text{V(III)}][\text{Cr(II)}]}{[\text{H}^+]}$$

The right-hand side is now a simple product of concentrations, without sums in the numerator or the denominator, and can be attributed to a single activated complex of the form

$$[\text{V(OH)Cr}]^{4+} \qquad\qquad 3.22$$

To this composition we can add or subtract water molecules.

The question then arises: Is this hydrolyzed dimer formed from hydrolyzed monomers (solvent molecules coordinated to the metal ion are not shown)

$$V(III) + H_2O \rightleftharpoons V(OH)^{2+} + H^+$$

or

$$Cr(II) + H_2O \rightleftharpoons Cr(OH)^+ + H^+$$

followed by

$$V(OH)^{2+} + Cr(II) \longrightarrow [V(OH)Cr]^{4+}$$

or

$$Cr(OH)^+ + V(III) \longrightarrow [V(OH)Cr]^{4+}$$

or is formed from nonhydrolyzed monomers that form an intermediate that then hydrolyzes?

$$V(III) + Cr(II) \rightleftharpoons [VCr]^{5+}$$

$$[VCr]^{5+} + H_2O \longrightarrow [V(OH)Cr]^{4+} + H^+$$

In the first case, the formation will depend inversely on acid, and the composition 3.22 may also correspond to the composition of a real intermediate.

In the second case, the formation of the intermediate will be acid-independent. Subsequent hydrolysis is, of course, acid-dependent.

If the investigation of the reaction were confined to highly acidic solutions, only simple inverse acid dependence would be observed, and we would perhaps be tempted to accept the simplest explanation—formation from hydrolyzed species. Moreover, if hydrolysis were important at high acid concentrations, it would seem reasonable to assume that hydrolysis is also important at low acid and to expect similar behavior. In fact, at low acid, the hydrogen ion concentration dependence becomes more complicated and the mechanism is clearly more complex than the high-acid results would lead us to believe.

Additional complications arise if we also consider possible parallel paths for the decomposition of $[V(OH)Cr]^{4+}$, for example, an acid-independent path and an acid-catalyzed path. Generally, a number of different mechanisms can be written on the basis of the following considerations:

With or without intermediate(s)

By assuming steady state or fast preequilibria

By assuming intermediate(s) formation from hydrolyzed species or hydrolysis after dimerization

By considering acid-independent and acid-catalyzed paths of decomposition

This corresponds to quite a large number of alternatives, even for just one activated complex of composition $[V(OH)Cr]^{4+}$!

The choice *can* be narrowed down, but on nonkinetic grounds. Yet even after considerable speculation, the ambiguity is scarcely removed.

3.8. A QUALITATIVE COMPARISON OF DIFFERENT ACTIVATION ENERGIES

Molecular and Statistical Activation Energies

As was noted earlier, the potential energy barrier V_B is the point of highest potential energy on the minimum energy pathway from reactant to product molecules. Barrier heights can vary from zero to ~400 kJ mol^{-1}. The quantity ΔE_0 obtained from V_B after correcting for zero-point energy is important in determining isotope effects. A schematic representation of different kinds of activation energies is given in Figure 3.12.

The quantities V_B and ΔE_0 are true molecular properties. The other activation energies in Figure 3.12, namely, $\Delta E_T^{\ddagger}$, $\Delta H_T^{\ddagger}$, and E_a, are statistical averages. Their numerical values may be larger or smaller than V_B. The value of $\Delta E_T^{\ddagger}$ differs from ΔE_0 by the thermal energy, $\int C_v^{\ddagger} \, dT - \int C_v \, dT$. For a bimolecular reaction, this difference is likely to be negative ($\int C_v^{\ddagger} \, dT < \int C_v \, dT$), because in the activated complex some translational degrees of freedom are lost, which means that the heat capacity decreases ($C_v^{\ddagger} < C_v$).

For an ordinary reaction, the enthalpy is obtained from the internal energy by adding the factor PV. Accordingly, the change in PV between the activated complex and the reactants connects the activation enthalpy with the internal energy of activation. For reactions in the condensed phase, the difference is small.

The value of E_a in Figure 3.12 is the experimentally determined Arrhenius activation energy. The upper tip does not coincide with that of $\Delta E_T^{\ddagger}$. The factors contributing to that difference are not well understood. It suffices to say that some of these factors tend to make E_a larger than $\Delta E_T^{\ddagger}$, and some operate in the opposite direction. At any rate, it should not be forgotten that the Arrhenius activation energy is the fact; the other energies are theoretical, and it is not always simple to match the two exactly. Estimates of $\Delta H^{\ddagger}$ and $\Delta S^{\ddagger}$ can, of course, be made using the Eyring equation, but this is not an independent way of testing this equation. In (simple) cases in which $\Delta H_T^{\ddagger}$ can be estimated independently, there may be some discrepancy with the experimental values.

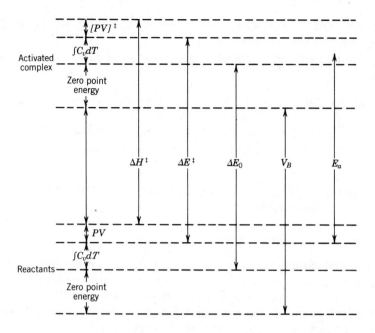

Figure 3.12. Schematic representation of a qualitative comparison between different kinds of activation energies.

In solution chemistry, $\Delta H_T^{\ddagger}$ is often simply written as $\Delta H^{\ddagger}$ and is estimated from the empirical E_a using the equation

$$\Delta H^{\ddagger} = E_a - RT$$

Practical Considerations

Generally, the value of E_a is obtained by plotting $\ln k_{obs}$ as a function of $1/T$. On the other hand, $\Delta H^{\ddagger}$ is obtained by plotting $\ln (k_{obs}/T)$ as of function of $1/T$. Alternatively, the appropriate values of E_a and A or of $\Delta H^{\ddagger}$ and $\Delta S^{\ddagger}$ can be obtained by numerical analysis.

The Gibbs free energy of activation for two opposing reactions can be written

$$\ln k_1 = \frac{-\Delta G_1^{\ddagger}}{RT} + \text{constant} \qquad 3.23$$

$$\ln k_{-1} = \frac{-\Delta G_{-1}^{\ddagger}}{RT} + \text{constant} \qquad 3.24$$

At equilibrium,

$$\ln K = \ln \left(\frac{k_1}{k_{-1}}\right) = \frac{-\Delta G^{\circ}}{RT} \qquad 3.25$$

By combining 3.23, 3.24, and 3.25 we obtain

$$\Delta G_1^{\ddagger} - \Delta G_{-1}^{\ddagger} = \Delta G^{\circ}$$

3.9. PRESSURE EFFECTS AND THE VOLUME OF ACTIVATION

Definitions and Concepts

The effect of pressure, P, on the rate constant of an elementary reaction at constant temperature is given by the equation

$$\left(\frac{\partial \ln k}{\partial P}\right)_T = \frac{-\Delta V^{\ddagger}}{RT}$$

The volume of activation $\Delta V^{\ddagger}$ is the difference in partial molar volume between the transition state and the reactants.

For $\Delta V^{\ddagger} > 0$, k decreases with pressure.
For $\Delta V^{\ddagger} < 0$, k increases with pressure.

In practice, pressures up to 2 or 3 kbar are used. Plots of $\ln k$ as a function of P are often nonlinear, which is interpreted to mean that $\Delta V^{\ddagger}$ is pressure-dependent. Linear $\ln k$ versus P plots imply independence of $\Delta V^{\ddagger}$ from pressure, and the value of the volume of activation can be obtained directly from the slope. For nonlinear plots, empirical relations are used, such as

$$\ln k_p = \ln k_0 - \frac{P\,\Delta V_0^{\ddagger}}{RT} + \Delta\beta^* \frac{P^2}{2RT}$$

where k_p and k_0 are the rate constants at pressure P and zero, respectively, $\Delta V_0^{\ddagger}$ is the volume of activation at zero pressure, and $\Delta\beta^*$ is the pressure-independent compressibility of activation.

By convention, in nonlinear cases, only the volume of activation at zero pressure ($\Delta V_0^{\ddagger}$) is quoted.

Interpretation of the Volume of Activation

The experimentally determined volume of activation can be thought of as arising from an intrinsic part and a solvation part:

$$\Delta V_{obs}^{\ddagger} = \Delta V_{intr}^{\ddagger} + \Delta V_{solv}^{\ddagger}$$

TABLE 3.1. Kinetic Data for the Fading of Bromophenol Blue as a Function of Pressure

Pressure atm	1.0	270	540	820	1090
Rate constant k, $M^{-1}\,s^{-1}$	9.58×10^{-4}	11.1×10^{-4}	13.1×10^{-4}	15.3×10^{-4}	17.9×10^{-4}

Thus, $\Delta V_{intr}^{\ddagger}$ reflects changes in size when going from the reactants to the activated complex, and in principle it provides information about the molecular (intrinsic) mechanism. On the other hand, $\Delta V_{solv}^{\ddagger}$ reflects reorganization of the solvent.

The contribution of each of these parts to the observed volume of activation is not always easy to determine. However, if in the transformation

$$\text{Reactants} \longrightarrow \text{transition state}$$

there is no change in formal charge, $\Delta V_{solv}^{\ddagger}$ is expected to be small and $\Delta V_{obs}^{\ddagger} \simeq \Delta V_{intr}^{\ddagger}$. When there are changes in charge, $\Delta V_{solv}^{\ddagger}$ becomes important and cannot be neglected.

An Example

The fading of the indicator bromophenol blue[2] in alkaline solution is a second-order process between hydroxide ion and the quinol form of the dye.

$$\text{Quinol form (blue)} + \text{OH}^- \longrightarrow \text{carbinol form (colorless)} + H_2O$$

In a series of experiments at 25°C, second-order rate constants were determined as a function of the hydrostatic pressure. The observed rates are given in Table 3.1.

In order to compute the volume of activation, $\Delta V^{\ddagger}$, the data are plotted as in Figure 3.13. The transition state (i.e., the activated complex) has a smaller volume than the reactants.

---

[2]Bromophenol blue (3',3",5',5"-tetrabromophenolsulfonephthalein) in the water-soluble form

has the formula

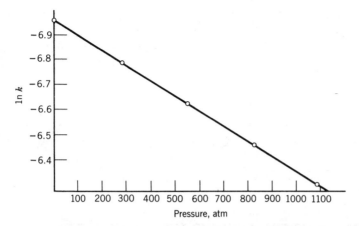

Figure 3.13. Dependence on pressure of the second-order rate constant of the reaction between the quinol form of bromophenol blue and hydroxide ion. (The slope corresponds to 0.000578 atm^{-1}.)

Thus, in summary,

$$\frac{-\Delta V^{\ddagger}}{RT} = 0.000578 \text{ atm}^{-1}$$

$$\Delta V^{\ddagger} = -0.01413 \text{ L mol}^{-1} = -14.13 \text{ mL mol}^{-1}$$

Interpretation of these effects in terms of possible mechanisms is discussed in Chapter 6. ☐

3.10. THE "STRUCTURE" OF THE ACTIVATED COMPLEX

If for a chemical compound information is available (*a*) about its energy, (*b*) about its composition, and (*c*) about its structure, a great deal is known about its identity and reactivity. This general premise also applies to the activated complex. We have talked at length about the energy of the activated complex and about its composition. Now we will deal with its structure.

The structure of an activated complex is generally studied in order to correlate it with behavior (reactivity). The behavior is complicated, and its variety practically unlimited. The structure is much easier to grasp, classify, and use for explaining behavior, or even for predicting behavior. "Structureless" matter (e.g., the soil, the environment) is more difficult to describe concisely and rationally.

By definition, the activated complex is transient, undergoing drastic changes in interatomic distances and angles. Hence, the best methods for structure determination, notably x rays and spectroscopy, cannot be used. Clearly, any structural information is necessarily indirect.

The structural elements for stable compounds are interatomic distances and angles. In the activated complex the changing distances and angles are obviously not "static" structural features. Yet, even then, one can speculate about relative positions, which atoms are closer to each other and in what direction, and whether these positions are taken only part of the time or throughout the activation process. In addition, other interatomic distances or angles may change within only a limited range, as in stable species.

In conclusion, in the case of the activated complex, structure is still relevant, but it is now understood to be dynamic, with some of the geometrical elements missing.

3.11. METHODS FOR STUDYING THE STRUCTURE OF ACTIVATED COMPLEXES

Information on the structure of the activated complex is obtained by measuring the effect of various parameters (e.g., temperature), by comparing to other processes, and/or by simulating the activated complex to stable compounds or to intermediates.

(a) *Parametric methods.* A study of the effect of *parameters* such as temperature and pressure. Concentration, too, can be conveniently considered a parameter. The quantities obtained by varying these parameters (composition of the activated complex, energy of activation, volume of activation, and the like) contain considerable structural information.

(b) *Competition methods.* The process of interest competes with or is compared to better-known processes. The results are interpreted on a structural basis. Dependence on various acidity functions can conveniently be included in this category.

(c) *Simulation methods.* Intermediate or stable compounds modeling the activated complex are isolated or synthesized under special, better controlled conditions and are characterized—structurally and otherwise—as thoroughly as possible. Competition methods involve comparisons of processes. Simulation methods involve comparison to other (stable or metastable) structures and modeling.

The best tactic is to use combinations of all these methods. A few examples may help illustrate their application.

EXAMPLE I

The rate law for the reaction

$$BrO_3^- + 5Br^- + 6H^+ \longrightarrow 3Br_2 + 3H_2O$$

is given by the equation

$$\text{Rate} = k[BrO_3^-][Br^-][H^+]^2$$

The coefficients in the stoichiometric equation are large. This necessarily means a complex reaction, simply because multiple encounters are highly unlikely. Yet the rate law is quite simple. This should probably be interpreted to mean that one of the steps dominates the overall rate.

From the rate law it may be concluded that the simplest empirical formula for the composition of the activated complex is $[BrO_3^- \cdot Br^- \cdot H^+ \cdot H^+]$ or $[H_2Br_2O_3]$. Possible structures are the trigonal pyramidal and planar structures illustrated here. In the former, there is no electrostatic obstacle, whereas in the latter, more repulsion is expected. In the trigonal pyramidal structure a Br—Br bond has already been formed.

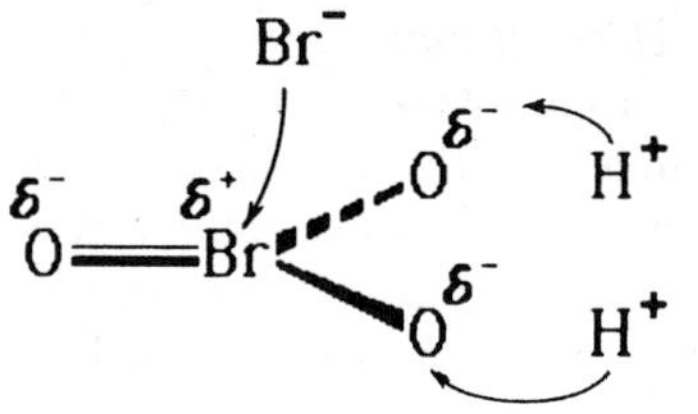

Trigonal pyramidal
structure

(a)

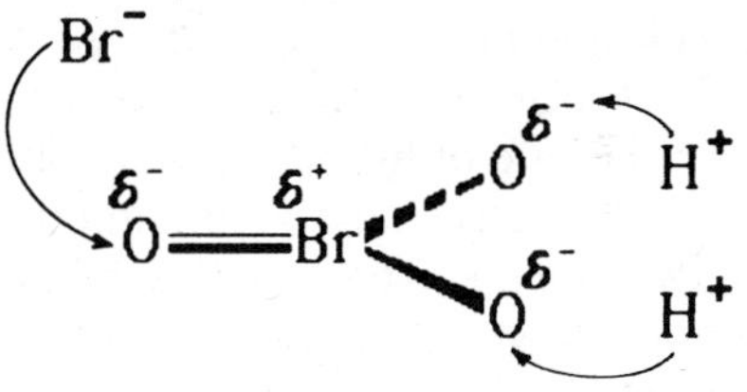

Planar structure

(b)

Removal of a water molecule (solvent) from the two structures yields, respectively,

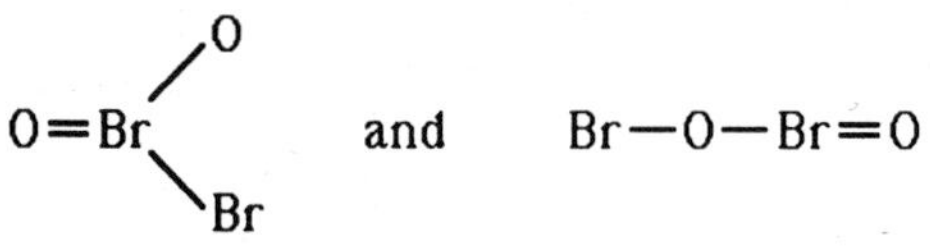

and $Br-O-Br=O$

The first of these structures resembles BrO_3^-. Stable structures like the second one are not known.   □

EXAMPLE II

In the exchange reaction

$$Mn(CN)_6^{3-} + {}^*CN^- \rightleftharpoons Mn(CN)_5({}^*CN)^{3-} + CN^-$$

the entropy of activation is found to be -4.4 eu. This negative value is interpreted to mean that there is more order in the activated complex than in the reactants, which is most consistent with a 7-coordinated activated complex.   □

EXAMPLE III

In the decomposition of dinitrogen tetroxide in carbon tetrachloride solutions, the volume of activation was found equal to about 10 cm³ mol⁻¹. The difference in molar volume between products and reactants is 20 cm³ mol⁻¹. It has been argued that the increase in molar volume is due to the increasing length and eventually breaking of the N—N bond, and that the transition state is midway along in this path □

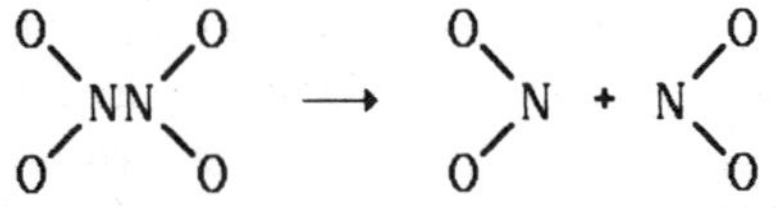

EXAMPLE IV

Outer- and inner-sphere activated complexes in electron-transfer reactions. If the rate for electron transfer between metal ion complexes is much faster than the rate of ligand substitution in these complexes, it can be argued that the electron is transferred without ligand removal, via an *outer-sphere activated complex.* The complexes $Fe(CN)_6^{4-}$ and $Fe(CN)_6^{3-}$ are inert to substitution. Replacement of a cyanide ion ligand by cyanide ion from the bulk of the solution or by a solvent molecule is slow relative to the rate of electron exchange:

Electron transfer must then necessarily take place with both coordination spheres intact. Schematically,

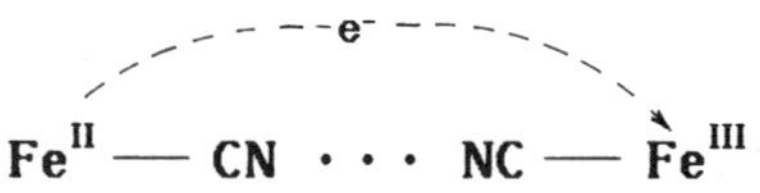

If the rate of ligand substitution for at least one of the reacting partners is fast compared to electron transfer, the reaction *can* proceed by an *inner-sphere activated complex* (*bridged activated complex*). Water exchange in $Cr(H_2O)_6^{2+}$ is fast ($t_{1/2} \simeq 10^{-9}$ s). Usually, in reactions with $Cr(H_2O)_6^{2+}$ the electron is donated *after* removal of one water molecule and formation of the $Cr(H_2O)_5^{2+}$ fragment. In Taube's first classic example,

$$Co^{III}(NH_3)_5X^{2+} + Cr^{2+}(aq) \longrightarrow CrX^{2+}(aq) + Co^{2+} + 5NH_3$$

the activated complex has the bridged structure

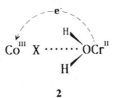

1

The lability of $Cr(H_2O)_6^{2+}$ is not proof for the bridged structure. It is a necessary condition but not a sufficient one. Substitution may be fast, but electron transfer could still take place by an outer-sphere path:

$$Co^{III} \quad X \cdots\cdots OCr^{II}$$

2

Fortunately, in this case, interconversion for the products is slow; water exchange in $Cr(H_2O)_6^{3+}$ has $t_{1/2} \simeq 10^6$ s, that is, it is 10^{15} times slower than water exchange in $Cr(H_2O)_6^{2+}$. Thus the choice of structure **1** over **2** is based on the composition of the Cr^{III} products. Structure **1** leads to Cr^{III}—X, whereas structure **2** leads to Cr^{III}—OH_2. ☐

EXAMPLE V

Isotopic labeling. In Example IV, the conclusions about the activated complex were based on the position of X before and after the reaction. In this sense, the method can be characterized as "chemical labeling." By analogy, *isotopic labeling* could also supply useful information.

Isotopic labeling (using a *tracer*) is used in the determination of the rates of reactions in which there is no net chemical change. An example would be the exchange between $Fe(CN)_6^{4-}$ and $Fe(CN)_6^{3-}$ discussed above.

Another use of isotopic labeling is for obtaining information about the site of attack. By labeling chlorate ion with oxygen-18 it can be shown that its reaction with sulfite ion follows the path

$$SO_3^{2-} + {}^{18}OClO_2^- + 2H^+ \longrightarrow$$

$$[O_2S^{18}OClO_2^- \cdot H_2O] \longrightarrow O_2S^{18}O + ClO_2^- + H_2O$$

The activated complex could be more explicitly represented by a structure similar to that of the activated complex in the BrO_3^-—Br^- reaction.

From the position of ^{18}O before and after the reaction conclusions are drawn about the intermediate stages. The application of the method rests again on specific requirements for the rates of substitution. ☐

EXAMPLE VI

Diffusion-controlled reactions. Solution reactions in which every encounter of reactant molecules leads to products are said to be *diffusion-controlled*. Their rate is limited only by diffusion, which has an activation energy of 10–20 kJ mol^{-1}. The rate constants for diffusion-controlled reactions range from 10^9 to 10^{10} M^{-1} s^{-1} and depend on the nature of the solvent. The nature of the reactants has little effect, but even small differences may supply precious information, for example, about the distance of closest approach.

As an example, consider the reaction

$$H_3O^+ + OH^- \longrightarrow 2H_2O$$

From the rate constant it was concluded that the reacting species, H_3O^+ and OH^-, do not really come in direct contact with each other, but that "coordinated" water molecules are interposed (hydrogen-bonded).

$$H_3O^+ \cdot 3H_2O + OH^- \cdot 3H_2O$$

Eigen based this conclusion on a comparison with the theoretical diffusion rate constant, which depends on the distance of closest approach.

A more elaborate comparison with diffusion is made in the reaction between iron(II) and chlorine(III), which fits the rate law

$$\text{Rate} = k_1[\text{Fe}^{II}][\text{Cl}^{III}] + \frac{k_2[\text{Fe}^{II}][\text{Cl}^{III}]}{[\text{H}^+]}$$

where $[\text{Cl}^{III}]$ represents the total analytical concentration of $HClO_2$ and ClO_2^-. The values of the rate constants are $k_1 = 1.9 \times 10^3$ M^{-1} s^{-1} and $k_2 = 58$ s^{-1}.

Since the predominant species in highly acidic solutions are $Fe(H_2O)_6^{2+}$ and $HClO_2$, the first term is most likely a bimolecular reaction between these species. The inverse hydrogen ion dependence in the second term indicates reaction of hydrolyzed species. Then

$$k_2 = k_2'K$$

where K is the hydrolysis constant of $Fe(H_2O)_6^{2+}$:

$$Fe(H_2O)_6^{2+} \underset{}{\overset{K_h}{\rightleftharpoons}} Fe(H_2O)_5OH^+ + H^+$$

or the dissociation constant of $HClO_2$:

$$HClO_2 \underset{}{\overset{K_a}{\rightleftharpoons}} H^+ + ClO_2^-$$

Both K_h and K_a are known independently. At 25°C and unit ionic strength ($\mu = 1.0\ M$), $K_h = 3.2 \times 10^{-10}$. If $K = K_h$, we obtain

$$k_2' = \frac{k_2}{K_h} = 1.8 \times 10^{11}\ M^{-1}\ s^{-1}$$

which is a value higher than the maximum values (10^9–$10^{10}\ M^{-1}\ s^{-2}$) for diffusion-limited reactions. Since this is not reasonable, the pathway involving hydrolysis of $Fe(H_2O)_6^{2+}$ is ruled out.

A more reasonable value for k_2' is obtained if the dissociation constant for $HClO_2$ is used. Thus, $K = K_a \simeq 10^{-2}$, and

$$k_2' = 5.8 \times 10^3\ M^{-1}\ s^{-1}$$

Mechanistically, then, k_2' corresponds to the reaction between $Fe(H_2O)_6^{2+}$ and ClO_2^-. It is clear that this information is not enough to allow us to comment on the structures of the activated complexes that correspond to the two terms. Evidence is also needed as to whether the reaction goes by an inner-sphere mechanism or whether it goes through a Cl^{II} or an Fe^{IV} intermediate, and so forth. A comparison with the diffusion-controlled limit contributes toward specifying the structure, but by itself it does not give the final answer.   □

EXAMPLE VII

Thermal or photochemical oxidations by $S_2O_8^{2-}$ are believed to proceed via formation of the anion radical SO_4^- by means of the dissociation process

$$^-O_3SOOSO_3^- \longrightarrow 2SO_4^-$$

Under normal conditions SO_4^- reacts rapidly and is difficult to detect. However, sufficiently high concentrations can be built up by flash photolysis. It can then be argued, by analogy, that SO_4^- is also formed thermally.   □

EXAMPLE VIII

Carbon dioxide reduction is a process of the utmost importance. However, it is a very complicated process with many stages that are difficult even to detect. So, the trick is simulation. By using, for example, analogues of the form

$$X{=}C{=}Y \qquad \text{or} \qquad \begin{matrix} X \\ \diagdown \\ Y \diagup \end{matrix} C{=}Z$$

and by trying to isolate intermediates that, hopefully are true analogues to those in the "real" system, some inferences can be gleaned.

A brief outline of some highlights follows:

1. Low-valence transition metal ions promote C—C bond formation, for example,

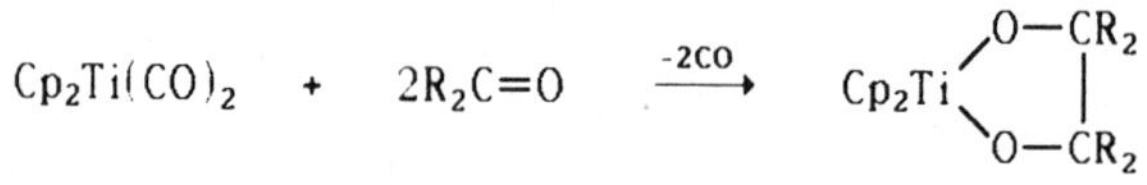

$$Cp_2Ti(CO)_2 \quad + \quad 2R_2C=O \quad \xrightarrow{\ -2CO\ } \quad Cp_2Ti \begin{matrix} O-CR_2 \\ | \\ O-CR_2 \end{matrix}$$

A free radical mechanism has been proposed for this reaction:

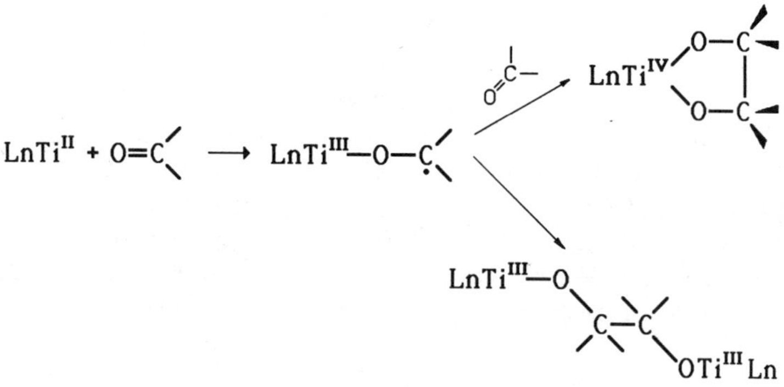

2. The reaction between $Cp_2Ti(CO)_2$ and diphenylketene gives an η^2-bonded complex,

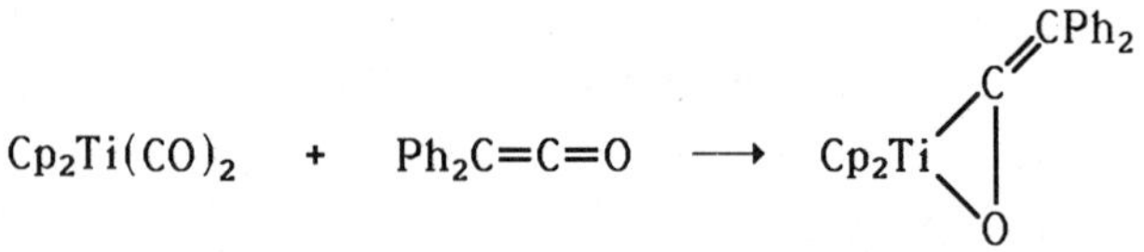

$$Cp_2Ti(CO)_2 \quad + \quad Ph_2C=C=O \quad \longrightarrow \quad Cp_2Ti \begin{matrix} C=CPh_2 \\ | \\ O \end{matrix}$$

which is formally reduced by a two-electron change to the corresponding dianion and adds to another diphenylketene to form a metallocycle without a C—C bond:

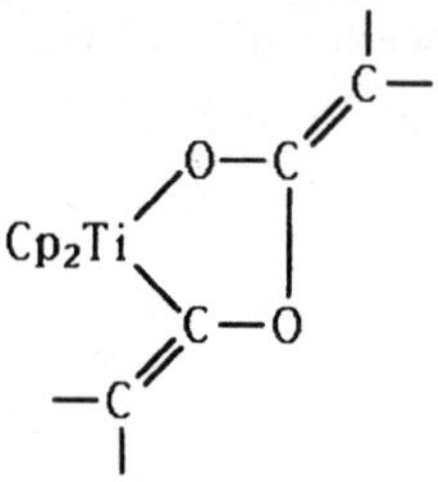

Similar products are formed by low-valence group VIII metal ions. The products mentioned here were isolated and characterized, and they give guidelines for some of the conditions required for the commercially important C—C bond formation. They apply to carbon dioxide by analogy.

3. Pyruvic acid,

$$\begin{array}{c} \text{HOOC} \\ \diagdown \\ \text{C=O} \\ \diagup \\ \text{H}_3\text{C} \end{array}$$

can be regarded as a simulation of the carbonate ion,

$$\left[\begin{array}{c} \text{O} \\ \diagdown \\ \text{C=O} \\ \diagup \\ \text{O} \end{array} \right]^{2-}$$

which is the form carbon dioxide assumes in aqueous alkaline solution.

Reduction of pyruvic acid by $V^{2+}(aq)$ or $Ti^{3+}(aq)$ leads also to C—C bond formation. In contrast, reduction by Cr^{2+} or Eu^{2+} leads to C—H bond formation (hydrogenation of the carbonyl group). The corresponding products from carbonates would have been oxalic and formic acids.   □

GENERAL REFERENCES

E. A. Boucher, Stochastic Approach to Reaction and Physico-Chemical Kinetics, *J. Chem. Educ.* **51**, 580 (1974).

R. K. Boyd, Some Common Oversimplifications in Teaching Chemical Kinetics, *J. Chem. Educ.* **55**, 84 (1978).

P. R. Brooks, Reactions of Oriented Molecules, *Science* **193**, 11 (1976).

A. I. Burshtein, Molecular-Kinetic Aspects of the Chemical Physics of the Condensed State, *Russ. Chem. Rev.* **47**, 120 (1978).

F. R. Cruickshank, A. J. Hyde, and D. Pugh, Free Energy Surfaces and Transition State Theory, *J. Chem. Educ.* **54**, 288 (1977).

J. H. Espenson, *Chemical Kinetics and Reaction Mechanisms*, McGraw-Hill, New York, 1981.

H. Eyring and E. M. Eyring, *Modern Chemical Kinetics*, Reinhold, New York, 1963.

S. J. Formosinho and M. G. M. Miguel, Markov Chains for Plotting the Course of Complex Reactions, *J. Chem. Educ.* **56**, 582 (1979).

A. Haim, The Ambiguity of Mechanistic Interpretations of Rate Laws: Alternate Mechanisms for the Vanadium(III)–Chromium(II) Reaction, *Inorg. Chem.* **5**, 2081 (1966).

J. E. Hulse, R. A. Jackson, and J. S. Wright, Energy Surfaces, Trajectories, and the Reaction Coordinate, *J. Chem. Educ.* **51**, 78 (1974).

H. S. Johnston, *Gas Phase Reaction Rate Theory*, Ronald, New York, 1966.

T. W. Newton and S. W. Rabideau, A Review of the Kinetics of the Aqueous Oxidation-Reduction Reactions of Uranium, Neptunium and Plutonium, *J. Phys. Chem.* **63**, 365 (1959).

P. D. Pacey, Changing Conceptions of Activation Energy, *J. Chem. Educ.* **58**, 612 (1981).

L. M. Raff, Illustration of Reaction Mechanism in Polyatomic Systems via Computer Movies, *J. Chem. Educ.* **51**, 712 (1974).

H. Taube, The Role of Kinetics in Teaching Inorganic Chemistry, *J. Chem. Educ.* **36**, 451 (1959).

R. E. Weston and H. A. Schwarz, *Chemical Kinetics,* Prentice-Hall, Englewood Cliffs, NJ, 1972.

PROBLEMS

1. Show that the maximum change of the rate constant with temperature is equal to $(4AR/E_a)e^{-2}$, where E_a and A are the Arrhenius activation energy and the preexponential factor, respectively, and show that this maximum is achieved at $T_{max} = E_a/2T$.

2. The rate of an elementary reaction is doubled if the temperature is increased from 27 to 37°C. Assume all other conditions remain constant. What is the energy of activation?

3. The two opposing reactions

$$2AB \underset{k_{-1}}{\overset{k_1}{\rightleftharpoons}} A_2B_2$$

are elementary. The reaction from left to right is exothermic. Which of the two rate constants, k_1 and k_{-1}, increases faster with temperature? Justify your answer.

4. Can the energy of activation of an endothermic elementary reaction be smaller than the heat of the reaction? Explain.

5. Suppose that compound A reacts by two parallel paths:

$$A \left\{ \begin{array}{l} \overset{k_1}{\longrightarrow} B + C \\ \overset{k_2}{\longrightarrow} D + E \end{array} \right.$$

Assume that each of these reactions is first-order in A and that the two preexponential factors are equal, but that the activation energy of the first reaction is larger than that of the second. The desired products are B and C; D and E are unwanted impurities. Is it possible, by changing only the temperature, to make the rate of the first reaction faster than that of the second? Explain.

6. Design a series of experiments for determining the rate law of the reaction A + B $\rightleftharpoons$ C and for measuring the rate constant and activation parameters. Consider, if you desire, a specific example.

7. Give an approximate (order-of-magnitude) estimate for the concentration of a solution in which the average distance between solute molecules is the same as the average distance in a gas at 25°C and 1 atm. How many solvent molecules correspond to each solute molecule? What is the magnitude of the average distance compared to the size of the molecules?

8. Does the structure of the activated complex depend on temperature? Explain.

9. In the reaction

$$(NH_3)_5Co\!-\!O\!-\!CO_2^+ + 2H_3O^+ \longrightarrow$$

$$(NH_3)_5Co\!-\!OH_2^{3+} + 2H_2O + CO_2$$

either of the bonds Co—O or O—CO$_2$ could break. Propose experiments to determine which bond breaks. What is the difference in the geometry of the corresponding activated complexes?

10. Imagine searching for an effective catalyst of an elementary reaction that is thermodynamically unfavorable, which makes it difficult experimentally to follow the formation of products in a closed system. Suggest ways to overcome the difficulties.

4

EXPERIMENTAL METHODS AND HANDLING OF THE DATA

4.1. TIME SCALE FOR CHEMICAL REACTIONS AND THE CHOICE OF EXPERIMENTAL TECHNIQUE

Time Scale

It is customary to divide the techniques used in mechanistic studies into two categories: those appropriate for "slow" reactions and those appropriate for "fast" reactions. The distinction between "slow" and "fast" is quite arbitrary: A reaction is characterized as fast if it is completed in less than a few seconds; otherwise it is characterized as slow.

An upper limit to the time required to complete a reaction can be thought to be set by the reactions associated with the nuclear decay of ^{238}U: $t_{1/2} = 4.5 \times 10^9$ yr or 1.4×10^{17} s. This is in fact close to the "age" of the archaeal rocks (3.85×10^9 yr). However, the rate determining step in this case is nuclear rather than chemical.

The lower limit is set by the physics of molecular movement and the time period of the individual vibrations. A chemical bond cannot break, and the atoms cannot get away from each other, in a time less than it takes these same atoms to undergo one vibration—which involves only short distances about the equilibrium position. For a bond of length ~0.1 nm, the time period for one vibration is of the order of 10^{-12}–10^{-13} s. Techniques to reach, or even exceed, this limit are now available (femtosecond spectroscopy). Thus, the chemist deals with a time range of 10^{32} s, from 10^{-15} to 10^{17} s.

The lower time limit in solution chemistry does not exceed 10^{-10}–10^{-11} s— the time needed for the molecules to diffuse and meet each other. The maximum value for a second-order rate constant, when every encounter of two reacting spherical neutral species leads to reaction, is given by

$$k_D = \frac{4\pi N_0 (D_A + D_B)(r_A + r_B)}{1000}$$

where r_A, r_B are the molecular radii, D_A, D_B the diffusion coefficients, and N_0 Avogadro's number. (For example, if $r_A = r_B = 2 \times 10^{-8}$ cm, and $D_A = D_B = 10^{-5}$ cm^2 s^{-1}, then $k_D = 6 \times 10^9$ M^{-1} s^{-1}).

For ions of the same size and diffusion characteristics, diffusion-controlled rate constants are generally smaller if the charges are the same, and larger if the charges are opposite in sign.

Choice of Experimental Technique

The choice of a suitable experimental technique(s) is dictated largely by the rate but also by the physical properties of the reacting species. The objective of this chapter is to draw attention to the salient features of common experimental techniques that are relevant to kinetic and mechanistic studies. Some mathematical and/or experimental tricks for the acquisition and handling of data will also be discussed.

The descriptions given are intended only to achieve this limited objective. For comprehensive descriptions, which will be needed if a technique is actually going to be used, special publications and original papers should be consulted.

Generally speaking, two kinds of data are collected in mechanistic studies:

1. Structural or other data about the "stations" where the reacting system "stops" for a while or indefinitely, namely, information about the reactants, the products, and the intermediates.
2. Data on the way the system goes from "station" to "station."

The first category refers to species corresponding to minima in the potential energy surface. Structural information for such species can be obtained by spectroscopic methods, which are not always reliable but have the great advantage that they can be used without appreciably disturbing the system. More reliable structural data are obtained by x-ray measurements on crystals, but this requires stability, which intermediates seldom have. Moreover, the crystallization procedure cannot be carried out without interrupting the reaction. It may be assumed that a species characterized by x rays is the same or similar to that postulated to participate in the reaction, but it is good to remember that this assumption may not be more reliable than spectroscopy because of the drastic interferences involved.

The second kind of data refer to the dynamics of the reaction, and only kinetic measurements can provide such information. Mechanistic conclusions are sometimes drawn based only on structural data collected for reactants, products, and "stable" intermediates. It must be pointed out, however, that strictly speaking this is the approach of thermodynamics (no concern for the path) and that the most important part of the mechanism, its dynamics, is not tested.

4.2. EXPERIMENTAL METHODS FOR SLOW REACTIONS

It is obvious that the analytical method to be used must be faster than the reaction under investigation. The time it takes to carry out various manipulations, such as titrations, filtrations, and the like, and the time it takes the chemical processes of the analytical method to be completed must be short compared to the time needed for the reaction under investigation.

The reactants are often mixed in appropriate containers at zero time, and then analysis is performed at various time intervals. After mixing, exchange of matter with the environment usually does not take place, and during the reaction the temperature is usually kept constant. The examples that follow belong to this category.

4.3. APPLICATION OF A SIMPLE TITRATION TECHNIQUE

Europium(II), in acidic solutions, selectively reduces the carboxylic group of isonicotinic acid, giving the corresponding aldehyde:

Experimentally, several mixtures are prepared, all having the same initial concentrations of europium(II), isonicotinic acid, and perchloric acid (to supply the necessary protons), and the temperature is kept constant. During stirring, the reaction is interrupted at different times by adding a solution of iron(III), which reacts rapidly with europium(II):

$$Fe^{III} + Eu^{II} \longrightarrow Fe^{II} + Eu^{III}$$

Europium(III) and iron(II) are unreactive toward isonicotinic acid. Thus, iron(II) can be determined at leisure by titration with cerium(IV). The amount of cerium(IV) consumed is equivalent to the amount of europium(II) in the solution at the time the reaction was interrupted.

One of the limitations of the titration technique is the need for relatively high concentrations (usually higher than 0.01 M).

4.4. SPECTROPHOTOMETRIC METHODS

Spectrophotometric methods are very convenient. They are the most popular among the techniques used for studying kinetics and mechanisms. The time response of the optics and electronics of a spectrophotometer is fast compared to the rates of chemical processes. What may limit the use of a spectrophotometric method is the time required for handling the samples and/or mixing the solutions. Two general requirements are that the spectra of the products must be different from the spectra of the reactants and that all absorptivities, including those of the intermediates, must be known or measurable.

As a simple example, consider the stoichiometric hydrogenation in acidic solution of the carbon–carbon double bond of maleic acid by vanadium(II):

$$2V^{2+}(aq) + HO_2CCH{=}CHCO_2H + 2H^+ \longrightarrow$$

Maleic acid

$$2V^{3+}(aq) + HO_2CCH_2CH_2CO_2H$$

Succinic acid

Both vanadium(II) and vanadium(III) absorb in the visible; maleic and succinic acids do not. However, the spectra of V^{II} and V^{III} differ. Thus the reaction is followed by recording the change in absorbance at the appropriate wavelengths.

4.5. ISOTOPIC TRACERS

Exchange Reactions

The exchange of the coordinated water molecules in $Cr(H_2O)_6^{3+}$ with water molecules in the solvent can be studied by using water enriched in oxygen-18:

$$Cr(H_2O)_6^{3+} + H_2^{18}O \rightleftharpoons (H_2^{18}O)Cr(H_2O)_5^{3+} + H_2O \qquad 4.1$$

The oxygen symbol without an isotope number denotes the natural abundance of isotopes. In practice, ^{18}O does not mean "100% oxygen-18"; rather it means oxygen (or actually water) enriched in ^{18}O.

Since oxygen-18 is not radioactive, changes in isotopic composition are measured by using a mass spectrometer. If the isotopes used are radioactive, the reaction is followed by measuring radioactivity. Isotopes may also differ in their nuclear spins; in that case the change in isotopic composition can perhaps be followed by nuclear magnetic resonance.

It should be emphasized that reactions like the water exchange reaction noted above cannot be studied by ordinary chemical or physical methods. The products and the reactants are indistinguishable unless they are properly labeled.

McKay Plots

A general exchange reaction can be represented by the equation

$$A_\alpha + {}^*A_\beta \rightleftharpoons A_\beta + {}^*A_\alpha \qquad 4.2$$

The environment of the labeled species changes back and forth from α to β. Reaction 4.2 can be a ligand-exchange reaction such as reaction 4.1 or an electron-exchange reaction such as reaction 4.3.

$$A^{II} + {}^*A^{III} \rightleftharpoons A^{III} + {}^*A^{II} \qquad 4.3$$

McKay plots are obtained by using the equation

$$\ln (1 - f) = -R_{ex} \left(\frac{a + b}{ab} \right) t \qquad 4.4$$

where $f = [{}^*A_\alpha]_t / [{}^*A_\alpha]_\infty$, the fraction of the exchange at time t; $a = [A_\alpha] + [{}^*A_\alpha]$; $b = [{}^*A_\beta] + [A_\beta]$; and R_{ex} is the rate of exchange. Derivations of this exchange equation can be found in standard textbooks on kinetics. At $f = 0.5$ (half-exchange, $t = t_{1/2}$), $\ln (1 - f) = -\ln 2$, and equation 4.4 becomes

$$R_{ex} = \frac{ab}{a + b} \left(\frac{\ln 2}{t_{1/2}} \right) \qquad 4.5$$

The half-time $t_{1/2}$ is obtained from the slope of a $\ln (1 - f)$ versus time plot, and R_{ex} is determined using equation 4.5. Finally, the form of the rate law is established by varying a and b.

Nonlinear McKay Plots

Deviation of a McKay plot from linearity indicates that the exchange process is not as simple as presented above and that other reactions are involved, complicating the mechanism. As an example, consider the exchange reaction between the uranyl ion and water, which was studied using an oxygen-18 tracer:

$$UO_2^{2+} + H_2{}^*O \rightleftharpoons U^*O_2^{2+} + H_2O \qquad 4.6$$

The uranyl ion was sampled as a function of time by precipitation of uranyl ferrocyanide. A typical McKay plot is shown in Figure 4.1.

One plausible interpretation of the decrease in the exchange rate is that the reactants contain some "impurity" that catalyzes the exchange, but that the concentration of this catalyst gradually decreases. Additional experiments

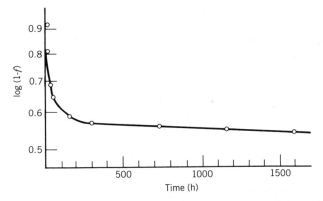

Figure 4.1. McKay plot for the oxygen exchange between uranyl ion and water.

suggest that the catalyst is indeed uranium(V) (UO_2^+). The proposed mechanism can be summarized as follows:

$$UO_2^{2+} + H_2^*O \rightleftharpoons U^*O_2^{2+} + H_2O \qquad \text{(very slow)}$$

$$UO_2^+ + H_2^*O \rightleftharpoons U^*O_2^+ + H_2O \qquad \text{(fast)}$$

$$U^*O_2^+ + UO_2^{2+} \rightleftharpoons UO_2^+ + U^*O_2^{2+} \qquad \text{(fast)}$$

but UO_2^+ is unstable and disproportionates:

$$UO_2^+ + UO_2^+ \overset{4H^+}{\rightleftharpoons} U^{4+} + UO_2^{2+} + 2H_2O \qquad\qquad 4.7$$

Thus, the UO_2^{2+} exchange process is catalyzed by UO_2^+, which slowly disappears, resulting in a nonlinear McKay plot.

4.6. GENERAL METHODS FOR HANDLING EXPERIMENTAL DATA FOR SLOW REACTIONS

The Method of Finite Differences

For slow reactions, the differential ratio dc_i/dt is replaced by the finite difference $\Delta c_i/\Delta t$, where Δc_i is the measurable finite change in the concentration of component i during a finite change in time Δt. This substitution constitutes an approximation that is justified only if the slope dc_i/dt does not change appreciably with time.

The slope can be determined more accurately by differentiating an empirical relation, such as

$$c_i = at + bt^2 + ct^3 + \cdots$$

in which the concentration of the ith component at time t is expressed as a power series of time.

If the rate depends on the mth power of a single concentration c and is determined at times t_1 and t_2, then

$$\frac{\Delta c_1}{\Delta t} = k c_1^m \qquad \text{4.8}$$

$$\frac{\Delta c_2}{\Delta t} = k c_2^m \qquad \text{4.9}$$

Taking the logarithms of these two equations and subtracting, we obtain

$$\log\left(\frac{\Delta c_1}{\Delta t}\right) - \log\left(\frac{\Delta c_2}{\Delta t}\right) = m(\log c_1 - \log c_2) \qquad \text{4.10}$$

from which the order m with respect to component c is calculated.

An expression of a similar form is obtained if the rates are measured at zero time (initial rates) but at two different initial concentrations:

$$\left(\frac{\Delta c_0}{\Delta t}\right)_1 = k(c_0)_1^m$$

$$\left(\frac{\Delta c_0}{\Delta t}\right)_2 = k(c_0)_2^m$$

and

$$\log\left(\frac{\Delta c_0}{\Delta t}\right)_1 - \log\left(\frac{\Delta c_0}{\Delta t}\right)_2 = m(\log (c_0)_1 - \log (c_0)_2)$$

The Method of Integration

Integration of the differential equations gives each concentration as a function of time: $c_i = c_i(t)$. Simple examples of such functions are included in Table 4.1.

Straight lines are obtained if we plot

For zero-order reactions, c versus time.

For first-order reactions, $\log c$ versus time.

For ($n \geq 2$)-order reactions, $1/c^{n-1}$ versus time.

TABLE 4.1. Some Common Rate Laws[a]

Order	Differential Rate Law	Integrated Rate Law	Half-Lives	Usual Units
0	$-\dfrac{dc_1}{dt} = k$	$c_1 = (c_1)_0 - kt$	$\dfrac{(c_1)_0}{2k}$	$M\,s^{-1}$
1	$-\dfrac{dc_1}{dt} = kc_1$	$\ln \dfrac{(c_1)_0}{c_1} = kt$	$\dfrac{0.693}{k}$	s^{-1}
2 (simple)	$-\dfrac{dc_1}{dt} = kc_1^2$	$\dfrac{1}{c_1} - \dfrac{1}{(c_1)_0} = kt$	$\dfrac{1}{k(c_1)_0}$	$M^{-1}\,s^{-1}$
2 (mixed)	$-\dfrac{dc_1}{dt} = kc_1c_2$	$\dfrac{\alpha \ln [c_1/(c_1)_0]\,[c_2/(c_2)_0]}{\beta(c_1)_0 - \alpha(c_2)_0} = kt$	$\dfrac{\alpha}{k[\beta(c_1)_0 - \alpha(c_2)_0]} \ln \dfrac{\alpha(c_2)_0}{2\alpha(c_2)_0 - \beta(c_1)_0}$	$M^{-1}\,s^{-1}$
$n \geqslant 2$ (simple)	$-\dfrac{dc_1}{dt} = kc_1^n$	$\dfrac{1}{c_1^{n-1}} - \dfrac{1}{(c_1)_0^{n-1}} = (n-1)kt$	$\dfrac{2^{n-1} - 1}{(n-1)k(c_1)_0^{n-1}}$	$M^{1-n}\,s^{-1}$

[a] c_1 and c_2 represent the concentrations of substances 1 and 2, respectively, at time t. $(c_1)_0$ and $(c_2)_0$ are the corresponding concentrations at zero time. α and β are the coefficients for compounds 1 and 2 in the chemical equation. The half-lives refer to substance 1.

Half-lives can also be used (Table 4.1). In testing the form of the rate law it is important that data points for *at least* 75% conversion be included. When experimental errors are taken into account, it is not unusual for typical plots of log c or c^{-1} versus time to appear "equally" linear during initial stages of the reaction—and in some cases about up to one half-life. There are statistical criteria to help decide the best fit. However, it is most important to cover a large percentage conversion of the reaction and to allow as wide a variation of the initial reactant concentrations as possible.

The Guggenheim Method

This method is applied to first-order kinetic data whenever the final reading is unknown. Suppose that λ_0 and λ_t are known but λ_∞ is unknown, where λ represents a physical property that is proportional to the concentration, such as light absorbance.

At time t_1,

$$\lambda_{t_1} - \lambda_\infty = (\lambda_0 - \lambda_\infty)e^{-kt_1} \qquad 4.11$$

At time $t_1 + \delta$,

$$\lambda_{t_1+\delta} - \lambda_\infty = (\lambda_0 - \lambda_\infty)e^{-k(t_1+\delta)} \qquad 4.12$$

Subtracting equation 4.12 from 4.11,

$$\lambda_{t_1} - \lambda_{t_1+\delta} = (\lambda_0 - \lambda_\infty)e^{-kt_1}(1 - e^{-k\delta}) \qquad 4.13$$

or, assuming $\lambda_{t_1} - \lambda_{t_1+\delta} > 0$,

$$kt_1 + \ln(\lambda_{t_1} - \lambda_{t_1+\delta}) = \ln[(\lambda_0 - \lambda_\infty)(1 - e^{-k\delta})] \qquad 4.14$$

If the time changes $(t_1, t_2, t_3, \ldots)$ but δ is kept constant, then the right-hand side of equation 4.14 is also constant. Therefore,

$$kt_1 + \ln(\lambda_{t_1} - \lambda_{t_1+\delta}) = \text{constant}$$

At any time t,

$$\ln(\lambda_t - \lambda_{t+\delta}) = -kt + \text{constant} \qquad 4.15$$

A graph of $\ln(\lambda_t - \lambda_{t+\delta})$ versus time should be linear with a slope of k. For the best accuracy, the value of δ must be as large as possible.

4.7. HANDLING COMPLICATED RATE LAWS

The rate laws in Table 4.1 are relatively simple. In more complicated cases, straightforward integration may not be feasible or practical. Numerical or other methods should then be used for plotting (generally handling) the data, or special conditions must be chosen for doing the experiments. Some of these methods are described briefly in the following sections.

$$* \quad * \quad * \quad * \quad *$$

Warning! Rate laws like those included in Table 4.1 are valid provided:

The system is closed; the reaction takes place in a "batch reactor" without mass flow in or out.

There is good stirring (homogeneity), and there is uniformity in temperature and pressure.

There is no change in volume.

If these conditions are not met, the expressions must be modified accordingly.

$$* \quad * \quad * \quad * \quad *$$

Ostwald's Method of Isolation

For component i, the concentration of which changes according to the equation

$$\frac{dc_i}{dt} = kc_1^m c_2^n c_3^l \cdots \qquad\qquad 4.16$$

the order with respect to c_1 can be determined if it is possible to keep the other components c_2, c_3, . . . in excess so that there is a minimal change in their concentration during the time period under investigation. This is the method known as "Ostwald's method of isolation" or the "flooding technique." Under these conditions, equation 4.16 becomes

$$\frac{dc_1}{dt} = k'c_1^m$$

where k' is a new constant that includes the concentrations of all species held constant raised to the appropriate power. In other words, the reaction becomes *pseudo m-order.*

As an example, consider the reaction between bromate ion and bromide ion, which at constant hydrogen and bromide ion concentrations has been shown to have the rate law

$$-\frac{d[BrO_3^-]}{dt} = k_1[BrO_3^-]$$

The pseudo-first-order rate constant k_1 is related to the actual rate constant by the expression

$$k_1[BrO_3^-] = k[BrO_3^-][Br^-][H^+]^2$$

The latter expression was in fact determined by following the reaction only under pseudo-first-order conditions but over a wide variation in initial concentrations.

By studying the reaction with excess—and virtually constant—bromide ion and hydrogen ion concentrations, we reduce the complexity of a fourth-order expression to that of a first-order expression.

For rate data of the customary precision, a tenfold excess usually suffices to consider the concentration of a component to be constant.

Inhibition

EXAMPLE

In the reaction

$$V^{3+} + Cr^{2+} \rightleftharpoons V^{2+} + Cr^{3+}$$

which is first-order in both V^{3+} and Cr^{2+}, the data shown in Table 4.2 have been reported.

TABLE 4.2. Observed Second-Order Rate Constants for the Reaction between V^{3+} and Cr^{2+} at Various Hydrogen Ion Concentrations

H^+, M	k_{obs}, $M^{-1}\,s^{-1}$
0.027	5.00
0.040	4.32
0.115	2.66
0.155	2.29
0.192	2.03
0.250	1.78
0.269	1.63
0.365	1.25
0.500	1.04

The reaction is obviously inhibited by an increase in the hydrogen ion concentration. A plot of log k versus log [H$^+$] is "roughly" linear with an apparent slope of -0.5, which is consistent with a rate law of the form

$$\frac{k_{obs}[V^{3+}][Cr^{2+}]}{[H^+]^{0.5}}$$

However, the data can be better represented and mechanistically interpreted by expressing the rate law as in equations 4.17 or 4.18.

$$k_{obs} = -\frac{d[Cr^{2+}]/dt}{[V^{3+}][Cr^{2+}]} = \frac{k_1}{k_2 + [H^+]} \qquad\qquad 4.17$$

or

$$k_{obs} = -\frac{d[Cr^{2+}]/dt}{[V^{3+}][Cr^{2+}]} = k_2 + \frac{k_1}{[H^+]} \qquad\qquad 4.18$$

At low hydrogen ion concentrations, the [H$^+$] in the denominator of equation 4.17 can be neglected, and the rate becomes independent of [H$^+$]. The data (pluses in Fig. 4.2) show a tendency of leveling off in this [H$^+$] region. In constrast, at low hydrogen ion concentrations, equation 4.18 predicts an order of -1. At high hydrogen ion concentration, the order with respect to [H$^+$] predicted by equation 4.17 is -1, and it is again supported by the data shown in Figure 4.2. The order predicted from equation 4.18 at high hydrogen

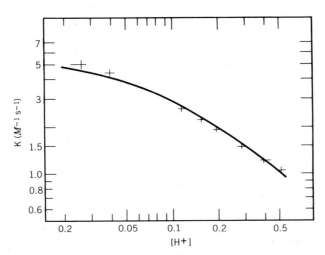

Figure 4.2. A plot of log k_{obs} versus log [H$^+$] for the data in Table 4.2. The data behave as predicted by equation 4.17.

ion concentration is zero (leveling off in the log–log plot), which is clearly inconsistent with the data.

Two-Term Rate Laws

EXAMPLE I

The rate of the electron exchange reaction between Sn^{II} and Sn^{IV} is increased in the presence of sulfate ion but inhibited by increases in the concentration of hydrogen ion. Data as a function of hydrogen and sulfate ion concentrations are given in Table 4.3.

The rate law can be written in the form

$$k_{obs} = \frac{rate}{[Sn^{II}][Sn^{IV}]} = k_2[SO_4^{2-}]^a[H^+]^b \qquad 4.19$$

The points at constant sulfate ion concentration allow evaluation of $b = -1.75$, and the points at constant hydrogen ion give the value of $a = 0.10$ and a calculated value of $k_2 = 0.700$. However, the data, instead of being expressed by equation 4.19 with fractional orders, are better expressed as a sum of two terms. The expression

$$k_{obs} = \frac{0.67}{[H^+]^2} + \frac{0.074[SO_4^{2-}]}{[H^+]} \qquad 4.20$$

can be shown to be a good approximation by plotting $k_{obs}[H^+]$ versus $[H^+]^{-1}$ at constant $[SO_4^{2-}]$ (the resulting slope is 0.67) and by plotting $k_{obs}[H^+]$ versus $[SO_4^{2-}]$ at constant $[H^+]$ (the resulting slope is 0.074).   □

TABLE 4.3. Dependence of the Observed Second-Order Rate Constant for the Sn^{II}–Sn^{IV} Electron-Exchange Reaction on Hydrogen Ion and Sulfate Ion Concentrations

$[H^+]$, M	$[SO_4^{2-}]$, M	k_{obs}, $M^{-1}\,s^{-1}$
4.96	0.99	0.042
3.99	0.99	0.060
3.60	0.99	0.072
2.85	0.99	0.109
1.99	3.00	0.280
1.49	0.99	0.348
0.99	0.99	0.760
0.99	1.40	0.791
0.99	0.70	0.739
0.99	0.20	0.702

EXAMPLE II

The data in Table 4.4 were collected for reactions of the type

$$(NH_3)_5CoO_2CRNH_2^{2+} + Cr^{2+} + 5H^+ \rightleftharpoons CrO_2CRNH_2^{2+} + Co^{2+} + 5NH_4^+$$

In this equation, coordinated and bulk water molecules have been omitted. The two data sets correspond to similar reactions, but the hydrogen ion dependence is markedly different. What rate laws will best fit the data?

A plot of log k_{obs} as a function of log $[H^+]$ is very instructive. As can be seen from Figure 4.3, data set 1 is almost linear, whereas data set 2 is definitely dish-shaped. Both curves are quite typical of parallel reactions. In the case of data set 1, two terms should be sufficient to fit the data. A minimum of three terms may be needed for data set 2. For data set 1, an increase in the hydrogen ion concentration by a factor of 10 increases the rate by more than a factor of 10. The slope of the log–log plot is between 1.5 and 1.6. A good guess at the rate law is

$$k_{obs} = k_1[H^+] + k_2[H^+]^2$$

Thus the overall order would be between 1 and 2. A plot of $k_{obs}/[H^+]$ versus $[H^+]$ gives $k_1 = 37 \, M^{-2} \, s^{-1}$ and $k_2 = 116 \, M^{-3} \, s^{-1}$, but the points corresponding to very large rate constants are weighted much too heavily. Since each point is known experimentally to about the same precision (± 3–4%), a more appropriate weighting factor would be $1/k_{obs}^2$. In this case, rather than minimize

TABLE 4.4. Rate Constants for the Reaction between Cr^{2+} and $(H_3N)_5Co^{III}O_2CRNH_2^{2+}$

[H+], M	Data Set 1 k_{obs}, $M^{-1} \, s^{-1}$	Data Set 2 k_{obs}, $M^{-1} \, s^{-1}$
2.50	900	–
2.00	500	180
1.80	414	170
1.60	–	161
1.50	300	–
1.30	–	148
1.20	204	–
1.00	150	135
0.80	104	128
0.60	66	122
0.50	–	120
0.40	36	120
0.30	–	123
0.20	14	135
0.10	–	180

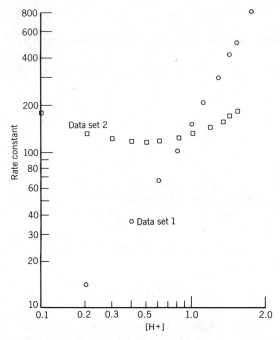

Figure 4.3. Rate data as a function of hydrogen ion concentration.

$(k_{obs} - k_{calc})^2$, the percent error [i.e., $(1/k_{obs}^2)(k_{obs} - k_{calc})^2$] is minimized. The results are $k_1 = 49.0 \ M^{-2} \ s^{-1}$ and $k_2 = 102 \ M^{-3} \ s^{-1}$. The exact method by which individual data points are weighted can generally make an important difference.

For the second data set, a more complicated rate law is necessary. Two of the most straightforward possibilities are

$$k_{obs} = \frac{k_a}{[H^+]} + k_b[H^+] \qquad 4.21$$

and

$$k_{obs} = \frac{k_c}{[H^+]} + k_d + k_e[H^+] \qquad 4.22$$

According to equation 4.21, a plot of $k_{obs}/[H^+]$ versus $[H^+]^{-2}$ should be linear. Such a treatment results in $k_a = 20.5 \ s^{-1}$ and $k_b = 9.75 \ M^{-2} \ s^{-1}$, but the fit is rather poor.

On the other hand, the fit to equation 4.22 results in $k_c = 9.99 \ s^{-1}$, $k_d = 75.1 \ M^{-1} \ s^{-1}$, and $k_e = 49.9 \ M^{-2} \ s^{-1}$, and it can be characterized as excellent (better than $\pm 1\%$ fit to every data point).

Obtaining Rate Constants by Using Concentration–Time Integrals

The concentration–time integral method uses graphically determined concentration–time integrals as variables instead of time itself. The integral

$$\int_0^t [B]\, dt$$

is obtained directly from a graph of the concentration of a particular species as a function of time and is used as a variable. Some simple examples will show the utility of the method.

The integral of the rate expression for the second-order reaction

$$A + B \xrightarrow{k_2} \text{products}$$

can be obtained by routine methods unless B also disappears by some concomitant parallel or consecutive reaction that does not consume A. However,

$$-\frac{d[A]}{dt} = k_2[A][B] \qquad\qquad 4.23$$

always correctly describes the rate, and the relation

$$-\int_{A_0}^{A} \frac{d[A]}{A} = k_2 \int_0^t [B]\, dt \qquad\qquad 4.24$$

is always valid. Thus, by evaluating the concentration–time integral for B, denoted by B_t, k_2 can be estimated from the equation

$$\ln \frac{[A]_0}{[A]} = k_2 B_t \qquad\qquad 4.25$$

provided that the disappearance of A and B can be followed independently. An interesting feature is that the graph of ln [A] as a function of B_t is linear with a slope of $-k_2$, independent of the other decomposition pathways for B.

Another example involves the reactions

$$A \xrightarrow{k_1} B$$

$$A + B \xrightarrow{k_2} C$$

or the reactions

$$A \xrightarrow{k_1} C$$

$$A + B \xrightarrow{k_2} D$$

In both cases the rate of dissappearance of A is given by the equation

$$-\frac{d[A]}{dt} = k_1[A] + k_2[A][B] \qquad\qquad 4.26$$

This equation can be rewritten in terms of the concentration–time integral for B as

$$\ln\frac{[A]_0}{[A]} = k_1 t + k_2 B_t$$

Thus, k_1 and k_2 can be estimated directly from a plot of $\ln([A]_0/[A])/B_t$ as a function of t/B_t.

The method can also be applied to simple sets of consecutive reactions such as

$$A + B \xrightarrow{k_1} C + E$$

$$A + C \xrightarrow{k_2} D + E$$

The integrated form of the rate equation for this set can be written in terms of the time integrals as

$$\ln\frac{[A]_0}{[A]} = k_1 B_t + k_2 C_t$$

Simplification of the System of Differential Rate Equations

Even though the system of differential equations for a particular mechanism may be nonlinear, sometimes it can be put into a simpler, easier-to-manipulate form.

As an example, consider a mechanism consisting of four elementary bimolecular reactions, involving seven components:

$$A + B \xrightarrow{k_1} C$$

$$C + B \xrightarrow{k_2} D$$

$$E + B \xrightarrow{k_3} F$$

$$F + B \xrightarrow{k_4} G$$

For this mechanism, in which four substances (A, C, E, and F) compete for B, the full system of differential equations is as follows:

$$\frac{d[A]}{dt} = -k_1[A][B] \qquad\qquad 4.27$$

$$\frac{d[B]}{dt} = -k_1[A][B] - k_2[C][B] - k_3[E][B] - k_4[F][B] \qquad\qquad 4.28$$

$$\frac{d[C]}{dt} = k_1[A][B] - k_2[C][B] \qquad\qquad 4.29$$

$$\frac{d[D]}{dt} = k_2[C][B] \qquad\qquad 4.30$$

$$\frac{d[E]}{dt} = -k_3[E][B] \qquad\qquad 4.31$$

$$\frac{d[F]}{dt} = k_3[E][B] - k_4[F][B] \qquad\qquad 4.32$$

$$\frac{d[G]}{dt} = k_4[F][B] \qquad\qquad 4.33$$

The goal is to obtain a simplified equivalent system. This can be achieved by taking the following combinations:

Combination I: (4.27)
Combination II: (4.27) + (4.29)
Combination III: (4.31)
Combination IV: (4.31) + (4.32)
Combination V: (4.27) + (4.29) + (4.30)
Combination VI: (4.28) − 2(4.27) − (4.29) − 2(4.31) − (4.32)
Combination VII: (4.31) + (4.32) + (4.33)

Each of the original equations appears in these combinations at least once. In choosing the combinations, the right-hand side is made as simple as possible. Thus, in V, VI, and VII, the right-hand sides are equal to zero. In I, II, III, and IV, the right-hand sides have only one term.

The simplified system is the following:

$$\frac{d[A]}{dt} = -k_1[A][B] \qquad 4.27'$$

$$\frac{d([A] + [C])}{dt} = -k_2[C][B] \qquad 4.28'$$

$$\frac{d[E]}{dt} = -k_3[E][B] \qquad 4.29'$$

$$\frac{d([E] + [F])}{dt} = -k_4[F][B] \qquad 4.30'$$

$$\frac{d([A] + [C] + [D])}{dt} = 0 \qquad 4.31'$$

$$\frac{d([B] - 2[A] - [C] - 2[E] - [F])}{dt} = 0 \qquad 4.32'$$

$$\frac{d([E] + [F] + [G])}{dt} = 0 \qquad 4.33'$$

The solution of the last three equations (4.31', 4.32', 4.33') is straightforward:

$$[A] + [C] + [D] = c_1$$

$$[B] - 2[A] - [C] - 2[E] - [F] = c_2$$

$$[E] + [F] + [G] = c_3$$

where c_1, c_2, and c_3 are constants related to the initial conditions; for example,

$$c_2 = [B]_0 - 2[A]_0 - [C]_0 - 2[E]_0 - [F]_0$$

Any of the initial concentrations except $[A]_0$ and $[B]_0$ may in fact be zero.

The remaining system of equations, 4.27', 4.28', 4.29', and 4.30', can be solved by numerical methods after substituting $[B]$ by $2[A] + [C] + 2[E] + [F] + c_2$. However, certain useful relations can be obtained more easily by simply taking ratios. Thus dividing equation 4.29 by equation 4.27, we obtain

$$\frac{d[C]}{d[A]} = \frac{k_2[C]}{k_1[A]} - 1 \qquad 4.34$$

Similarly, from equations 4.31 and 4.32,

$$\frac{d[F]}{d[E]} = \frac{k_4[F]}{k_3[E]} - 1 \qquad\qquad 4.35$$

Both of these equations are ordinary simple first-order differential equations, and they can be solved readily, giving the ratios k_2/k_1 and k_4/k_3, respectively:

$$\frac{k_2}{k_1} = 1 - \frac{[A]_0([A]/[A]_0)^{k_2/k_1} - [A]}{[C] - [C]_0} \qquad\qquad 4.34'$$

$$\frac{k_4}{k_3} = 1 - \frac{[E]_0([E]/[E]_0)^{k_4/k_3} - [E]}{[F] - [F]_0} \qquad\qquad 4.35'$$

A third ratio of rate constants, k_1/k_3, can be obtained from equations 4.27 and 4.31

$$\frac{k_1}{k_3} \frac{d[E]}{[E]} = \frac{d[A]}{[A]} \qquad\qquad 4.36$$

which gives

$$\frac{k_1}{k_3} = \frac{\ln[A] - \ln[A]_0}{\ln[E] - \ln[E]_0} \qquad\qquad 4.36'$$

Laplace Transforms for Solving Differential Rate Equations

The general Laplace transform of the function $c_i(t)$, (e.g., the concentration of a component) is defined as

$$\phi(p) = T(c_i) = \int_0^\infty c_i(t) \exp(-pt)\, dt \qquad\qquad 4.37$$

where t and p are variables, $c_i(t)$ is a function of t, and $\phi(p)$ is a function of p.

The following "properties" of the Laplace transform are very useful in treating systems of kinetic differential equations:

1. The transform of a sum of several terms is the sum of the transforms of each term.
2. The transform of a constant multiplied by a variable is the constant multiplied by the transform of the variable.
3. The transform of the derivative dc_i/dt is given by

$$T\left(\frac{dc_i}{dt}\right) = pT(c_i) - c_i(0)$$

where $c_i(0)$ is the value of c_i at $t = 0$ (the initial concentration). In the case when $c_i(0) = 0$,

$$T\left(\frac{dc_i}{dt}\right) = pT(c_i)$$

By applying these simple rules, a system of kinetic differential equations is transformed into a system of algebraic equations, which are easier to solve by exact or approximate methods.

Retransforming the solution of the transformed algebraic system, we obtain the concentration of each component as a function of time, and the appropriate rate constants.

A simple example will be used as an illustration. A general solution will be obtained for the system of differential equations corresponding to a mechanism consisting of two opposing elementary reactions:

$$A \underset{k_2}{\overset{k_1}{\rightleftharpoons}} B$$

The system is

$$\frac{dc_A}{dt} = -k_1 c_A + k_2 c_B \qquad 4.38$$

$$\frac{dc_B}{dt} = k_1 c_A - k_2 c_B \qquad 4.39$$

According to the rules stated, the transforms of equations 4.38 and 4.39 are

$$pT(c_A) - c_A(0) = -k_1 T(c_A) + k_2 T(c_B) \qquad 4.40$$

$$pT(c_B) - c_B(0) = k_1 T(c_A) - k_2 T(c_B) \qquad 4.41$$

Thus, a system of two linear algebraic equations in two unknowns, $T(c_A)$ and $T(c_B)$, is obtained. The solution of this system is simple:

$$T(c_A) = \frac{k_2\{c_A(0) + c_B(0)\}}{p(k_1 + k_2 + p)} + \frac{c_A(0)}{k_1 + k_2 + p} \qquad 4.42$$

$$T(c_B) = \frac{k_1\{c_A(0) + c_B(0)\}}{p(k_1 + k_2 + p)} + \frac{c_B(0)}{k_1 + k_2 + p} \qquad 4.43$$

The Laplace transforms of the right-hand sides of 4.42 and 4.43 are the sums of the transforms of the corresponding terms, and they can be found in tables (consult the *Handbook of Chemistry and Physics*). This new transformation yields the concentrations as functions of time. Rearranging and taking the natural logarithm, we have

$$\ln \left[\frac{k_2\{c_A(0) + c_B(0) - c_A\} - k_1 c_A}{k_2 c_B(0) - k_1 c_A(0)} \right] = -(k_1 + k_2)t \qquad 4.44$$

After some mathematical manipulations, and taking into account material balance,

$$c_A(0) + c_B(0) = c_A + c_B$$

we finally obtain

$$\ln \frac{k_2 c_B - k_1 c_A}{k_2 c_B(0) - k_1 c_A(0)} = -(k_1 + k_2)t$$

4.8. EXPERIMENTAL METHODS FOR FAST REACTIONS

The most important methods for fast reactions are listed in Table 4.5, where a typical half-time that can be determined in each case is also indicated. Some of these methods are applicable when the system is close to equilibrium (e.g., temperature jump) or practically at equilibrium (e.g., NMR). For systems far from equilibrium, flow methods are used or the system is perturbed drastically. Some of the methods for fast reactions are "close-to-equilibrium" methods; others (e.g., flow methods, flash photolysis) are "far-from-equilibrium" methods.

The time period of the electromagnetic waves used in certain techniques must be short compared to the time scale of the chemical change under

TABLE 4.5. Experimental Methods for Fast Reactions

Method	$t_{1/2}$, s
Flow techniques	10^{-3}
Pressure jump	10^{-6}
NMR	10^{-6}
Temperature jump	10^{-7}
Acoustical methods	10^{-9}
Electrochemical methods	10^{-8}
ESR	10^{-9}
Fluorescence	10^{-10}
Flash photolysis, picosecond and femtosecond spectroscopy, pulse radiolysis	10^{-12}

investigation. Microwave spectroscopy, for example, is a relatively "fast" technique with periods of the order of 10^{-10} s, and it is appropriate for studying processes with energy barriers in the range of 0–20 kJ mol^{-1}. In contrast, NMR is a slower technique (periods 10^{-1} to 10^{-9} s), and it is appropriate for processes with energy barriers in the range of 35–100 kJ mol^{-1}.

4.9. FLOW METHODS

General Description

A simplified schematic presentation of a flow setup is given in Figure 4.4. Solutions of A and B enter the mixing chamber M, where the reaction

$$A + B \longrightarrow C$$

is initiated. As the mixture moves away from M, the concentrations of A and B decrease, whereas the concentration of C increases.

Treatment of the Data

Consider the volume element dV between cross sections S_1 and S_2 (Fig. 4.4). If at the surface S_1 the concentration of C is x, the corresponding concentration at S_2 will be $x + dx$. Assuming that C is formed by a second-order reaction

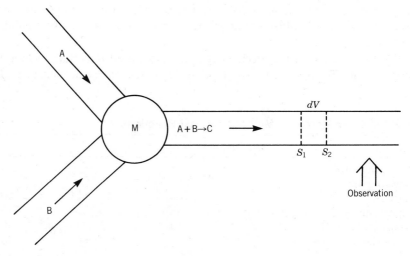

Figure 4.4. Simplified schematic presentation of a flow setup. Measurements are performed at various points along the tube containing the mixture.

(first-order each in A and B), we have

$$\frac{dx}{dt} = k(a - x)(b - x) \qquad 4.45$$

where a and b are the initial concentrations of A and B at the time of mixing ($t = 0$). To simplify the analysis further, suppose that temperature and pressure do not change, that the flow is smooth, and that the tube is filled with solution.

Mass balance requires that the number of molecules of C coming out of the volume element dV in time dt be equal to the number that entered plus the number formed within this element. Hence,

$$u(x + dx)\, dt \quad = \quad ux\, dt \quad + k(a - x)(b - x)\, dV\, dt$$

| Number of C molecules coming out of dV in time dt | Number of C molecules coming into dV in time dt | Number of C molecules produced within dV in time dt |

where u is the constant flow rate. Upon rearrangement we obtain

$$\frac{dx}{(a - x)(b - x)} = \frac{k}{u}\, dV$$

and by integrating over the total volume V_0, from the point the reaction starts until the point the observation is made,

$$\int_0^{x_f} \frac{dx}{(a - x)(b - x)} = \int_0^{V_0} \frac{k}{u}\, dV$$

and

$$\frac{1}{a - b} \ln \frac{b(a - x_f)}{a(b - x_f)} = \frac{k}{u} V_0 \qquad 4.46$$

where x_f is the concentration of C at the observation point. Equation 4.46 is derived assuming that the initial concentration of C (at the mixing chamber) is zero.

If the experimental data fit equation 4.46, it is concluded that the rate law is given by 4.45. The rate constant is obtained from the slope of the plot of $\ln [b(a - x_f)]/[a(b - x_f)]$ versus $1/u$.

The mathematical treatment applied here can also be used for the calculation of the output of the cylindrical chemical flow reactors used in industry. In industry the ratio V_0/u, which has dimensions of time, is known as "contact time," because it represents the averge time it takes a molecule to pass through a reactor of volume V_0.

Stopped Flow

In a variation of the above method, the flow is stopped suddenly, and then the rate at which the system reaches equilibrium is followed. This is the "stopped flow method." In commercially available instruments the flow is started and stopped using power-driven syringes. Some instruments also have provision for multimixing (e.g., with a third reagent).

Observation Techniques

The reactions studied by flow methods are fast; hence the method of analysis (observation) must also be fast. The most widely used method is UV-visible spectroscopy. The signal from the photomultiplier is often transmitted to a storage oscilloscope. Alternatively, the signal can be transferred to a transient recorder coupled to a programmable calculator or computer for online data treatment. The information processed can be absorbance as a function of time or absorbance as a function of distance from the mixing point.

UV-visible spectroscopy is a sensitive method; concentrations as low as 10^{-4}–10^{-6} M or even lower can be detected. Spectroscopy is also quite accurate. However, it is not "diagnostic" enough; the structural information it provides does not differentiate sufficiently between various substances.

Other popular detection methods include fluorescence and conductivity as well as NMR and EPR. A combination of such methods is also possible. Furthermore, with multimixing devices a rapid quencher can be introduced into the reacting mixture, in order to stop the reaction at a present time or study the reactions or properties of the intermediates themselves.

EXAMPLE

One of the early reactions studied using flow techniques was the decomposition of carbonic acid, H_2CO_3:

$$H_2CO_3 \longrightarrow H_2O + CO_2 \qquad\qquad 4.47$$

Carbonic acid is formed by mixing a solution of $NaHCO_3$ with a solution of HCl in a flow apparatus:

$$NaHCO_3 + HCl \rightleftharpoons H_2CO_3 + NaCl$$

The decomposition reaction 4.47 is followed by measuring the pH (pH < 8). The initial solutions contain an indicator (e.g., bromophenol blue), the color of which changes because during the reaction the concentration of hydrogen ion changes. The rate law is

$$-\frac{d[H_2CO_3]}{dt} = \frac{d[CO_2]}{dt} = k[H_2CO_3]$$

Generally, if there are no intermediates or if the concentration of the intermediates can be considered constant, the rate of disappearance of the reactants multiplied by the appropriate stoichiometric factor (1 in this example) equals the rate of product formation.

At 18°C the rate constant k for the decomposition of carbonic acid is 12.3 s^{-1}, which corresponds to a half-life of 0.056 s.

4.10. FLASH PHOTOLYSIS—PULSE RADIOLYSIS

The chemical system under investigation is illuminated with an intense light flash of short duration, and the concentrations of free radicals, atoms, and excited species formed during the flash are followed spectrometrically. The method is applicable only if one or more of the components can be excited or decomposed photochemically. In pulsed radiolysis the decomposition is caused by a pulsed beam of high-energy electrons.

EXAMPLE I

The determination of the rate of the reaction

$$I + I + M \longrightarrow I_2 + M \qquad\qquad 4.48$$

is a simple example of applying flash photolysis. A glass vessel containing diodine, I_2, is flashed with light of the appropriate frequency, which splits the I_2:

$$I_2 + h\nu \longrightarrow 2I$$

The reaction 4.48 is followed spectroscopically.   □

EXAMPLE II

The reaction of Cr^{2+} with O_2 in aqueous solutions is too fast to be followed by flow techniques. By the time the separate solutions are mixed, the reaction is over. However, it can be studied by generating chromium(II) *in situ* by pulsed radiolysis from chromium(III) in a solution already containing O_2.

It is well known that irradiation of water with ionizing radiation leads to formation of hydrated electrons, hydrogen atoms, hydroxyl radicals, hydrogen peroxide, and dihydrogen. In the presence of formate ion and oxygen, H and

OH are quenched, and hydrated electrons react with chromium(III) to give chromium(II):

$$Cr^{3+} + e^-(aq) \longrightarrow Cr^{2+}$$

After the pulse is completed, chromous ion reacts further with O_2 via a quite stable intermediate, CrO_2^{2+}. □

4.11. PICOSECOND AND FEMTOSECOND SPECTROSCOPY

Picosecond (10^{-12} s) and femtosecond (10^{-15} s) spectroscopy were made possible by the development of laser systems that can give intense picosecond or shorter pulses at various wavelengths, along with the simultaneous measurement of time, wavelength, and intensity. How this is achieved technically is not within the scope of this book.

EXAMPLE I

Excitation of the stretching vibration of the C—H bond in ethanol occurs at 2900 cm^{-1}. Picosecond spectroscopy shows that during relaxation this vibration is transformed into two bending vibrations of the C—H bond at 1450 cm^{-1}. □

EXAMPLE II

The first stage in the process of vision is the excitation of rhodopsin, which subsequently is partially deactivated and forms the first intermediate, prelumirhodopsin or bathorhodopsin. It was shown by using picosecond spectroscopy that prelumirhodopsin results from an intramolecular proton transfer—a "jump" of a proton from one position to another. □

4.12. TEMPERATURE JUMP AND ACOUSTICAL METHODS

A system at equilibrium can be perturbed by a sudden change in temperature or pressure or by a periodic change, such as a change in pressure or temperature brought about by the passage of sound waves through the system. For a finite but small temperature jump δT, the equilibrium constant changes according to the equation

$$\ln \frac{K_1}{K_2} = - \frac{\Delta H^\circ \delta T}{RT^2}$$

where ΔH° is the standard enthalpy difference between products and reactants.

Relaxation to the new equilibrium state is controlled by the rates of the chemical reactions involved. The development of the reaction is usually followed spectrophotometrically.

The jump in temperature can be accomplished by discharging a high-voltage (kV) capacitor through a small volume of solution. In order to decrease the resistance, an inert salt is also added. Typical rise times in 0.1 M solutions are on the order of 1–5 μs.

Treatment of Relaxation Data: Simple Relaxation

Consider the reaction

$$A + B \underset{k_2}{\overset{k_1}{\rightleftharpoons}} C \qquad 4.49$$

If the concentrations in the new equilibrium position are symbolized by $\bar{c}_A$, $\bar{c}_B$, and $\bar{c}_C$, and the deviations at time t from these final values by $\Delta\bar{c}_A$, $\Delta\bar{c}_B$, and $\Delta\bar{c}_C$, the concentrations at time t are $(\bar{c}_A - \Delta\bar{c}_A)$, $(\bar{c}_B - \Delta\bar{c}_B)$, and $(c_C - \Delta\bar{c}_C)$. The rate of formation of c_C is

$$\frac{d(\bar{c}_C - \Delta\bar{c}_C)}{dt} = k_1(\bar{c}_A - \Delta\bar{c}_A)(\bar{c}_B - \Delta\bar{c}_B) - k_2(\bar{c}_C - \Delta\bar{c}_C)$$

from which it follows that

$$\frac{d\bar{c}_C}{dt} - \frac{d(\Delta\bar{c}_C)}{dt} = k_1\bar{c}_A\bar{c}_B - k_2\bar{c}_C - k_1\bar{c}_B\,\Delta\bar{c}_A$$

$$- k_1\bar{c}_A\,\Delta\bar{c}_B + k_1\,\Delta\bar{c}_A\,\Delta\bar{c}_B + k_2\,\Delta\bar{c}_C \qquad 4.50$$

At equilibrium the rates of the two opposing reactions become equal, that is,

$$\frac{d\bar{c}_C}{dt} = k_1\bar{c}_A\bar{c}_B - k_2\bar{c}_C = 0 \qquad 4.51$$

In view of 4.51, equation 4.50 becomes

$$-\frac{d(\Delta\bar{c}_C)}{dt} = -k_1\bar{c}_B\,\Delta\bar{c}_A - k_1\bar{c}_A\,\Delta\bar{c}_B + k_1\,\Delta\bar{c}_A\,\Delta\bar{c}_B + k_2\,\Delta\bar{c}_C \qquad 4.52$$

For the techniques discussed here, deviations from equilibrium are generally small and the term $k_1\,\Delta\bar{c}_A\,\Delta\bar{c}_B$ can be neglected. Moreover, from the stoichiometry it follows that $\Delta \bar{c}_C = -\Delta \bar{c}_A = -\Delta \bar{c}_B$. Thus equation 4.52 becomes

$$-\frac{d(\Delta \bar{c}_C)}{dt} = k_1(\bar{c}_A + \bar{c}_B) \Delta \bar{c}_C + k_2 \Delta \bar{c}_C$$

or

$$-\frac{d(\Delta \bar{c}_C)}{dt} = k_{obs} \Delta \bar{c}_C \qquad\qquad 4.53$$

where k_{obs} is the observed rate constant $[k_1(\bar{c}_A + \bar{c}_B) + k_2]$ and equals the reciprocal of the relaxation time, τ^{-1}. Integration of equation 4.53 yields

$$\Delta c_C = (\Delta \bar{c}_C)_0 \exp(-k_{obs}t)$$

The plot of the observed rate constant k_{obs} versus $\bar{c}_A + \bar{c}_B$ gives a straight line with slope k_1 and intercept k_2.

For reactions other than equation 4.49, the general expression 4.53 is still valid, but the reciprocal of the relaxation time is different (Table 4.6). In all these cases, the inverse of the relaxation time—the observed rate constant for the reestablishment of the equilibrium—depends not only on the rate of the forward reaction but also on the rate of the back-reaction.

As a numerical example, consider the data shown in Table 4.7 for an equilibrium of the form

$$2A \underset{k_2}{\overset{k_1}{\rightleftharpoons}} B + H_2O$$

The reciprocal of the relaxation time is

$$\tau^{-1} = 4k\bar{c}_A + k_2(\bar{c}_B + \bar{c}_{H_2O})$$

TABLE 4.6. Expressions for the Relaxation Times of Various Reactions

Reaction	Relaxation Time
$A + B \underset{k_2}{\overset{k_1}{\rightleftharpoons}} C + D$	$\tau^{-1} = k_1(\bar{c}_A + \bar{c}_B) + k_2(\bar{c}_C + \bar{c}_D)$
$A \underset{k_2}{\overset{k_1}{\rightleftharpoons}} B$	$\tau^{-1} = k_1 + k_2$
$2A \underset{k_2}{\overset{k_1}{\rightleftharpoons}} B$	$\tau^{-1} = 4k_1\bar{c}_A + k_2$
$A + B + C \underset{k_2}{\overset{k_1}{\rightleftharpoons}} D$	$\tau^{-1} = k_1(\bar{c}_A\bar{c}_B + \bar{c}_B\bar{c}_C + \bar{c}_C\bar{c}_A) + k_2$

TABLE 4.7. Relaxation Data

τ^{-1}, s^{-1}	$\bar{c}_A$, M
0.0342	0.0010
0.0377	0.0015
0.0413	0.0020
0.0450	0.0025
0.0488	0.0030
0.0524	0.0035
0.0557	0.0040
0.0595	0.0045
0.0630	0.0050
0.0700	0.0060
0.0774	0.0070
0.0810	0.0075

However,

$$\bar{c}_{H_2O} \gg \bar{c}_B.$$

Therefore,

$$\tau^{-1} \simeq 4k\bar{c}_A + k_2\bar{c}_{H_2O}$$

It should be noted that $\bar{c}_{H_2O}$ is a constant. A plot of the experimental τ^{-1} versus $\bar{c}_A$, as shown in Figure 4.5, gives a straight line. From the slope, $k_1 = 1.8\ M^{-1}\ s^{-1}$; and from the intercept, $k_2 = 4.86 \times 10^{-4}\ M^{-1}\ s^{-1}$.

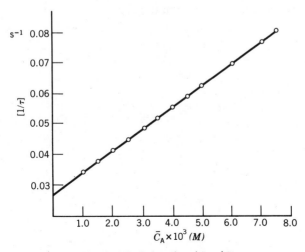

Figure 4.5. Relaxation time plot.

Multiple Relaxations

Consider the simple case of two consecutive first-order equilibria:

$$A \underset{k_2}{\overset{k_1}{\rightleftharpoons}} B \underset{k_4}{\overset{k_3}{\rightleftharpoons}} C$$

The complete system of rate equations is the following:

$$\frac{dc_A}{dt} = -k_1 c_A + k_2 c_B \qquad 4.54$$

$$\frac{dc_B}{dt} = k_1 c_A - k_2 c_B - k_3 c_B + k_4 c_C \qquad 4.55$$

$$\frac{dc_C}{dt} = k_3 c_B - k_4 c_C \qquad 4.56$$

Any two of these equations are independent; the third can be eliminated by using mass balance. By considering equations 4.54 and 4.56 and by keeping the notation used in the previous example, we obtain

$$\frac{d\Delta \bar{c}_A}{dt} = -k_1 \, \Delta \bar{c}_A + k_2 \, \Delta \bar{c}_B \qquad 4.57$$

$$\frac{d\Delta \bar{c}_C}{dt} = k_3 \, \Delta \bar{c}_B - k_4 \, \Delta \bar{c}_C \qquad 4.58$$

With the aid of the mass balance relation,

$$\Delta c_B = -\Delta \bar{c}_A - \Delta \bar{c}_C$$

equations 4.57 and 4.58 are put into the form

$$-\frac{d\Delta \bar{c}_A}{dt} = (k_1 + k_2) \, \Delta \bar{c}_A + k_2 \Delta \bar{c}_C$$

$$= a_{11} \Delta \bar{c}_A + a_{12} \Delta \bar{c}_C$$

$$-\frac{d\Delta \bar{c}_C}{dt} = k_3 \Delta \bar{c}_A + (k_3 + k_4) \, \Delta \bar{c}_C$$

$$= a_{21} \Delta \bar{c}_A + a_{22} \Delta \bar{c}_C$$

where $a_{11} = (k_1 + k_2)$, $a_{12} = k_2$, $a_{21} = k_3$, and $a_{22} = k_3 + k_4$, or in matrix form,

$$
\begin{bmatrix} \dfrac{d\Delta\bar{c}_A}{dt} \\[2ex] \dfrac{d\Delta\bar{c}_C}{dt} \end{bmatrix}
=
\begin{bmatrix} a_{11} & a_{12} \\[2ex] a_{21} & a_{22} \end{bmatrix}
\begin{bmatrix} \Delta\bar{c}_A \\[2ex] \Delta\bar{c}_C \end{bmatrix}
\qquad 4.59
$$

The solutions of this homogeneous system of linear equations with constant coefficients are of the form

$$
\Delta\bar{c}_A = m_1 \exp\left(-\frac{t}{\tau}\right)
$$

and

$$
\Delta\bar{c}_B = m_2 \exp\left(-\frac{t}{\tau}\right)
$$

The constants m_1 and m_2 are obtained by solving the algebraic system

$$
(a_{11} - \tau^{-1})m_1 + a_{12}m_2 = 0
$$
$$
a_{21}m_1 + (a_{22} - \tau^{-1})m_2 = 0
$$

which has nontrivial solutions only if

$$
\begin{bmatrix} a_{11} - \tau^{-1} & a_{12} \\[2ex] a_{21} & a_{22} - \tau^{-1} \end{bmatrix} = 0
\qquad 4.60
$$

The reciprocal times obtained from the vanishing determinant 4.60 are the eigenvalues of the matrix in equation 4.59.

The expressions for τ^{-1} obtained from 4.60 are

$$
\tau_1^{-1} = \tfrac{1}{2}(a_{11} + a_{22}) + \tfrac{1}{2}[(a_{11} + a_{22})^2 - 4(a_{11}a_{22} - a_{12}a_{21})]^{1/2}
$$

$$
\tau_2^{-1} = \tfrac{1}{2}(a_{11} + a_{22}) - \tfrac{1}{2}[(a_{11} + a_{22})^2 - 4(a_{11}a_{22} - a_{12}a_{21})]^{1/2}
$$

If $a_{11} \gg a_{22}$, two separate relaxations are observed, corresponding to two equilibria with the following inverse relaxation times:

$$\frac{1}{\tau_1} \simeq a_{11} = k_1 + k_2$$

$$\frac{1}{\tau_2} \simeq a_{22} - \frac{a_{12}}{a_{11}} a_{21} = \frac{k_3}{1 + k_2/k_1} + k_4$$

If $a_{22} \gg a_{11}$, again two relaxations are observed, with

$$\frac{1}{\tau_1} \simeq a_{22} = k_3 + k_4$$

$$\frac{1}{\tau_2} \simeq a_{11} - \frac{a_{21}}{a_{22}} a_{12} = \frac{k_2}{1 + k_3/k_4} + k_1$$

Graphical methods can be used to resolve differences of about a factor of 4 or 5, and computer methods can be used to resolve τ values differing by a factor of about 2 or less.

Generalization

By manipulations such as those just described, differential equations *for small deviations from equilibrium* can be placed in the following linear form:

$$-\frac{d\Delta c_i}{dt} = \sum_{j=1}^{n} a_{ij} \Delta \bar{c}_j$$

This can be done even if the original equations in terms of concentrations are nonlinear, because for small differences, terms of second or higher order are neglected. In matrix form:

$$
\begin{bmatrix}
\dfrac{d\Delta \bar{c}_1}{dt} \\[2ex]
\dfrac{d\Delta \bar{c}_2}{dt} \\[2ex]
\vdots \\[2ex]
\dfrac{d\Delta \bar{c}_n}{dt}
\end{bmatrix}
=
\begin{bmatrix}
a_{11} & a_{12} & \cdots & a_{nn} \\
a_{21} & a_{22} & \cdots & a_{2n} \\
 & & \vdots & \\
a_{n1} & a_{n2} & \cdots & a_{nn}
\end{bmatrix}
\begin{bmatrix}
\Delta \bar{c}_1 \\[2ex]
\Delta \bar{c}_2 \\[2ex]
\vdots \\[2ex]
\Delta \bar{c}_n
\end{bmatrix}
$$

or, more compactly,

$$\frac{d\overline{C}(t)}{dt} = \overline{A}\,\overline{C}(t)$$

The solution is

$$\overline{C}(t) = \overline{C}_0 \exp(-\overline{A}t)$$

The eigenvalues of matrix $\overline{A}$ are the reciprocal relaxation times.

The Use of Indicators in Relaxation Studies

The study of many relaxation processes is difficult because the absorbance changes are small. The solution to the problem in such cases may be the use of indicators. The reaction

$$(H_2O)_3Ni(terpy)^{2+} + NH_3 \underset{k_r}{\overset{k_f}{\rightleftharpoons}} (H_2O)_2NH_3Ni(terpy)^{2+} + H_2O$$

where terpy is 2,2':6':2"-terpyridine*, for example, was followed (in fact, the change in acidity associated with the reaction was followed) by using phenol red as an indicator. The following coupled reactions were taking place:

$$NH_3 + H^+ \rightleftharpoons NH_4^+ \qquad K_H = 1.6 \times 10^9$$

$$In^- + H^+ \rightleftharpoons HI_n \qquad K_I = 8 \times 10^7$$

The relaxation time is given by the equation

$$\frac{1}{\tau} = k_f \left(\frac{M_{eq}}{1 + \alpha} + L_{eq} \right) + k_r \qquad 4.61$$

where

$$\alpha = \frac{K_H[H^+](K_I[In^-] + K_I[H^+] + 1)}{K_H[NH_3](K_I[H^+] + 1) + K_I(In^-] + [H^+]) + 1} \qquad 4.62$$

and M_{eq}, L_{eq} are the equilibrium concentrations of the complex and ammonia, respectively, at the higher temperature.

---

*2,2':6',2"-terpyridine

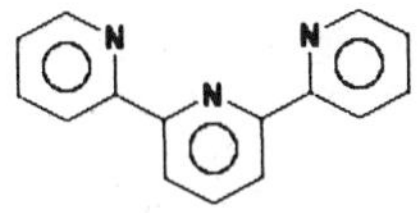

Most of the measurements were made with $6 \times 10^{-3}\ M$ ammonia at pH 7.4, which corresponds to a hydrogen ion concentration of $4 \times 10^{-8}\ M$, and with $5 \times 10^{-6}\ M$ indicator. By carefully choosing an indicator with the correct dissociation constant and molar absorptivity, $\alpha \ll 1$. Under these conditions, equation 4.61 simplifies to

$$\frac{1}{\tau} = k_f(M_{eq} + L_{eq}) + k_r$$

Fitting the Relaxation Times to a Mechanism

The reaction between oxovanadium(IV) and glycine has been studied by a large number of experimental techniques including rapid mixing, stopped-flow, temperature-jump, and proton NMR line broadening.

The following equations depict the equilibria in the system under consideration:

$$\text{VOHGly}^{2+} \underset{k_{-1}}{\overset{k_1}{\rightleftharpoons}} \text{VOGly}^+ + \text{H}^+ \qquad K_1 \qquad\qquad 4.63$$

$$\text{VOGly}^+ \underset{k_{-2}}{\overset{k_2}{\rightleftharpoons}} \text{VO(Gly)}^+ \qquad\qquad K_2 \qquad\qquad 4.64$$

In VOGly$^+$, glycine is coordinated as a monodentate ligand and VO(Gly)$^+$ is a bidentate glycine complex. The relaxation times are summarized in Table 4.8.

On the basis of these data, an expression for τ can be derived:

$$\frac{1}{\tau} = \frac{k_2 K_1([\text{H}^+] + [\text{VOHGly}^{2+}])}{[\text{H}^+]^2 + K_1([\text{H}^+] + [\text{VOHGly}^{2+}])} + k_{-2}$$

TABLE 4.8. Typical Relaxation Times in the Reaction between Oxovanadium(IV) and Glycine

pH	$[\text{VO}^{2+}]_{eq}$, $\times 10^3\ M$	$[\text{VOHGly}^{2+}]_{eq}$, $\times 10^3\ M$	$[\text{VOGly}^+]_{eq}$, $\times 10^3\ M$	$m\tau$, s
3.879	6.30	6.63	2.68	16.7
3.368	6.66	10.19	1.26	21.8
3.408	6.52	10.05	1.36	21.9
3.458	6.33	9.85	1.50	18.9
3.542	6.02	9.46	1.75	17.9
3.642	5.62	8.94	2.08	18.0
3.512	5.02	10.36	1.78	19.4
3.509	7.28	11.36	1.94	18.2

Since K_1 is known independently to have a value of 5.30×10^{-5}, and since the concentrations of $[H^+]$ and $[VOHGly^{2+}]$ are also known, the graphically evaluated values of k_2 and k_{-2} are

$$k_2 = 37 \text{ s}^{-1} \qquad k_{-2} = 21 \text{ s}^{-1}$$

and

$$K_2 = \frac{k_2}{k_{-2}} = 1.8$$

4.13. EMISSION SPECTROSCOPY

Figure 4.6 graphically represents a transition from the ground state to an electronically excited state. The transition is to higher vibrational levels of the electronically excited state.

Excitation is followed first by a return to the lower vibrational levels of the electronically excited state, and then, with the emission of a photon, the system returns to the electronic ground state but to higher vibrational levels of this state. The last stage is relaxation to the equilibrium distribution of vibrational levels.

In Figure 4.6 the minimum of the curve for the excited state is shifted to the right relative to the corresponding minimum of the ground state. In physical terms this means that the geometry of the excited state is different from that of the ground state; some bond lengths and/or bond angles have changed.

If the spin multiplicities of the ground and excited states are the same, the emission is called *fluorescence*. If they are different, it is called *phosphorescence*. Usually lifetimes of excited states are shorter in fluorescence than in

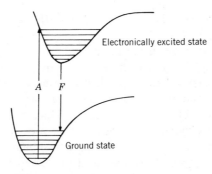

Figure 4.6. Absorption of light, A, and fluorescence, F. The energy differences (and photon frequencies) are not the same.

phosphorescence. The study of the competition between photon emission and other processes involving the excited state (such as chemical reactions, intersystem crossing, and thermal deactivation) supplies information about these processes.

Identification of the excited states participating in these processes is by no means simple, and it is frequently the subject of investigation. It took a lot of effort, for example, to establish that in the photochemically induced substitution of the complex $Cr(NH_3)_2(NCS)_4^-$ the excited state involved is the $^4L_1$ state, resulting from the lowest energy d–d transition.

More details on the reactions of excited states will be presented in Chapter 9 on inorganic photochemistry.

4.14. ELECTROCHEMICAL METHODS

Parameters and Processes

The uses of electrochemical methods in mechanistic studies can be classified into three categories:

1. Analysis of reactant and/or products
2. Study of the mechanisms of the processes taking place in electrochemical cells
3. Use of an electrochemical method to generate and study species that are important in other reactions

The first application is straightforward. The electrochemical method is used simply as an analytical tool, provided, of course, that the corresponding properties of the products differ from those of the reactants and that sufficient time is available to perform the measurements.

As a simple example consider the reaction

$$Co(en)_2Cl_2^+ + H_2O \longrightarrow Co(en)_2Cl(H_2O)^{2+} + Cl^-$$

which involves a net increase in charge and therefore a change in conductivity.

The transport of charge across the electrode interfaces, as reflected in the current–voltage characteristics of an electrochemical cell, is intimately connected to the mechanism of the processes taking place at these interfaces and also in the bulk of the electrolyte solution. In fact, it must be recalled that the flow of charge is a dynamic, nonequilibrium process. The use of electrochemical cells in obtaining thermodynamic data should not allow this fact to be forgotten. The study of electrochemical processes is essentially a special type of mechanistic study involving the flow of charges over long distances, not just a transfer between molecular species.

The potential–current characteristics of an electrochemical cell generally depend on *parameters* such as temperature, concentration, and solvent, which affect the rates of the processes taking place in the cell. More specifically, these *processes* are:

Diffusion to and from the electrodes.

Electron transfer and other chemical reactions on the surface (including adsorption and desorption).

Chemical reactions taking place in the bulk of the electrolyte solution.

Thus the potential–current characteristics at various temperatures and concentrations will in principle supply information on all these processes. In some cases, however, the situation is simpler, because some processes may not take place or may be too fast compared to the others. One such simple case is the competition between diffusion of charges and the neutralization reaction:

$$CH_3COCOO^- + H^+ \xrightarrow{k} CH_3COCOOH$$

In the polarographic measurement of the rate ($k = 1.3 \times 10^{10} \, M^{-1} \, s^{-1}$ at 25°C), this rate is compared to the rate of diffusion of charges toward the dropping mercury electrode.

One variation of the third category of applications of the electrochemical methods is to simply use the cell as a generator of unstable species that are suspected intermediates in a reaction and study their behavior (including the way they decay) either electrochemically or in combination with other techniques, such as UV-visible spectroscopy.

Time Scale of Electrochemical Methods

The time scale of the various electrochemical techniques is summarized in Table 4.9.

Special Characteristics

Some features are peculiar to electrochemical methods:

They are restricted to systems containing electrolytes (liquid solutions, fused salts, conducting solids).

The cell contains more than one interface. Homogeneous and heterogeneous phenomena are intermingled. The systems are inherently complicated.

The situation is somewhat simplified if one of the electrodes (the reference electrode) has relatively simple and well-characterized behavior.

Electrochemical methods are especially useful in determining the re-

TABLE 4.9. Time Scale of Various Electrochemical Techniques

Technique	$t_{1/2}$, s
AC polarography[a]	2×10^{-4} to 10^{-1}
Rotating disc electrode voltammetry[a]	10^{-3} to 3×10^{-1}
Chronopotentiometry[a] Chronoamperometry[a] Chronocoulometry[a]	10^{-3} to 50
Voltammetry, linear and cyclic[a]	10^{-4} to 1
DC polarography[a]	1 to 5
Coulometry[b] Macroscale electrolysis[b]	100 to 3000

[a]Small A/V conditions. The electrochemical measurement does not appreciably perturb the system; it does not alter bulk concentrations.

[b]The composition of the system is altered drastically.

versibility or irreversibility of the reactions taking place in the bulk of the solution or the overall reversibility of the cell (of all the processes taking place in the cell).

4.15. NUCLEAR MAGNETIC RESONANCE (NMR)

Resonance and Relaxation

Let us first recall the basic principles of the NMR phenomenon. In the absence of a magnetic field, the nuclear spins I are randomly oriented (Fig. 4.7a). If, however, a strong magnetic field $\mathbf{B}_0$ directed along the z axis is applied, the magnetic vectors of the nuclei take $2I + 1$ quantized orientations. For a proton, $I = \frac{1}{2}$ and the number of orientations is 2 (Fig. 4.7b). The term "orientation" here does not mean a certain direction; it means precession around axes parallel to $\mathbf{B}_0$ with the tops of the precession cones up or down (Fig. 4.7b). Spin is a quantum-mechanical concept. Hence "classical" pictures for individual spins such as that of Figure 4.7b are not strictly correct. They are appropriate only for overall macroscopic magnetization. However, chemists frequently use "inaccurate" illustrations (e.g., the dual particle–wave picture for the electron) in order to gain a better understanding.

The frequency of precession, the so-called Larmor (resonance) frequency, is given by

$$\omega_0 = 2\pi\nu_0 = \gamma\mathbf{B}_0 \qquad\qquad 4.65$$

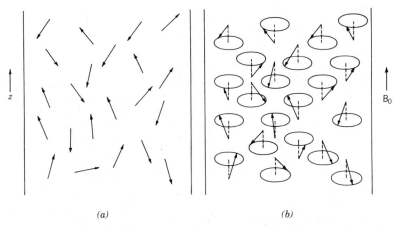

Figure 4.7. (*a*) Random orientation of nuclear spins in the absence of a magnetic field. (*b*) Orientation and precession in the presence of a strong magnetic field for $I = \frac{1}{2}$.

where γ is the magnetogyric ratio (2.675×10^8 rad s^{-1} T^{-1} for the proton), ν_0 is the frequency, and ω_0 is the corresponding angular velocity. Some of the commonly used nuclei are tabulated in Table 4.10.

The number of ^{1}H nuclei oriented in the direction of the field (sometimes called "parallel," although it is not really parallel) is larger than the number of nuclei having the opposite ("antiparallel") orientation. Parallel orientation corresponds to lower energy. The difference in energy equals $h\nu_0$. Because of the excess population in the lower level (and provided the excess is sustained), a net absorption of a radiofrequency ν_0 will be observed.

In the *continuous slow passage NMR* experiment, the radiofrequency is

TABLE 4.10. Nuclei Commonly Used in NMR Studies

Nucleus	Spin	Resonance[a] Frequency	Natural Abundance	Relative Sensitivity
^{1}H	$\frac{1}{2}$	100	99.985	1.000
^{2}D	1	15.4	0.015	0.0245
^{13}C	$\frac{1}{2}$	25.2	1.108	0.0317
^{14}N	1	7.2	99.635	0.0374
^{17}O	$\frac{5}{2}$	13.6	0.037	0.0789
^{19}F	$\frac{1}{2}$	94.0	100.0	0.858
^{31}P	$\frac{1}{2}$	40.5	100.0	0.104
^{35}Cl	$\frac{2}{2}$	9.8	75.4	0.0150
^{55}Mn	$\frac{5}{2}$	24.9	100.0	0.357
^{63}Cu	$\frac{2}{2}$	26.5	69.1	0.181
^{79}Br	$\frac{2}{2}$	25.2	50.52	0.157

[a]Megacycles, for a 1 Tesla (10 kG) field.

changed slowly through the resonance frequency v_0. In *pulsed NMR*, a radiofrequency equal to v_0 is suddenly applied in the form of a pulse of short duration.

The orientation effects of the two fields—the strong static field $\mathbf{B}_0$ and the small field that oscillates with the radiofrequency—are opposed by *relaxation,* which tends to randomize the spins. Under the conditions of the NMR experiment, directions are not all equivalent: the direction along z is drastically different from directions in the xy plane because of the strong magnetic field $\mathbf{B}_0$. Relaxation (randomization) is differentiated accordingly. The corresponding first-order rate constants for randomization along $\mathbf{B}_0$ (T_1^{-1}, "longitudinal") and perpendicular to $\mathbf{B}_0$ (T_2^{-1}, "transverse") are different. The *relaxation times T_1 and T_2* are also called spin–lattice and spin–spin relaxation times, respectively. These terms, however, may be somewhat misleading, as they can give the impression that the processes causing relaxation in the two directions are completely different. The fact is, the main interactions leading to relaxation operate in all directions, but with different effectiveness along $\mathbf{B}_0$ and perpendicular to it.

NMR Parameters

The parameters that can be obtained from an NMR experiment are the *relaxation times,* the *chemical shifts,* and the *coupling constants.* The observed resonance frequency is not exactly that predicted from equation 4.65 on the basis of only the magnitude of the externally applied field, $\mathbf{B}_0$. It is shifted because of the local fields created by the chemical environment around the nucleus.

Interactions (coupling) with other nearby nuclei lead to a splitting of the observed peaks. With paramagnetic species there are also interactions between electronic and nuclear spins.

Relaxation times depend on the interactions with other nuclei and electrons, but they also depend on chemical processes, such as intra- or intermolecular exchange. In an exchange process the magnetic nuclei change sites, that is, they go from one kind of minimum in the potential energy profile to another. These minima generally correspond to magnetically different environments. In the proton exchange process, for example,

$$CrNH + OH^- \rightleftharpoons CrN^- + H_2O$$

the change in the magnetic environment of the proton is caused by a corresponding change in the chemical environment. However, chemical equivalence of two sites does not necessarily also mean magnetic equivalence. In the compound $(phCH_2)_3AsF_aF_b$ there are three different sites for the $-CH_2-$ protons arising from the magnetic coupling with various combinations of the two fluorine spins (^{19}F, $I = \frac{1}{2}$):

	F_a	F_b
First combination	$\frac{1}{2}$	$\frac{1}{2}$
Second combination $\left\{\begin{array}{@{}c@{}} \\ \\ \end{array}\right.$	$\frac{1}{2}$	$-\frac{1}{2}$
	$-\frac{1}{2}$	$\frac{1}{2}$
Third combination	$-\frac{1}{2}$	$-\frac{1}{2}$

It should also be noted that the exchange of sites may involve substantial differences in chemical shift, but this is not necessary. Complex formation, for example, of the general form

$$A(solv) + B(solv) \rightleftharpoons AB(solv)$$

may not involve substantial changes in the chemical shift of B, but it will most certainly affect the relaxation times.

Mechanistic Information from NMR

The relaxation times provide both structural and kinetic information. In fact, this combination is often mechanistically very instructive. Chemical shifts and coupling constants provide structural information.

The special significance of NMR in mechanistic studies is that over a large temperature range (between ~100 and ~500 K) it provides information about species in solution. Low temperatures are appropriate for detection and structural characterization of relatively unstable intermediates that cannot be isolated or that do not live long enough at higher temperatures to be detected.

Other useful aspects of NMR are the structural information it can provide on labile systems and the information on exchange processes under equilibrium conditions.

One of the disadvantages (with presently available NMR techniques) is the inability to observe very unstable intermediates present at very low concentrations. The concentrations required are quite high. Repeated scanning and time averaging of the signal improve the sensitivity, but what is gained in sensitivity is to a large extent lost in time, because it may take several minutes to carry out these operations.

Another disadvantage of NMR is that it cannot be used for some paramagnetic species, although in such cases it may be possible to use EPR, since there is some complementarity between the two techniques.

Chemical reactions that can be studied by NMR may have rate constants in the range from 5×10^{-3} to $10^6 \, \text{s}^{-1}$. The samples needed are small (fractions of a milliliter), but the concentrations are somewhat high (~10^{-3} M with modern Fourier transform instruments). Another characteristic of NMR is that it is usually applied to equilibrium systems. However, techniques have also been developed for applications in nonequilibrium systems, for example, in combination with flow, stopped-flow, or flash photolysis. NMR–flow or NMR–stopped-flow combinations cover a range of reaction lifetimes from 5 min to 50 ms. The usefulness of NMR for the study of nonequilibrium systems is limited by its inherent disadvantages, namely, its insensitivity and its time characteristics.

An NMR Application: Exchange Processes by ^{17}O NMR

As an example of an NMR application in mechanistic studies, consider the exchange reactions of the general form:

$$M(OH_2)_n^{m+} + H_2^{17}O \underset{k'_{ex}}{\overset{k_{ex}}{\rightleftharpoons}} M(OH_2)_{n-1}(^{17}OH_2)^{m+} + H_2O \qquad 4.66$$

where M is a paramagnetic ion. Oxygen-17 NMR line-broadening techniques are applicable to such reactions if $10^3 \text{ s}^{-1} \leqslant k_{ex} < 10^8 \text{ s}^{-1}$.

The magnitude of the linewidth at half-height of the ^{17}O signal with a paramagnetic ion dissolved in ^{17}O-enriched water is affected by the spin relaxation rates of bulk and coordinated water, the difference in chemical shift

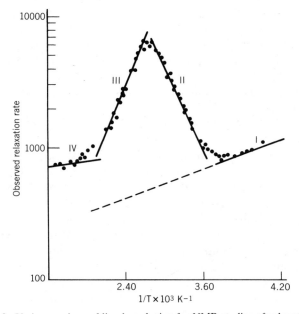

Figure 4.8. Various regions of line broadening for NMR studies of solvent exchange.

in these two environments, and the rate of the chemical exchange process represented by equation 4.66.

At low temperatures, long-range interactions between bulk solvent and the paramagnetic metal ion dominate (region I in Fig. 4.8). Region I is frequently called the *outer-sphere broadening region.*

Region II is dominated by the exchange process (equation 4.66). The observed half-linewidth in this region (which is also called the *exchange-controlled* region) *increases* rapidly with temperature.

In regions III and IV, the exchange is already very fast. The lifetime of water in the first coordination sphere is too short for complete relaxation to occur each time water is coordinated to the metal ion. Several encounters between solvent molecules and metal ion occur, and the linewidth is no longer directly proportional to the exchange rate. Region III is usually called the *fast-exchange* or *relaxation-controlled region.* The temperature dependence in region III is non-Arrhenius, that is, the linewidth decreases with increasing temperature.

4.16. ELECTRON PARAMAGNETIC RESONANCE (EPR)

A Brief Comparison of NMR and EPR

Nuclear spin is a nondisturbing probe for studying structure and following the course of chemical reactions; its effect on chemical properties is negligible. In contrast, the spin of an unpaired electron is not only a probe; it is also related to chemical behavior.

During a chemical reaction the nuclei generally change positions, but there is no change in their "shape" or "size." In contrast, during a chemical reaction the "shape" and "size" of the electron "cloud" change.

Since the nuclei occupy a negligible part of the atomic or molecular volume, it is meaningful to talk about the effect on their spin of local fields (chemical shift) or about the shielding from the external field provided by the immediate environment. On the other hand, an unpaired electron occupies a large part of the atomic or molecular volume, and it makes no sense to speak of local fields created by the rest of the chemical environment. There is no well-defined boundary between the system considered and its environment. Accordingly, in the basic EPR equation,

$$\Delta E = h\nu = g\beta H \qquad 4.67$$

H symbolizes the applied field, not the effective field (applied plus local) as in NMR. The effect of the environment is incorporated into the values of g.

Energy separations in EPR (equation 4.67) are generally larger than in NMR. The frequencies used for EPR lie in the microwave region. Also note that whereas for nuclear spin, $I = +\frac{1}{2}$ corresponds to lower energy and $I = -\frac{1}{2}$ to higher energy, for electron spin the signs are reversed. In equation 4.67, β is the Bohr magneton and equals 9.2741×10^{-17} erg T^{-1}. The value of g for a free electron is $g_e = 2.00232$.

The EPR experiment is usually performed by keeping the frequency constant and changing the field until the resonance condition is reached. The microwave field is applied continuously, but instruments with pulsed microwave generators are also available.

Parameters of EPR

EPR provides both structural and mechanistic information about free radicals and paramagnetic metal ions. The *relevant parameters* are *the values of g, their isotropy or anisotropy, hyperfine and fine splitting constants*, and *relaxation times*.

The hyperfine splitting of EPR peaks results from interactions of the electronic spin with nearby nuclear spins. Nuclei with nonzero spin and unpaired electrons interact with the externally applied magnetic field, but they also interact with each other. This additional interaction leads to splitting of the levels and of the EPR peaks.

Spin–spin interaction depends, of course, on the magnitude of the nuclear and electron spin magnetic moments, but it also depends on how strongly these spins are coupled—on the *hyperfine coupling constant* α. Hyperfine splitting constants are obtained from the corresponding separation of the peaks, which equal $\alpha/g\beta$.

The *number of peaks* ("lines") resulting from the interaction of an unpaired electron with a nucleus of spin I is $2I + 1$; for example, $2(\frac{1}{2}) + 1 = 2$ lines result from the interaction of the spin of the electron with a proton ($I = \frac{1}{2}$). With many (n) equivalent nuclei, the number of lines is $2nI + 1$. In the case of the methyl radical, CH_3, the unpaired electron interacts with three protons. Hence, $2 \times 3(\frac{1}{2}) + 1 = 4$ lines are expected, corresponding to different combinations of the orientations of the nuclear spins. For two sets of nuclei, that is, n nuclei with spins I_i and m nuclei with spins I_j, the number of lines is $(2nI_i + 1)(2mI_j + 1)$. A special case exists for n nonequivalent protons ($I = \frac{1}{2}$), where the number of peaks is 2^n.

Another coupling is also possible; coupling of the electron spin angular momentum with that of the orbital motion, which leads to *anisotropy* in the values of g, that is, to different values in different directions. Anisotropy can also appear in the electron–nuclear spin coupling constant α. Because of averaging, anisotropy cannot be detected in solution, but it can be detected in powders or glasses and in single crystals.

For more than one unpaired electron, there is also coupling between electron spins, which results in a *zero field splitting* and in the so-called *fine splitting* of the peaks.

EPR of Transition Metal Ions

With many transition metal ions, all these interactions may be present. In addition there are interactions between the orbital motions of the electrons. In Russell–Saunders coupling, electron spin–electron spin and orbital–orbital interactions are stronger than spin–orbital interactions. In j–j coupling, spin–orbital interaction is stronger (heavy elements).

In a free ion, both spin and orbital motions tend to become oriented if a magnetic field is applied. This is reflected in the value of g:

$$g = 1 + \frac{J(J + 1) + S(S + 1) - L(L + 1)}{2J(J + 1)} \tag{4.68}$$

where S, L, and J are the magnitudes of total spin, total orbital angular momentum, and their resultant, respectively.

If the ion is not free but surrounded by ligands, the spin is again freely oriented in the magnetic field—it is not inhibited by the ligands. However, the extended-over-space orbital motion is severely affected. The ligands act as obstacles to this motion, or, as we say, *the orbital motion is quenched.* If quenching is complete, L becomes effectively zero (even if for the free ion $L \neq 0$), and the complexed ion behaves as if $J = S$ and $g = 2$. In this case there is no sense in talking about spin–orbital coupling; decoupling is complete. It is like having zero spin–orbital coupling.

Most of the time, however, decoupling is not complete. The quenching of the orbital motions by the ligands is not 100% effective; spin–orbital coupling does not vanish. The value of g then is not 2, but it is not given by equation 4.68 either. Resonance does not appear where one would have expected for the spin-only case, nor does it appear where one would have expected if the orbital motions were also freely oriented in the field.

Mechanistic Information from EPR

In mechanistic studies, EPR is useful in detecting and characterizing paramagnetic species and in following their fate. EPR is especially useful for studying reactions of short-lived paramagnetic intermediates such as free radicals present at low concentrations. The rate constants that can be measured are as large as 5×10^7 s^{-1}. The amount of paramagnetic species that can be detected can be as low as 1 pg.

Chemical processes such as electron transfer from a free radical to a diamagnetic species (e.g., naphthalene$^-$ + naphthalene $\rightarrow$ naphthalene + naphthalene$^-$) or electron (spin) exchange between paramagnetic species (e.g., $W(CN)_8^{3-}$ + $W(CN)_8^{4-}$ $\rightarrow$ $W(CN)_8^{4-}$ + $W(CN)_8^{3-}$) contribute to the relaxation times. As in NMR, the spins tend to get aligned with the applied field, and thermal motion tends to oppose this alignment by randomizing the spin vectors. Likewise, with EPR, randomization along the field is characterized by the spin–lattice relaxation time T_1, and perpendicular to the field by the spin–spin relaxation time T_2. Again, as in NMR, these times can be obtained from pulsed experiments; T_2 is also obtained from linewidth analysis.

One of the limitations of EPR is that it is restricted to paramagnetic species, but even paramagnetic species do not always give an EPR spectrum in solution, because of fast relaxation and extensive broadening of the peaks, especially at ordinary temperatures. On the other hand, if relaxation is too slow, the rate of (thermal) growth of magnetization is also slow, because this rate is also controlled by the spin–lattice relaxation time. Thus, for a species to be detectable by EPR, it must have short enough T_1 for fast growth of magnetization but not so short as to cause too much broadening.

A special case where the limitation of the rate of growth of magnetization (polarization) does not exist is the so-called *chemically induced dynamic electron polarization* (CIDEP). Polarization in this case results from intersystem crossing of photoexcited molecules (see Chapter 7) and from chemical reactions between free radicals.

EPR is also used after rapidly freezing the reacting mixture and/or in combination with flow or stopped-flow techniques. The slow step in such combinations is freezing or mixing, which requires times of the order of milliseconds.

GENERAL REFERENCES

A. Abragam and B. Bleany, *Electron Paramagnetic Resonance of Transition Ions,* Oxford Clarendon, London, 1970.

W. F. Ames, *Non-Linear Ordinary Differential Equations in Transport Processes,* Academic, New York, 1968.

C. F. Bernasconi, *Relaxation Kinetics,* Academic, New York, 1976.

J. R. Bolton, *Electron Spin Resonance Theory,* McGraw-Hill, New York, 1972.

E. Breitmaier, K. H. Spohn, and S. Berger, ^{13}C Spin-Lattice Relaxation Times and the Mobility of Organic Molecules in Solution, *Angew. Chem. Intern. Ed.* **14**, 144 (1975).

E. F. Caldin, *Fast Reactions in Solution,* Blackwell's Scientific Publications, London, 1964.

R. S. Drago, *Physical Methods in Chemistry,* Saunders, Philadelphia, 1977.

D. N. Hague, *Fast Reactions,* Wiley-Interscience, London, 1971.

G. G. Hammes, *Techniques of Chemistry,* Vol. VI, *Investigation of Rates and Mechanisms of Reactions,* Part II, 3rd ed., Wiley, New York, 1974.

P. T. Kissinger and W. R. Heineman, *Laboratory Techniques in Electroanalytical Chemistry,* Dekker, New York, 1984.

H. Kruger, Techniques for the Kinetic Study of Fast Reactions in Solution, *Chem. Soc. Rev. London* **11**, 227 (1982).

P. Laszlo, Fast Kinetics Studied by NMR, *Progr. NMR Spectr.,* **13**, 257 (1979).

S. F. Lincoln, Kinetic Applications of NMR Spectroscopy, *Progr. Reaction Kinetics* **9**, 1 (1977).

E. McLaughlin and R. W. Rozett, Kinetics of Complex Systems Tending to Equilibrium, *Chem. Technol.* **5**, 120 (1971).

R. Pople, F. Schneider, and B. Bernstein, *High Resolution Nuclear Magnetic Resonance,* McGraw-Hill, New York, 1959.

G. Porter, Molecules in Microtime, *Proc. Roy. Inst. Gt. Brit.* **47**, 143 (1974).

B. Saville, On the Use of "Concentration–Time" Integrals in the Solutions of Complex Kinetic Equations, *J. Phys. Chem.* **75**, 2215 (1971).

T. R. Stengle and C. H. Langford, The Uses of Nuclear Magnetic Resonance in the Study of Ligand Substitution Processes, *Coord. Chem. Rev.* **2**, 349 (1967).

H. Strehlow and W. Knoche, *Fundamentals of Chemical Relaxation* (*Monographs in Modern Chemistry*, Vol. 10), Verlag Chemie, Weinheim, 1977.

T. J. Swift and R. E. Connick, NMR-Relaxation Mechanisms of ^{17}O in Aqueous Solutions of Paramagnetic Cations and the Lifetime of Water Molecules in the First Coordination Sphere, *J. Chem. Phys.* **37**, 307 (1962).

F. Wilkinson, *Chemical Kinetics and Reaction Mechanisms,* Van Nostrand-Reinhold, New York, 1980.

PROBLEMS

1. Experimentally, the following data were obtained:

Time, min	Concentration, M
0	2.34
3.08	2.10
5.32	1.90
8.76	1.66
14.4	1.36
20.0	1.12

Show that the statistical criteria (e.g., linearity) indicate a "good" fit for *both* first-order and second-order processes. What is wrong?

2. It has been found empirically that the decomposition of HI,

$$2HI(g) \rightleftharpoons H_2(g) + I_2(g)$$

follows second-order kinetics. The value of the rate constant at 600 K is $4.0 \times 10^{-6}\ M^{-1}\,s^{-1}$. Determine (1) the half-life, (2) the initial rate, and (3) the rate after one half-life. Assume that the vessel initially contained pure HI at a pressure of 1 atm. Neglect the reverse reaction.

3. In sulfur tetrafluoride, all four fluorine atoms are not equivalent, yet the ^{19}F NMR spectrum at room temperature shows only one peak. Which are the nonequivalent fluorine atoms in SF_4? Why is only one peak observed? What happens on a molecular scale?

4. Show that for a first-order reaction,

$$A + H_2O \longrightarrow B + C$$

the equation

$$\ln \left[\frac{A_t - A_\infty}{A_0 - A_\infty} \right] = -kt$$

is generally valid. A_t, A_0, and A_∞ are the absorbance values at time t, zero time, and infinity, respectively. Assume that A and B absorb only in the region of interest and that they have different absorptivities.

5. For the reactions

$$C_1 + C_2 \underset{k_2}{\overset{k_1}{\rightleftharpoons}} C_3 \underset{k_4}{\overset{k_3}{\rightleftharpoons}} C_4 + C_5$$

set up the system of differential equations for small deviations from the equilibrium concentration values and the secular equation for relaxation. Give the expressions for the reciprocal relaxation times.

6. Tracer experiments were performed on the reaction

$$A + {}^*B \rightleftharpoons {}^*A + B$$

and the following data were obtained:

Experiment	Initial A, M	Initial B, M	$t_{1/2}$, s
1	0.10	0.05	100
2	0.05	0.20	120
3	0.10	0.10	75

Estimate the rates of exchange, and determine the order with respect to A and B. Evaluate the rate constant.

7. In the reaction between hydroquinone and chlorine dioxide, the following transmittance data were collected at two different wavelengths as a function of time:

t, s	T_1	T_2
0	0.465	0.200
0.6	0.405	—
1.0	0.373	0.245
1.4	0.348	—
2.0	0.308	0.289
3.0	0.263	0.334
4.0	0.227	0.379
6.0	0.177	0.470
8.0	0.144	0.561
10.0	0.122	0.641
14.0	0.097	0.765
∞	0.065	1.00

The initial concentrations are $[\text{Hydroquinone}]_0 = 0.010\ M$ and $[\text{ClO}_2]_0 = 0.420\ M$. Determine the order of the reaction with respect to the hydroquinone, and calculate the corresponding rate constant. Can the transmittance data be used directly, or should they first be transformed into absorbances? Is it possible on the basis of these data alone to determine the order with respect to ClO_2?

8. The following data were obtained experimentally with initial concentrations $[\text{A}]_0 = 0.0156\ M$, $[\text{B}]_0 = 0.00119\ M$:

Time, ms	Absorbance	Time, ms	Absorbance
0	0.186	20.0	0.0463
2.2	0.135	40.0	0.0396
2.4	0.129	60.0	0.0335
2.6	0.124	80.0	0.0280
2.8	0.118	100.0	0.0241
3.4	0.105	120.0	0.0197
4.2	0.089	140.0	0.0162
4.8	0.081	160.0	0.0133
5.4	0.076	180.0	0.0104
6.6	0.064	∞	0.0000

Does the reaction take place in one step or in two consecutive steps? Can the rate constant(s) be calculated from these data, or should some additional information be given? Explain.

9. The carbonic acid equilibrium has been studied by means of pulsed high field conductivity. The equilibrium

$$H_2CO_3 \rightleftharpoons H^+ + HCO_3^- \tag{1}$$

is affected by the high field strength, but the equilibrium

$$H_2CO_3 \rightleftharpoons H_2O + CO_2 \tag{2}$$

is not. It was found that at 25°C

$$K_{\text{diss}} = \frac{[H^+][HCO_3^-]}{[H_2CO_3]} = 1.3 \times 10^{-4}$$

By other methods it is known that

$$K = \frac{[H^+][HCO_3^-]}{[CO_2] + [H_2CO_3]} = 4.5 \times 10^{-7}$$

Calculate the ratio of $[H_2CO_3]$ to $[CO_2]$. Propose two molecular paths by which the equilibrium between CO_2 and H_2CO_3 is established.

10. From the temperature dependence of the ^{19}F NMR spectrum of

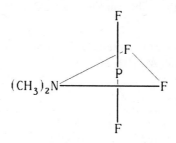

it was shown that there is an exchange between the axial fluorine atoms and the equatorial fluorine atoms. Give an interpretation of these results.

11. Calculate the "contact time" for a cylindrical flow reactor, assuming that the reaction that takes place is first-order.

5

MECHANISM AND STRUCTURE

There is in general a multiple correlation between geometrical structure, electronic configuration, composition, and properties. Here we examine in particular the relation between geometry, electronic configuration, and mechanism.

It would be ideal if one could predict the course of the reaction from the structure of the reactants, but unfortunately it is by no means easy to find general rules. However, progress has been made, and we will try to outline the state of the art.

5.1. GEOMETRICAL STRUCTURE AND MECHANISM

The Steric Factor

Perhaps the most general demonstration of the central role the geometrical structure plays in determining the course of a reaction is the necessity to introduce a *steric factor p* in the expression for the rate constant of a bimolecular reaction in the kinetic theory of gases:

$$k^{(2)} = pZe^{-E_a/RT}$$

The relative geometrical arrangement of the colliding molecules is one of the factors affecting the magnitude of p.

The Size

The size of atoms or ions is the simplest geometrical parameter known to affect rates and the mechanism. A very well known example is provided by the comparison of the reactivities of carbon tetrachloride and silicon tetrachloride. In spite of the similarities in geometrical and electronic structure between the two, carbon tetrachloride is inert to water, whereas silicon tetrachloride reacts vigorously with water, and, in fact, in a moist atmosphere it fumes. The hydrolytic reactions

$$CCl_4 + 6H_2O \longrightarrow CO_2 + 4H_3O^+ + 4Cl^-$$

$$SiCl_4 + 6H_2O \longrightarrow SiO_2 + 4H_3O^+ + 4Cl^-$$

have standard free energy values of -377 and -276 kJ mol^{-1}, respectively. Thus, the hydrolysis of CCl_4 is more favorable thermodynamically, but $SiCl_4$ reacts much faster. This example is by no means unique.

Perchlorates and periodates are (thermodynamically) strong oxidizing agents, but usually periodates react much faster.

Methane and silane are thermodynamically unstable toward oxidation by air, but silane will burn immediately, whereas methane can be kept "for ever." The same is true for ammonia and phosphine.

Of the reactions in aqueous solutions,

$$H_2O_2 \longrightarrow H_2O + \tfrac{1}{2}O_2$$

$$H_2S_2 \longrightarrow H_2S + S$$

the first is energetically more favorable, but the second reaction is considerably faster.

In all these and many other cases the differences in rates may reflect differences in the mechanism (see Chapter 6, "Group and Atom Transfer Reactions"). It can be argued in particular that for heavy central atoms the mechanism involves the empty d orbitals of the valence shell. The valence shell of a light central atom ($n = 2$) has no d orbitals. However, an equally plausible explanation is that the reactive molecules contain large central atoms, and the unreactive, small ones. The four chlorine atoms, for example, around the small carbon atom may leave very little space for another molecule to approach and attack, but they cannot effectively shield the larger silicon atom.

The Shape

Another geometrical factor that plays an important role in determining the rate and the mechanism is the *shape* of the molecule. The hydrolysis of sulfur tetrafluoride (SF_4) to hydrogen fluoride and sulfur dioxide is fast. In contrast, the hydrolysis of SF_6 is very slow, in spite of the fact that the S—F bonds in this molecule are weaker. The difference can be attributed to the difference in shielding by the ligands.

The size and shape of the ligands themselves are also important. The rates of substitution in the square planar complexes[1] $Pt(Et_4dien)X^+$ are orders of magnitude smaller than the rates of the corresponding reaction of $Pt(dien)X^+$. The main reason seems to be the presence of the four ethyl groups, which inhibit approach from either side of the square.

Stereospecific Effects

There are cases in which two diastereoisomers exhibit a completely different chemical behavior. The complex $(+)Co(en)_2(+T)^+$, where en = ethylene-diamine and T = tartaric acid, reacts quite fast with NO_2^-, whereas its diastereoisomer $(-)Co(en)_2(+T)^+$ reacts very slowly.

5.2. HOMO AND LUMO

When two molecules come to within interaction distance from each other, there is generally a tendency for electrons to move from the *h*ighest *o*ccupied *m*olecular *o*rbital (HOMO) of one to one of the *l*owest *u*noccupied *m*olecular *o*rbitals (LUMO) of the other. The HOMO and LUMO are collectively called *f*rontier *m*olecular *o*rbitals (FMO). The first of these molecules can act as an electron *donor,* and the second as an electron *acceptor.* In this respect the introduction of the HOMO and LUMO can be regarded as the contemporary way of handling the old[2] donor–acceptor concept.

The terms "molecule" and "molecular orbitals" are used here in a generalized sense. If the donor or acceptor is an atom, the corresponding orbitals will, of course, be atomic. On the other hand, a molecular orbital may be localized mainly on a single atom in the molecule; that is, it may have strong atomic character.

It is well known that many molecules act in some cases as donors and in other cases as acceptors. For example, water can contribute its nonbonding electrons, which are polarized on the oxygen, but it can also accept electrons into its LUMO.

Water acts either as donor or as acceptor, not both simultaneously. Other species, such as CN^-, CO, or N_2, can act as donors and acceptors simultaneously.

Complicated molecules may have more than one position for accepting or giving electrons, either simultaneously or not.

---

[1]dien = diethylenetriamine, $H_2NCH_2CH_2NHCH_2CH_2NH_2$; and Et_4dien = tetraethyldiethyl-enetriamine, $(C_2H_5)_2NCH_2CH_2NHCH_2CH_2N(C_2H_5)_2$.

[2]"The opposites agree with each other and the differences create the most beautiful harmony, and all results from strife." (Heraclitus, 544–483 B.C.)

5.3. MINIMUM ENERGY PATHWAYS AND THE MAXIMUM OVERLAP CRITERION

The Overlap

Under the conditions at which chemical reactions usually take place, there is a statistical distribution in the relative orientations of nonspherical donor and acceptor molecules as they approach each other. However, some of these orientations (pathways) are energetically more favorable than others; they are called the *minimum energy pathways*.

Minimum energy pathways are related to orbital overlap. For a donor electron to be totally or partially transferred to the acceptor, there must be a *positive overlap* between the corresponding HOMO and LUMO; that is, the so-called *overlap integral* must be positive: $S = \int \psi_1\psi_2 \, dT > 0$, where ψ_1, ψ_2 are the overlapping orbitals. In simple words, the algebraic signs of the overlapping orbital functions must be the same. Examples of positive, negative, and zero overlaps are given in Figure 5.1.

Figure 5.1. Positive, negative, and zero overlap of atomic orbitals.

The larger the positive overlap, the stronger the bond formed between donor and acceptor. The positively charged donor and acceptor nuclei tend to move toward the region of positive overlap, where the electron density is enhanced.

Sigma and Pi Donors and Acceptors

The symbols σ and π denote the symmetry with respect to the donor–acceptor permanent or incipient bond. Examples of σ and π donor–acceptor interactions are given in Figure 5.2. Donors or acceptors from or to *n*onbonding orbitals are sometimes called *n*; those from or to *b*onding orbitals, *b*; and those from or to *a*ntibonding orbitals, *a*.

Asymmetric Donor–Acceptor Bonds

An example where the bond formed is neither σ nor π is the nucleophilic attack on an olefinic double bond by hydride ion (Fig. 5.3). The occupied *s* orbital of the donor overlaps with the empty π^* orbital of the acceptor in the manner shown in the figure. Because of the inherent lack of symmetry of the system, there is a local deviation from σ symmetry.

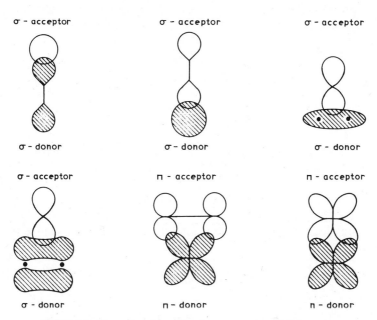

Figure 5.2. Examples of σ and π donor–acceptor orbital overlap.

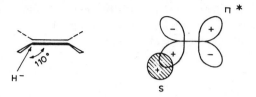

Figure 5.3. Minimum-energy pathway in the nucleophilic attack of an olefin by hydride ion.

5.4. WHICH IS THE DONOR AND WHICH THE ACCEPTOR?

Chemical Tendency for Electron Displacement

The criterion of the positive overlap can be regarded as a necessary condition, but it is not sufficient. For the electron(s) to be displaced there must also be what has been called a *chemical tendency* for the transfer. We examine here the origin of this tendency.

It is well known that Mulliken defined the electronegativity of an atom as the half-sum of its ionization potential and its electron affinity. This definition is based on the argument that in a compound A—B, in which A and B have the same electronegativity, the energy required to move an electron from A to B will be equal to the energy required to move an electron from B to A. In the first case this energy is proportional to $I_A - E_B$, and in the second case it is proportional to $I_B - E_A$, where I_A, I_B are the ionization potentials and E_A, E_B the electron affinities. For equal electronegativities,

$$I_A - E_B = I_B - E_A$$

or

$$I_A + E_A = I_B + E_B$$

If A is more electronegative than B, there is a tendency for charge to move from B to A, and

$$I_A + E_A > I_B + E_B$$

If B is more electronegative, there is a tendency for charge to move from A to B, and

$$I_A + E_A < I_B + E_B$$

By analogy, if

$$I_{H1} + E_{L1} > I_{H2} + E_{L2}$$

where I_{H1} and I_{H2} are the energies required to remove an electron from the HOMO of molecules 1 and 2, respectively, and E_{L1}, E_{L2} are the corresponding energies released when an electron enters the LUMO, there is an initial tendency for charge to move from molecule 2 to molecule 1. In other words, there is an initial tendency for molecule 2 to act as a donor and molecule 1 as an acceptor. Thus, which molecule is going to act as a donor and which as an acceptor will depend not only on the energy needed to remove an electron from the HOMO, but also on the energy released when an electron enters the LUMO.

EXAMPLES

Some examples will help illustrate these ideas. Consider first the reaction

$$N_2 + O_2 \longrightarrow 2NO$$

According to the criterion of positive overlap, electron transfer from the HOMO of N_2 (σ) to the LUMO of O_2 (π^*) is not favored (Fig. 5.4a). Transfer of an electron from O_2 to N_2 satisfies the criterion of maximum positive overlap (Fig. 5.4b), but it is ruled out because oxygen is more electronegative than nitrogen and the driving force for movement of the electrons is in the opposite direction.

The conclusion to be drawn from this description is that for the reaction between N_2 and O_2 to take place, higher-lying molecular orbitals must be used, not the HOMO and LUMO. The activation energy will then be high.

In contrast, the reaction

$$2NO + O_2 \longrightarrow 2NO_2$$

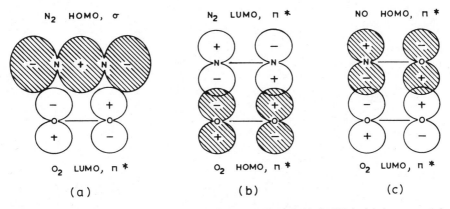

Figure 5.4. Schematic presentation of orbital overlap. The σ orbital of N_2 in (a) the oxygen π^* in (b) and the NO π^* in (c) are occupied. Partially occupied orbitals can act either as HOMO or as LUMO.

is fast. The first step in this reaction is probably

$$NO + O_2 \longrightarrow NO_3$$

and for this step it is predicted that the electron(s) should move from NO toward O_2. One should then use the HOMO of NO and the LUMO of O_2. The overlap will be positive (Fig. 5.4c). The reaction is "allowed" and fast.

The Polarization of the Orbitals

Compare the orbitals of dinitrogen with those of the isoelectronic carbon monoxide (Fig. 5.5). The occupied σ orbital of carbon monoxide is polarized on carbon, but the two occupied π orbitals (one of which is shown in Fig.

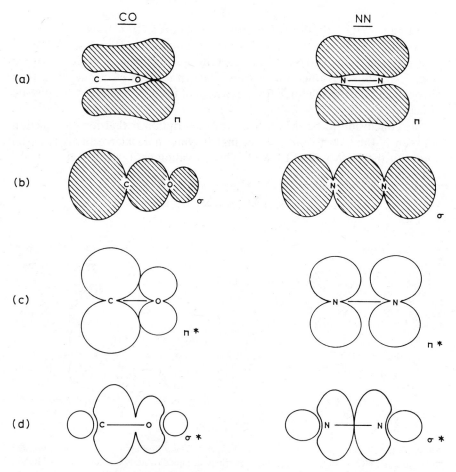

Figure 5.5. Comparison of CO and NN orbitals.

5.5) are polarized on oxygen. As a result, there is a net displacement of negative charge on the more electronegative oxygen.

With the empty σ^* and the two π^* orbitals, the situation is reversed. The first is polarized on oxygen, the others on the less electronegative carbon.

5.5 HOMO AND LUMO IN REACTIONS OF TRANSITION METAL COMPLEXES

In coordination compounds, the HOMO and LUMO are usually orbitals with d character and energetically lie relatively close to each other. It must also be recalled that the d orbitals of the transition metal ions are better shielded from the nucleus than the p and s orbitals, and their average distance from the nucleus is larger. This simply means that access to d orbitals is easier. It has been stressed several times that for a reaction to take place there must be the maximum possible overlap. However, d orbitals are "inner" orbitals, in the sense that they are surrounded by ligands. Therefore, in order to have the necessary overlap, one of the following must occur:

The d orbitals must be "diffuse" and extend beyond the ligands.

A ligand must somehow be displaced or removed in order to expose the d orbitals.

In a preceding step, use must be made of the ligand orbitals.

In any case, if the symmetries of the donor HOMO and acceptor LUMO do not match, the reaction can take place only through higher-lying unoccupied orbitals or lower-lying occupied orbitals, and more activation will be needed. Nevertheless, for transition metal complexes this is not necessarily forbidden, because the energy gaps between orbitals are small.

The examples given below illustrate the use of HOMO and LUMO in interpreting reactions of transition metal complexes.

In the reduction by Cr^{2+} the electron comes from an antibonding e_g^* orbital. In the reduction by V^{2+} the electron comes from a t_{2g} almost nonbonding orbital. Most probably because of this difference, and in spite of the fact that aqueous Cr^{2+} is thermodynamically a stronger reducing agent, aqueous V^{2+} reacts with $Cr^{III}(OAc)^{2+}$ at least a thousand times faster than Cr^{2+}. A plausible explanation is that $Cr^{III}(OAc)^{2+}$ acts as a π acceptor and can match better with V^{2+}, which can act as a π donor. The e_g^* orbital of Cr^{2+} cannot give π overlap; the overlap occurring in the case of Cr^{2+} is less effective, being nonsymmetric or involving an orbital of the acceptor other than its LUMO.

It is interesting to note that in cases such as the electron transfer to $Cr^{III}(NH_3)_6^{3+}$, in which the d orbitals do not extend beyond the ligands and where it is not easy to remove or displace the ligand and there are no good "auxiliary" orbitals on the ligands to be used in the first step, the reaction takes place by an outer-sphere mechanism, even with reductants like Cr^{2+}, which usually reacts by an inner-sphere mechanism (see Chapter 6).

Another example illustrating the importance of having easy access to the HOMO and LUMO is obtained by comparing the reactions

$$NaCN + RX \longrightarrow RCN + NaX$$

Nitrile

and

$$AgCN + RX \longrightarrow RNC + AgX$$

Isonitrile

In AgCN the cyanide group is bound to silver through the carbon, and it can be argued that the electrons of this bond are not readily available to combine with R. In contrast, NaCN is an ionic compound, and the carbon electrons of the cyanide ion are readily accessible.

5.6. HOMO AND LUMO IN REACTIONS OF COMPOUNDS OF REPRESENTATIVE ELEMENTS

In compounds of the lighter representative elements, the HOMO and LUMO are molecular orbitals obtained from s and p atomic orbitals, which are poorly shielded from the nucleus. By analogy, the molecular orbitals obtained from these atomic orbitals are also close to the nuclei and less accessible. Often, it also happens that the appropriate ligand-centered orbitals to be used instead of the HOMO or the LUMO of the whole molecule are not available. In order to achieve overlap it then becomes necessary to displace or remove a ligand. This seems to be the case in oxidation reactions by molecules such as H_2O_2, NO_3^-, or ClO_4^-, which are slow and are preceded by substitution. The same factor (the stronger attraction of the electron by the nuclei) is probably also responsible for the slowness of many reductions with such reactants as the halides, SH^-, Sn^{2+}, SO_3^{2-}, benzylamine, or phenol.

The heavier of the representative elements have empty d orbitals, which can conceivably act as LUMO, but even in this case activation may be high because of the larger energy gap between a HOMO with s, p character and a LUMO with d character.

5.7. WHAT HAPPENS AFTER THE OVERLAP

Positive overlap and a chemical tendency for the electrons to be displaced are the necessary "prerequisites," but most of the time the donor–acceptor system does not stop there. After the first interaction the donor and acceptor usually undergo rearrangements; bond lengths and bond angles change, even the number of bonds may change, until eventually the system develops into its final form (the products). So let us examine the salient features of this development. We examine first the gross effect of the HOMO–LUMO overlap on the other donor and acceptor orbitals, and later (Section 5.11) we attempt to follow the orbitals somewhat more systematically as they develop from orbitals of the reactants into orbitals of the products.

It must be emphasized again that the HOMO–LUMO positive overlap essentially leads to bond formation, no matter how weak this bond may be. In fact, the "construction" of bonding molecular orbitals from atomic orbitals involves similar overlap. The formation of this HOMO–LUMO bond gives the system the time required for transformations to occur. A simple collision in the gas phase lasts only $\sim 10^{-12}$ s. Conditions are more favorable in solution, where an encounter in a solvent cage may last as long as 10^{-10} s. However, these times are too short for the massive nuclei to move and for bonds to break or form. With the formation of even a weak bond, the time available is extended; in fact, sometimes it tends to infinity, that is, the bond becomes permanent.

If the donor and acceptor orbitals are nonbonding, the formation of the HOMO–LUMO bond has no appreciable effect on the other bonds. If the overlapping donor orbital is bonding or the acceptor orbital antibonding, the formation of the new bond causes a decrease in the strength of the preexisting bonds. If, on the other hand, the HOMO of the donor is antibonding or the LUMO of the acceptor is bonding, the other bonds are strengthened.

EXAMPLE I

In the reaction between NO and O_2, the transfer of electron density from the antibonding orbital of NO to the antibonding orbital O_2 results in a strengthening of the N—O bond and a weakening of the O—O bond. In fact, the bond O—O finally breaks, whereas simultaneously two new bonds are formed:

$$OO + NO \longrightarrow \begin{array}{c} O \\ \diagdown \\ \diagup \\ O \end{array} N\!\!-\!\!O \qquad \qquad \square$$

EXAMPLE II

The importance of the bonding or antibonding character of the overlapping HOMO and LUMO is seen clearly if a comparison is made in aqueous solution between Cr^{2+} and Cr^{3+}. One of the factors contributing to make Cr^{2+} a strong reducing agent is the fact that removal of the antibonding e_g^* electron is associated with a strengthening of the bonds. This electron also contributes in making $Cr(H_2O)_6^{2+}$ a highly labile system. Water exchange in $Cr(H_2O)_6^{3+}$ is 10^{15} times slower than in $Cr(H_2O)_6^{2+}$. (The water exchange half-life for

$Cr(H_2O)_6^{3+}$ is of the order of 10^6 s, and that for $Cr(H_2O)_6^{2+}$, of the order of 10^{-9} s.) The inertness of $Cr(H_2O)_6^{3+}$ can be attributed partly to the higher charge of the central metal ion, which makes it a better Lewis acid, but the fact that Cr^{3+} has no e_g^* electrons is also an important factor. □

EXAMPLE III

The reduction of inert Co^{III} complexes leads to labile Co^{II} species. The configuration of the low-spin Co^{III} is $(t_{2g})^6(e_g^*)^0$, whereas that of the high-spin Co^{II} is $(t_{2g})^5(e_g^*)^2$. In Co^{II}, two of the electrons occupy antibonding orbitals, and this leads to a facile breaking of the bonds. In contrast, the configuration of Ru^{III} in low-spin complexes is $(t_{2g})^5(e_g^*)^0$. In the reduction of these complexes, the electron enters a nonbonding t_{2g} orbital. Hence, in the Ru^{II} complex formed there is no weakening of the bonds. The result is that the low-spin Ru^{II} product may live long enough to dimerize. □

EXAMPLE IV

The reaction of carbon dioxide with water can be represented as follows:

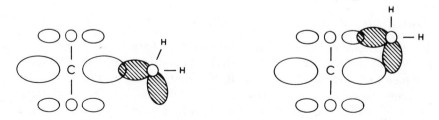

The first step is nonbonding electron donation from water to the antibonding LUMO of carbon dioxide (Fig. 5.6), resulting in a weakening of the CO_2 bonding system and eventual formation of H_2CO_3 or HCO_3^-, probably first H_2CO_3, because it involves intramolecular transfer of a proton. □

Figure 5.6. Two ways in which water may donate an electron to carbon dioxide.

5.8. ATTRACTIVE AND REPULSIVE INTERACTIONS

Attractive Interactions

Let us recall a very well known quantum-mechanical result: *The strongest interactions occur between orbitals of comparable energy,* provided, of course, that there is overlap. Thus, in a case like that depicted in Figure 5.7, the main interaction will be between the HOMO of the donor and the LUMO of the acceptor, because these orbitals have comparable energies. In the donor–acceptor complex, the pair of electrons will occupy a lower energy level; interaction between HOMO and LUMO is attractive.

Repulsive Interactions

Consider next the interaction between the two HOMOs. In the example given in Figure 5.8, this interaction is weaker because of larger differences in energy, and it results in some stabilization of the acceptor electrons and destabilization of the donor electrons. The donor–acceptor system as a whole is slightly destabilized, because the destabilization of one pair is slightly more than the stabilization of the other. The HOMO–HOMO interaction is therefore repulsive, but this repulsion is considerably smaller than the HOMO–LUMO attraction, because of the relative energies and also because HOMO–HOMO repulsion is the difference between the destabilization of one electron pair and the stabilization of the other whereas the HOMO–LUMO attraction is the result of the stabilization of only one pair.

From Figures 5.7 and 5.8 it can also be seen that two more interactions are possible: between the LUMO of the donor and the HOMO of the acceptor, and between the two LUMOs.

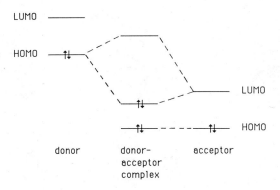

Figure 5.7. Attractive HOMO–LUMO interactions.

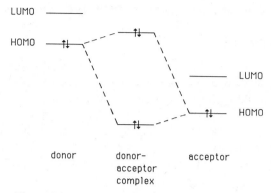

donor   donor–   acceptor
        acceptor
        complex

Figure 5.8. Repulsive HOMO–HOMO interactions.

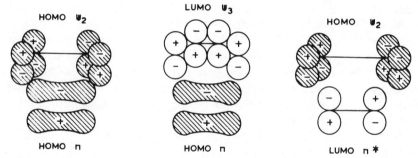

Figure 5.9. HOMO and LUMO in the formation of cyclohexane from ethylene and buta-1,3-diene.

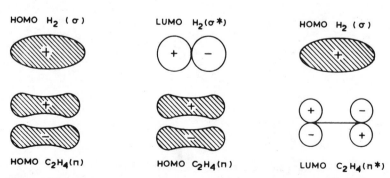

Figure 5.10. HOMO and LUMO for the reaction between ethylene and dihydrogen.

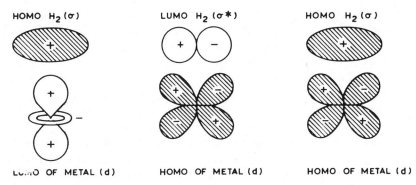

HOMO H$_2$ (σ)   LUMO H$_2$ (σ*)   HOMO H$_2$ (σ)

LUMO OF METAL (d)   HOMO OF METAL (d)   HOMO OF METAL (d)

Figure 5.11. Activation of dihydrogen by a transition metal.

Zero Overlap

In the analysis above it has been assumed that there is a net positive or net negative overlap. If there is no net overlap, there is no net interaction. Thus, in the reaction of ethylene with butadiene (Fig. 5.9), the HOMO–HOMO interaction is ignored (zero overlap). Only the attractive HOMO–LUMO interactions are important. The reaction (Diels–Alder) has a small activation energy, exactly because these dominant interactions are attractive.

In contrast, in the hydrogenation of a carbon–carbon double bond (Fig. 5.10) the attractive HOMO–LUMO interaction is the one that vanishes, and the repulsive HOMO–HOMO interaction is the one that remains. The reaction is thermodynamically favored ($\Delta G° = -102$ kJ mol^{-1}) but kinetically unfavored. In practice the reaction is carried out only in the presence of a catalyst.

In catalytic hydrogenation, the obstacles are removed. The catalyst (a transition element) accepts electron density from the σ bonding orbital of H$_2$, and simultaneously back-donates to the antibonding σ* orbital (Fig. 5.11). These interactions result in a weakening of the H—H bond, because electron "cloud" is taken from the bonding area and is transferred to the antibonding area. Similar interactions are possible with the double-bond system. In the presence of a catalyst the reaction takes place readily.

5.9. AN ALTERNATIVE APPROACH BASED ON THE CONCEPT OF FORCE

Stable Configurations

The structure of a molecule and the way it changes are usually discussed in terms of energy. In a bonding interaction, the potential energy is lowered; in an antibonding interaction, the potential energy is raised. A stable configuration corresponds to a minimum.

Alternatively, in describing chemical bonding and dynamics we can use the concept of *force*. Intuitively, this concept is perhaps better understood, and it is more directly related to human experience.

Consider as an example a square planar complex (Fig. 5.12). Two categories of forces can be distinguished: "central," that is, from and to the central ion, and "peripheral," that is, between ligands or electrons associated with ligands.

In the equilibrium position (stable configuration), *the resultant of all the forces acting on any of the nuclei of the system vanishes* (Fig. 5.12).

Usually, chemists are interested mainly in the central forces—those directly related to the chemical bond. However, from Figure 5.12 it is obvious that an equivalent way to describe the equilibrium configuration can be based on the *"peripheral"* forces.

The Hellmann–Feynman Theorem

According to the Helmann–Feynman theorem, the component of the force acting on the nucleus A in the x direction is given by the equation

$$F_{Ax} = \frac{\partial}{\partial x_A} \left[\sum_{B(\neq A)} \frac{Z_A Z_B e^2}{r_{AB}} \right] - \int \rho \, \epsilon_{Ax} \, dr \qquad 5.1$$

where Z_A, Z_B are the charges of nuclei A and B, respectively, r_{AB} the distance between them, ρ the electron density at the point (x, y, z), and ϵ_{Ax} the electric field produced by nucleus A along the x axis at the position of the ith electron.

The electron distribution $\rho(x, y, z)$ is calculated quantum-mechanically, possibly by using an approximate combination of orbitals. Apart from this, equation 5.1 itself is not an approximation; it is a rigorous result. It gives the force F_{Ax} exerted on nucleus A by all the other nuclei and the electrons of the system and implies that this force is simply the sum of all classical electrostatic forces operating between A and all the other nuclei and between A and all the electrons.

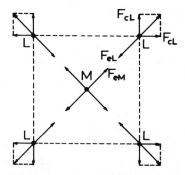

Figure 5.12. Forces in a square planar complex.

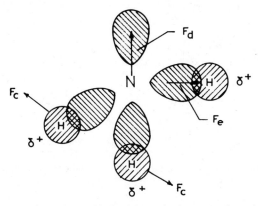

Figure 5.13. Different kinds of forces in the ammonia molecule.

The right-hand side of equation 5.1 can be transformed in such a way as to become the sum of three terms:

The dipole force, F_d
The exchange force, F_e
The effective charge force, F_c

An illustration is given in Figure 5.13.

The dipole force F_d shown in Figure 5.13 is exerted on the nitrogen atom by the hybrid lone electron pair, and the exchange forces F_e on nitrogen are exerted by the electron "cloud" in the overlap region. Effective charge repulsive forces, F_c on hydrogens are also shown.

With this clarification, the nature of the forces that have been depicted for a square planar complex should be obvious. In Figure 5.13, effective charge forces on M have been omitted.

Forces and Potential Energy

A simple demonstration of the relationship between the formalism based on potential energy and the formalism based on forces is given in Figure 5.14. In the attractive region (point A), $F_e > F_c$. In the repulsive region (point B), $F_e < F_c$. At the equilibrium position, $F_e = F_c$.

Application to Dynamics

Consider the simple reaction

$$H_\alpha + H_\beta{-}H_\gamma \longrightarrow H_\alpha{-}H_\beta + H_\gamma$$

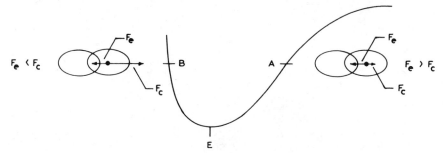

Figure 5.14. Correspondence between forces and potential energy. Only the forces acting on one of the nuclei are shown.

The forces acting on H_α as it approaches H_β—H_γ can be represented as in Figure 5.15. This picture is admittedly oversimplified, but it shows clearly that there is at first a repulsive force that should be overcome and that is related to the activation energy. A more involved application will be given in Chapter 8 when we examine the trans effect in substitution reactions of square planar complexes.

Forces in HOMO, LUMO Interactions

The correspondence between the HOMO–LUMO, HOMO–HOMO initial interactions and the attractive or repulsive forces is illustrated in Figure 5.16. The dipolar force F_d on the donor nucleus is created by the polarization of the electron cloud caused by the proximity of the acceptor nucleus. The acceptor LUMO is also polarized. There are no effective charge forces in HOMO–LUMO, but in HOMO–HOMO these forces dominate (Fig. 5.16).

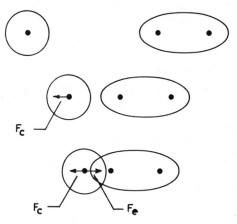

Figure 5.15. Sketch of the forces during the approach of a hydrogen atom to a dihydrogen molecule.

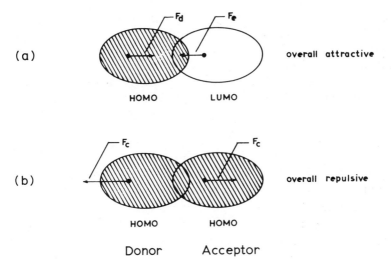

Figure 5.16. Relation between HOMO–LUMO and HOMO–HOMO interactions and the dipolar exchange and effective charge forces. The magnitudes of the various forces are not drawn to scale.

In the cases depicted in Figures 5.15 and 5.16 it is assumed that there is a positive overlap. If there is no overlap, there is no net force.

5.10. THE CONSERVATION OF ORBITAL SYMMETRY

Wigner–Witmer and Woodward–Hoffmann Rules

During a chemical reaction many quantities change: bonds, angles, electron distribution, and so on. However, there are also quantities that are conserved: The mass is conserved, the total energy in an isolated system is conserved, and *the total angular momentum is conserved.*

The total angular momentum can be analyzed into a nuclear spin part, an electron spin part, and an orbital part. It may be expected that under certain conditions not only the total angular momentum but also its individual components are conserved.

The study of the conservation of orbital angular momentum in particular led to the formulation of two sets of rules:

The Wigner–Witmer (1928) rules, which are based on the assumption of conservation of the *total orbital angular momentum* of all the electrons of the reacting system. This implies conservation of the overall symmetry of the reacting system.

The Woodward–Hoffmann (1965) rules, which are based on the assumption of conservation of the *orbital angular momentum of each electron separately*. This implies conservation of orbital symmetry.

These rules apply to elementary (conserted) reactions, which represent events at the molecular level, and in many cases they are extremely useful and provide a powerful framework for mechanistic interpretations. They are not universal, however. If certain conditions are not met, the rules are not applicable. The following possibilities can be distinguished:

Total symmetry, and necessarily the symmetry of the individual orbitals, is not conserved. The reaction is then characterized as *nonfeasible*.

Total symmetry is conserved, but not the symmetry of the individual orbitals. The reaction is then characterized as *feasible* (Wigner–Witmer rules) but *forbidden* (Woodward–Hoffmann rules).

Both total and individual symmetries are conserved. The reaction is then characterized as *feasible* and *allowed*.

From all of this it is obvious that the rules of the conservation of orbital symmetry correspond to a more detailed picture of the system. The example given below refers to these rules only.

Application of the Orbital Symmetry Conservation Rules

The simplest cycloaddition reaction is the dimerization of ethylene to form cyclobutane. Suppose that the two molecules approach each other with the highest possible symmetry, with the two olefinic planes parallel to each other, according to Figure 5.17. After the two molecules have come sufficiently close to each other, there is an interaction between the orbitals, and the electron distribution gradually changes until it finally becomes the distribution corresponding to the products. Throughout this process orbital symmetry is conserved. The correlation diagram is given in Figure 5.18. The interactions

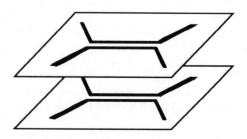

Figure 5.17.

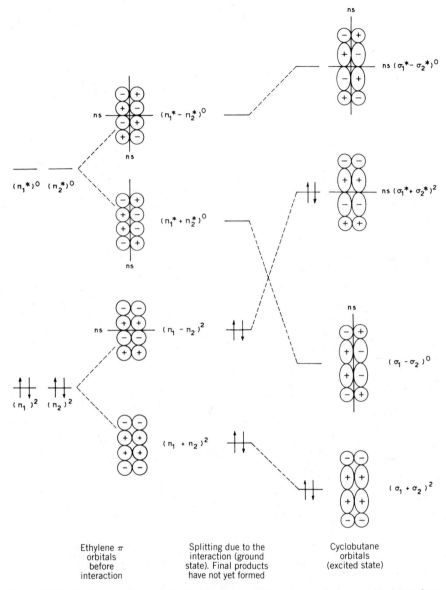

Ethylene π orbitals before interaction

Splitting due to the interaction (ground state). Final products have not yet formed

Cyclobutane orbitals (excited state)

Figure 5.18. Correlation diagram for the thermal reaction of two ethylene molecules to form cyclobutane. Nodal surfaces are denoted by ns. The relative positions of the second and third levels on the left depend on the relative positions of the original π and π^* levels and on the strength of the interaction between them. Here the level with the perpendicular nodal surface lies higher than the level with the horizontal nodal surface.

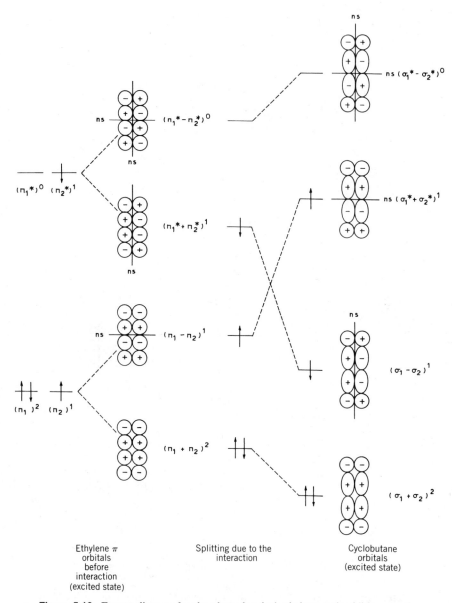

Ethylene π orbitals before interaction (excited state)

Splitting due to the interaction

Cyclobutane orbitals (excited state)

Figure 5.19. Energy diagram for the photochemical ethylene cycloaddition reaction.

considered in this case are not those between the HOMO and LUMO of the two identical molecules, because there is a large difference in energy between them. The interactions that are strong are the HOMO–HOMO and LUMO–LUMO interactions, for which energy matching is perfect.

A rule of thumb for remembering which orbital combination is higher in energy and which lower is to count new nodal surfaces (ns), that is, nodal surfaces that did not exist in the original molecules. On the left side in Figure 5.19, the combination with the lowest energy has no such node. The next combination has one such node between the two ethylene molecules (horizontal). The third (from the bottom) again has one node perpendicular to the original bonds. The highest has two new nodes. On the right-hand side, the lowest orbital again has no node. The next has one node perpendicular to the original ethylene bonds, like the third level on the left. The third on the right again has one horizontal nodal plane, like the second on the left. The highest level on the right has two nodal planes, like the highest level on the left.

It is seen from Figure 5.18 that the ground state of the reactants is correlated with an excited state of the product, not with its ground state. One of the electron pairs occupies the orbital $\sigma_1^* + \sigma_2^*$, not the σ_1–σ_2 orbital. Accordingly, the activated energy for the process is expected to be high, and indeed it is. In contrast, if one of the ethylenes is excited, for example, by light absorption, the situation is different (Fig. 5.19). The photochemical reaction is not forbidden; the sum of the energy of the four electrons has decreased.

5.11. WHAT IS THE CONNECTION BETWEEN HOMO AND LUMO, AND THE CONSERVATION OF ORBITAL SYMMETRY?

In the example of ethylene cycloaddition, we concentrated mainly on the occupied orbitals of the reactants. We essentially tried to follow the development of the shape of the electron cloud, as it is transformed continuously all the way from reactant cloud to product cloud.

However, there is also another perspective in looking at this development. During the reaction, the system passes through transient states, which are higher in energy than the reactants or the products (activation energy); in other words, it passes through states that can be considered to be excited.

In quantum-mechanical terms, the interaction between reactant molecules is equivalent to mixing the ground state with excited states, that is, it is equivalent to combining the wavefunction representing the ground state with the wavefunctions representing excited states.

For a donor–acceptor system, the simplest possible mixing is to combine the HOMO with the LUMO. If these two orbitals have comparable energies and the overlap is good, then the energy of the transition state will be comparable; there will be no need to mix in excited states. However, if there is

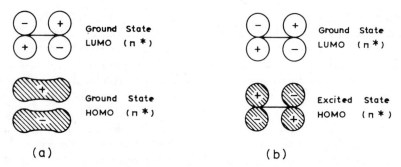

Figure 5.20. If HOMO and LUMO do not match (*a*) the transition state is obtained by using higher orbitals (*b*), which require activation.

no positive overlap between HOMO and LUMO, it will be necessary to use higher-lying states; the reaction will require activation.

In cycloaddition, HOMO and LUMO do not match (Fig. 5.20*a*). The transition state is obtained by using higher-lying orbitals (Fig. 5.20*b*).

5.12. MECHANISTIC CONSEQUENCES OF THE PAULI PRINCIPLE

To a first approximation, the Hamiltonian operator of a molecule includes terms corresponding to the kinetic energy of the electron–nuclei relative motion, and to the electron–nucleus, electron–electron, and nucleus–nucleus potential energies. In describing the salient features of the molecule (e.g., structure or reactivity), there is no need to include terms corresponding to the spin energy. The effect of spin energy on these features is negligible. Only if we were interested in fine and hyperfine coupling effects would this energy be important.

Yet it is well known that electronic configurations corresponding to different spin states often differ greatly in energy. It is also reasonable to suppose (in spite of the doubts sometimes expressed) that the spin of the activated complex is somehow related to the activation energy and plays a significant role in the mechanism.

There seems to be a contradiction: The spin energy is negligible compared to activation energy, and yet we insist that spin may have a serious effect on the mechanism.

The role of spin in the activation process is indirect. It is not related to its own energy but *originates in the role it plays—on how, in what orbitals, and at what energy levels the electrons are distributed.* In other words, the role of spin in the activation process is related to the Pauli principle. Electron distribution is affected by the electrostatic interactions, but it is also affected, independently, by the Pauli principle. This is well known from the aufbau principle. An electron will be placed in an orbital of lower energy if and only if the Pauli principle is not violated. For example, electrostatically (energetically), it is more favorable to put the third electron of the lithium atom into the same orbital with the other two, but this violates the Pauli principle. Thus, the lithium ground state is $1s^2 2s^1$ and not $1s^3$. The energy of the system has been influenced by spin, but not directly, not because of spin energy per se.

For dynamic systems like those examined in this book, the situation is analogous. During a reaction, at any point on its path, the Pauli principle must be obeyed.

For chemists, the most familiar formulation of the Pauli principle is that *electrons having all four quantum numbers the same cannot exist in an atom.* A more general formulation is that *the maximum number of electrons in an orbital, atomic or molecular, is two with antiparallel spins.*

An even more general formulation, which shows clearly the fundamental character of the principle, is that *the total wavefunction of a molecule, ion, or atom is antisymmetric.* The total wavefunction depends on space coordinates and on spin. For two electrons we can write

$$\psi(M_1, \omega_1, M_2, \omega_2)$$

where M_1 represents the space coordinates of electron 1, and M_2 those of electron 2. The quantities ω_1, ω_2 are the corresponding spins. According to the Pauli principle,

$$\psi(M_1, \omega_1, M_2, \omega_2) = -\psi(M_2, \omega_2, M_1, \omega_1) \qquad 5.2$$

That is, if electron 1 takes the position and spin of electron 2, or vice versa, the sign of the wavefunction changes. The probability dp that the two electrons occupy the same position in space ($M_1 = M_2 = M$) and have the same spin ($\omega_1 = \omega_2 = \omega$) is

$$dp = |\psi(M, \omega, M, \omega)|^2 \, du$$

with du being the volume element.

But according to 5.2,

$$\psi(M, \omega, M, \omega) = -\psi(M, \omega \, M, \omega)$$

or

$$\psi(M, \omega, M, \omega) = 0$$

and hence the probability is zero. There is no way to have two electrons with the same spin at the same place. (Occupancy of the same orbital in the last analysis means competition for the same space region.) Electrons with the same spin tend to move away from each other as much as possible, not only because of their mutual electrostatic repulsion but also because of the Pauli principle. It looks as if, in addition to the electrostatic forces, other repulsive forces operate too—the so-called Pauli forces. These are the forces dictating the structure $1s^2 2s^1$ for Li, instead of the electrostatically more favorable $1s^3$.

In the field of mechanisms, the application of the Pauli principle can lead to very useful interpretations. Take as one example (many such examples can be cited) the exchange of coordinated water in $V(H_2O)_6^{2+}$ or $V(H_2O)_6^{3+}$ and water in the bulk solution. The electron configuration of the two central ions can be presented as $(t_{2g})^3(e_g^*)^0$ and $(t_{2g})^2(e_g^*)^0$, respectively. Vanadium(III) has an empty t_{2g} orbital of relatively low energy, which can act as LUMO and form an incipient bond with the entering ligand. For vanadium(II), such a possibility does not exist, because the entering ligand donates an electron pair and the Pauli principle excludes use of t_{2g} orbitals already containing one electron. For vanadium(II), the higher-lying e_g^* orbitals must be used, or a change in configuration must take place first, for example,

$$\uparrow \quad \uparrow \quad \uparrow \qquad t_{2g}$$

to

$$\uparrow\downarrow \quad \uparrow \quad \bigcirc \qquad t_{2g}$$

which also requires considerable energy (~ 209 kJ mol^{-1}). Thus water exchange in $V(H_2O)_6^{2+}$ is expected to be slower than in $V(H_2O)_6^{3+}$, and indeed it is—inspite of the higher charge of V^{3+}, which makes the V^{3+}—OH_2 bonds stronger.

5.13. IS THE SPIN CONSERVED?

Spin Conservation Rules

In mechanistic studies, spin is of interest from two aspects:

1. Because, as we saw, it affects the activation energy in accord with the Pauli principle.
2. Unpaired electrons are readily detected (magnetic properties, via EPR); in this respect spin can be regarded as a "tracer."

In both cases it is necessary to examine the conditions under which spin is conserved or changes.

For elementary reactions of the type A + B $\rightleftharpoons$ AB, which develop on a single potential surface, the following rules (formulated by Wigner and

Witmer) apply:

$$S_A + S_B \rightleftharpoons S_{AB}$$

$$S_A + D_B \rightleftharpoons D_{AB}$$

$$S_A + T_B \rightleftharpoons T_{AB}$$

$$S_A + Q_B \rightleftharpoons Q_{AB}$$

$$D_A + D_B \rightleftharpoons S_{AB} \text{ or } T_{AB}$$

$$D_A + T_B \rightleftharpoons D_{AB} \text{ or } Q_{AB}$$

$$D_A + Q_B \rightleftharpoons T_{AB} \text{ or } Qi_{AB}$$

$$T_A + T_B \rightleftharpoons S_{AB} \text{ or } T_{AB} \text{ or } Qi_{AB}$$

$$T_A + Q_B \rightleftharpoons D_{AB} \text{ or } Q_{AB} \text{ or } Sx_{AB}$$

$$Q_A + Q_B \rightleftharpoons S_{AB} \text{ or } T_{AB} \text{ or } Qi_{AB} \text{ or } Sp_{AB}$$

where

$$S = \text{singlet}, \quad 0 \text{ unpaired electron}$$

$$D = \text{doublet}, \quad 1 \text{ unpaired electron}$$

$$T = \text{triplet}, \quad 2 \text{ unpaired electrons}$$

$$Q = \text{quartet}, \quad 3 \text{ unpaired electrons}$$

$$Qi = \text{quintet}, \quad 4 \text{ unpaired electrons}$$

$$Sx = \text{sextet}, \quad 5 \text{ unpaired electrons}$$

$$Sp = \text{septet}, \quad 6 \text{ unpaired electrons}$$

AB can be a final product or an intermediate, or even an activated complex. "Developing on a single potential surface" means that the system does not cross over to a higher-energy surface (Fig. 5.21). Such reactions are said to be adiabatic. The term "adiabatic" is also used in thermodynamics, but the meaning there is completely different.

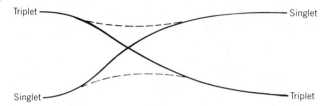

Figure 5.21. Schematic presentation of crossing singlet and triplet potential surfaces.

Two remarks on the spin selection rules have some special interest. From the first entry in the above list it is seen that if the two reactant molecules have all electrons paired, the electrons in the product will still be paired, and vice versa. The second remark is that the number of combinations allowed by the rules increases with the multiplicity. This means, for example, that for transition metal species with many unpaired electrons, spin conservation does not impose serious limitations; there are many alternatives.

Violation of the Spin Selection Rules

In polyelectron and polynuclear systems it may happen that the potential surface of the triplet state crosses that of the singlet state (Fig. 5.21). The distribution of the electrons in the triplet and singlet states are different everywhere, including the crossing point, but at the crossing point the positions of the nuclei and the energy of the system match. Consequently, in the region around the crossing point there is a nonzero probability of transition from one multiplicity to the other without absorption or emission of radiation (intersystem crossing). This means violation of the spin rules, but it is feasible if the spin interaction (coupling) with the environment is sufficiently strong, for example, if there is sufficient spin–orbit coupling or spin–lattice coupling. It is also reasonable to expect that the longer the system stays in the crossing region, the higher the probability for a change in multiplicity.

It must also be noted that the transition from one surface to the other provides a path with lower activation energy (the lower dashed line in Fig. 5.21). Often, however, reactions involving singlet–triplet transitions or vice versa are slow, because the probability for the transition is small.

EXAMPLE I

It is estimated that the reaction

$$SO_3^{2-} + O_2 \longrightarrow \underset{O}{\overset{O}{OSOO^{2-}}}$$
$$\quad (S) \qquad (T) \qquad\qquad (S)$$

is exothermic, at least by about -125 kJ mol^{-1}, but it has not been observed. It is difficult to decide whether SO_3^{2-} acts as a σ or π donor. The energy difference between the occupied orbitals that might be involved is small. It seems safer to say that SO_3^{2-} can act both as a σ donor and as a π donor with more or less the same efficiency. Overlap problems with the O_2 LUMO (π^*) do not exist. The fact that the reaction is not observed, presumably because it is slow, does not seem to be related to orbital symmetry, but it may very well be related to spin: The spin selection rules are not satisfied.   □

EXAMPLE II

The mechanism proposed for the isomerization of the maleic methyl ester to its fumaric isomer can be summarized as follows:

$$
\underset{\text{RC}=\text{CR}}{\overset{\text{H}\quad\text{H}}{|\quad\ |}} \longrightarrow \underset{\text{R}\dot{\text{C}}-\dot{\text{C}}\text{R}}{\overset{\text{H}\quad\text{H}}{|\quad\ |}} \longrightarrow \underset{\underset{\text{H}}{|}}{\overset{\text{H}}{|}}\text{R}\dot{\text{C}}-\dot{\text{C}}\text{R} \longrightarrow \underset{\underset{\text{H}}{|}}{\overset{\text{H}}{|}}\text{RC}=\text{CR}
$$

The spin rules are violated, and it can be argued that this is one of the factors contributing to the slowness of the reaction.   □

EXAMPLE III

The restrictions imposed by the spin selection rules may also be one of the reasons for the slowness of the reaction

$$N_2O_5 \longrightarrow N_2O_3 + O_2$$

    (S)   (S)   (T)   □

EXAMPLE IV

The spin and why butter gets rancid. Oxidation by dioxygen (air) of many commercial products, such as oils and fats, drugs, cosmetics, and rubber, often takes place by free radical mechanisms. Reactions such as

$$\dot{R} + \ddot{O}_2 \longrightarrow R\dot{O}_2$$

    (D)   (T)   (D)

do not violate the spin rules, and transition metal ions can accelerate these reactions because they have unpaired electrons, which can combine with the unpaired electron of the free radical or facilitate formation of such radicals. If we want to slow down the oxidation, a good idea is to remove or somehow deactivate these metal ions—for example, by adding appropriate chelating complexing agents.

The action of the so called antioxidants depends on their ability to remove the free radicals R· before they react with oxygen. The corresponding reaction then must be fast, without spin restrictions, and of course without any symmetry restrictions. Many antioxidants are free radical sources. Free radicals ·R' produced from antioxidants may remove R· by way of reactions such as

$$R· + ·R' \longrightarrow R:R' \qquad \square$$

5.14. A PARADIGM: THE REACTIONS OF DIOXYGEN

An attempt will now be made to apply what has been written in this chapter to the extremely important reactions of dioxygen (molecular oxygen). The following questions will be examined:

1. Thermodynamically the reactions of dioxygen with many of the materials with which it comes in contact (organic matter, humans) are favored. Why, then, is free molecular oxygen left in nature? What makes it inert?
2. How can the inert dioxygen be activated?
3. How can oxygen be transported, for example, by hemoglobin, without being destroyed and without acting as an oxidant during transportation?

Question 1. Why Is Dioxygen Inert?

The ground-state electronic configuration of dioxygen can be represented as follows:

$$KK(\sigma_{2s})^2(\sigma_{2s}^*)^2(\sigma_{2p_x})^2(\pi_{2y} = \pi_{2p_z})^4(\pi^*_{2p_y} = \pi^*_{2p_z})^2$$

The spin multiplicity is 3, and the total angular momentum around the internuclear axis vanishes ($L = 0$). This state is symbolized as $^3\Sigma_g^-$. The first two excited states differ from the ground state and from each other in the arrangement of the last two electrons in the two π^* orbitals:

$$\pi^* \uparrow \ \uparrow \qquad\qquad \pi^* \uparrow\downarrow \ \underline{\quad} \qquad\qquad \pi^* \uparrow \ \downarrow$$

Ground state   First excited state   Second excited state

$$^3\Sigma_g^- \qquad\qquad\qquad {}^1\Delta_g \qquad\qquad\qquad {}^1\Sigma_g^+$$

The Boltzmann distribution dictates that under ordinary conditions the ground state prevails. In trying, therefore, to interpret the inertness of dioxygen, let us concentrate on the ground state.

We first notice that dioxygen is a strong four-electron oxidizing agent:

$$4H^+ + O_2 + 4e^- \longrightarrow 2H_2O \qquad E^0 = 1.23 \ V$$

But, as we will see in Chapter 7, "Electron Transfer Reactions," simultaneous transfer of two electrons is a rather rare event. Simultaneous transfer of four electrons is practically impossible. Accordingly, we will confine our attention to:

The transfer of the four electrons one by one, in four successive steps
The less probable transfer in two steps, of two electrons each.

Transfer of the Electrons One By One

In acid aqueous solutions the relevant half-reactions and potentials[3] are

	$E^0 \ (V)$
$H^+ + O_2 + e^- \longrightarrow HO_2$	-0.32
$H^+ + HO_2 + e^- \longrightarrow H_2O_2$	1.68
$H^+ + H_2O_2 + e^- \longrightarrow OH + H_2O$	0.80
$H^+ + OH + e^- \longrightarrow H_2O$	2.74

This tabulation immediately reveals that although the reduction potential for the overall four-electron process is positive (1.23 V), the potential for the first step is negative. What this means is simply that the addition of the first electron is difficult—it requires a strong reducing agent.

One of the reasons for this barrier is probably the repulsion exerted on the entering electron by the electrons of the oxygen molecule itself. It can be said that all four entering electrons need to use the antibonding orbitals of the original molecule. This is not strictly correct, since after each addition the situation changes, but it can be accepted as a rough approximation. Addition of new electrons into antibonding orbitals means an increase of bond

---

[3]The potentials are different in neutral and alkaline solutions. In biological systems the hydrogen ion concentration and hence the potentials are controlled by buffers.

length, "swelling" of the electron cloud, a decrease in electron density, and a decrease in interelectronic repulsions. In the original neutral molecule, the charge density is higher, and the repulsive force on the entering first electron is stronger.

A second obstacle in the reaction of molecular oxygen is imposed by the spin conservation rules. Ordinary organic molecules usually have spin multiplicity 1. Hence, according to the spin rules, reactions of the form

$$\text{Organic molecule } + \text{ O}_2 \longrightarrow \text{ organic product}$$
$$\quad\quad\quad\text{(S)}\quad\quad\quad\quad\text{(T)}\quad\quad\quad\quad\quad\text{(S)}$$

are not allowed.

Of the spin-allowed reactions

$$\text{RH } + \text{ O}_2 \longrightarrow \text{ R } + \text{ HO}_2$$
$$\text{(S)}\quad\text{(T)}\quad\quad\text{(D)}\quad\text{(D)}$$

$$\text{R } + \text{ O}_2 \longrightarrow \text{RO}_2$$
$$\text{(D)}\quad\text{(T)}\quad\quad\text{(D)}$$

$$\text{RH } + \text{ O}_2 + \text{ HR} \longrightarrow \text{ R } + \text{ HOOH } + \text{ R}$$
$$\text{(S)}\quad\text{(T)}\quad\text{(S)}\quad\quad\text{(D)}\quad\quad\text{(S)}\quad\quad\text{(D)}$$

the first has a large activation energy, because it involves the formation of unstable free radicals. The second involves a free radical as a reactant, and the third involves a triple collision, which has a small probability of occurrence.

Transfer of the Electrons in Two Pairs

In this case a reduction potential barrier does not exist:

$$E^0(V)$$

$$2\text{H}^+ + \text{O}_2 + 2e^- \longrightarrow \text{H}_2\text{O}_2 \quad\quad 0.68$$
$$2\text{H}^+ + \text{H}_2\text{O}_2 + 2e^- \longrightarrow 2\text{H}_2\text{O} \quad\quad 1.77$$

Both steps have relatively large positive reduction potentials. But spin restrictions still exist, and in addition the quantum-mechanical probability for the simultaneous transfer of two electrons is small. Also, simultaneous transfer of two electrons requires simultaneous overlap of the two dioxygen π^* orbitals that are perpendicular to each other with two orbitals of the donor. The necessity of such a matching excludes all donors that do not have two occupied orbitals with the suitable symmetry and energy.

Question 2. How Is Dioxygen Activated?

We all know, of course, that the difficulties in reducing dioxygen are often surpassed. Burning with oxygen is certainly basic in our everyday activities and in life itself.

One way to overcome the inertness of dioxygen and make it react is simply to use brute force, for example, heating to a high enough temperature until all obstacles are overcome. This is done, for example, every time oxygen is used to burn fuels. A high enough temperature is used for free radicals to be formed and chain reactions to be started. In other cases, activation is more "refined," less violent. Two examples of such "gentle" activation are given below.

Reactions of Excited Dioxygen ($^1\Delta_g$)

State $^1\Delta_g$ lies 94 kJ mol^{-1} above the ground state; it does not have unpaired electrons, and its acid properties (electron acceptor) are enhanced because of the empty π^* orbital. This state can be generated photochemically or by microwave or radiowave excitation. It is also found as an intermediate in many reactions, such as the reaction of hydrogen peroxide with hypochlorite ion or bromine. Furthermore, many dye solutions absorb light and then transfer energy to dioxygen, forming $O_2(^1\Delta_g)$:

$$(\text{Dye}) \xrightarrow{h\nu} (\text{Dye})^*$$

$$(\text{Dye})^* + O_2(^3\Sigma_g^-) \longrightarrow (\text{Dye}) + O_2(^1\Delta_g)$$

$$O_2(^1\Delta_g) \longrightarrow \text{further reactions}$$

where (Dye)* is an excited state of the dye.

$O_2(^1\Delta_g)$ reacts with many organic molecules differently than $O_2(^3\Sigma_g^-)$, for example, in Diels–Alder addition:

Diels–Alder addition

It is left to the student to show that this addition, but not that of $O_2(^3\Sigma_g^-)$, is symmetry-allowed.

It is believed that $O_2(^1\Delta_g)$ participates in autooxidation reactions of fatty acids, fats, and oils, and in the formation of peroxides. It also plays an important role in atmospheric chemistry, particularly in the chemistry of the polluted atmosphere, and also in the photochemistry of some biological systems.

Activation by Transition Metals

The variety in the shapes and orientation of d orbitals, the small differences in energy between them, the existence of several oxidation states, and the high spin multiplicities allow transition metal species to react with many donors or acceptors without violating symmetry or spin conservation rules. The overlap of the d orbitals of a transition metal ion with dioxygen orbitals is quite complicated and depends on the number of d electrons and the geometry. Nevertheless, several of the possible combinations (Fig. 5.22 for

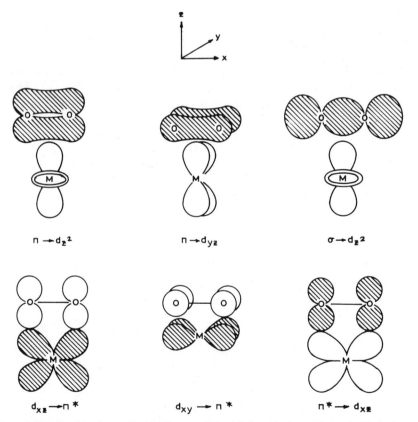

Figure 5.22. Combinations of transition metal d orbitals with the orbitals of η^2-coordinated dioxygen. Shading indicates occupied orbitals.

η^2 coordination) result in a decrease of electron density in the bonding region or an increase in the antibonding region. In either case the net result is activation of the dioxygen molecule.

If dioxygen receives electron density into one of its π^* orbitals, the resulting electron configuration will resemble that of the monoanion O_2^-. If, simultaneously, dioxygen donates electron density from the other π^* orbital, the electronic configuration will resemble that of the excited $O_2(^1\Delta_g)$.

Three possible geometrical structures for mononuclear dioxygen complexes are

There are also binuclear dioxygen complexes and complexes containing more than one oxygen molecule.

In the reactions of dioxygen that are mediated by transition metal compounds, complexation is the first step, but what happens next? How does coordinated oxygen react?

Two main types of reactions have been distinguished: those in which the substrate to be oxidized is also coordinated, and those in which only dioxygen is coordinated. In the first case, both dioxygen *and* the substrate are activated, and in the second case only dioxygen is activated.

Reactions like

must be classified in the second category. In the catalytic oxidation of tertiary phosphines to the corresponding oxides, it appears that both dioxygen and the phosphine become coordinated, and oxygen atom transfer takes place intramolecularly.

Question 3. How Is Dioxygen Transported?

In some cases dioxygen coordinated to a transition metal reacts further oxidatively, while in others it is simply transported. What are the factors determining this behavior?

Consider the mechanism

$$L_nM + O_2 \underset{}{\overset{K}{\rightleftharpoons}} L_nMO_2$$

$$L_nMO_2 \xrightarrow{k} L_{n-1}M + L_{oxidized} + H_2O_2$$

It is clear that if k has a large value the catalytic oxidation of L is favored. This, for example, is what happens with the enzyme oxidase. If, on the other hand, k is small and K relatively large, there is only transportation, as in the case of hemoglobin.

Heme itself combines with dioxygen strongly and irreversibly. This association is accompanied by oxidation of Fe^{II} to Fe^{III}. The presence of the protein prevents this oxidation and makes the coordination of dioxygen reversible. The irreversible oxidation of heme occurs through the intermediate formation of a compound, with an oxygen molecule bridging two iron ions belonging to different hemes. In myoglobin and in hemoglobin the "folding" of the polypeptide chain forms cavities where the hemes can fit, becoming inaccessible to each other. Thus, bridge formation is inhibited, and the oxidation of Fe^{II} to Fe^{III} does not take place. Furthermore, Fe^{II} coordinates to the histidene nitrogen, and this causes electron displacement from the nitrogen to the iron and subsequent weakening of the iron–dioxygen bond.

GENERAL REFERENCES

R. F. W. Bader, The Use of the Hellmann-Feynman Theorem to Calculate Molecular Energies, *Can. J. Chem.* **38**, 2117 (1960).

J. D. Bradley and G.C. Gerrans, Frontier Molecular Orbitals, A Link Between Kinetics and Bonding Theory, *J. Chem. Educ.* **50**, 463 (1973).

M. C. Caserio, Reaction Mechanisms in Organic Chemistry, *J. Chem. Educ.* **48**, 782 (1971).

R. P. Feynman, Forces in Molecules, *Phys.Rev.* **56**, 340 (1939).

K. Fukui and H. Fujimoto, Orbital Symmetry and Electronic Rearrangements, *Mech. Mol. Migr.* **2**, 117 (1969).

H. Hellmann, *Einfuhrung in die Quantenchemie,* Deuticke, Leipzig, 1937.

H. Nakatsuji, Electrostatic Force Theory for a Molecule and Interacting Molecules. I. Concept and Illustrative Applications, *J. Am. Chem. Soc.* **95**, 345 (1973).

R. G. Pearson, Symmetry Rules for Chemical Reactions, *Accounts Chem. Res.* **4**, 152 (1971).

R. G. Pearson, *Symmetry Rules for Chemical Reactions: Orbital Topology and Elementary Processes,* Wiley-Interscience, New York, 1976.

D. M. Silver, Hierarchy of Symmetry Conservation Rules Governing Chemical Reaction Systems, *J. Am. Chem. Soc.* **96**, 5959 (1974).

E. Wigner and E. E. Witmer, Molecular Spectra of Dialomic Molecules in Modern Quantum-Mechanics, *Z. Phys.* **51**, 859 (1928).

R. B. Woodward and R. Hoffmann, *The Conservation of Orbital Symmetry,* Academic, New York, 1970.

PROBLEMS

1. It is believed that in the hydrogen ion–catalyzed decomposition of oxalate ion, the following reactions occur:

$$C_2O_4^{2-} \longrightarrow CO + CO_3^{2-}$$

$$CO_3^{2-} \xrightarrow{2H^+} CO_2 + H_2O$$

   Predict the products of the analogous reaction of the isoelectronic oxide N_2O_4. Using this prediction, propose a mechanism for the nitration of aromatic compounds with a HNO_3–H_2SO_4 mixture.

2. In acid solution, hydrazine forms hydrazinium ions, $N_2H_5^+$, the same way ammonia forms ammonium ions, NH_4^+. But in contrast to NH_4^+, $N_2H_5^+$ is a strong and fast reducing agent. What, in your opinion, are the reasons?

3. Why are reductions by hydroxylamine, NH_2OH, usually fast?

4. The cyclic compound $P_3N_3Cl_6$ (hexachlorophosphazene(I)) resembles benzene. In benzene, π bonds are formed by overlap of p orbitals and in $P_3N_3Cl_6$ by overlap of the nitrogen p orbitals with phosphorus d orbitals. In benzene, π electrons form a full ring covering all carbon atoms (symbolized as a circle within the ring).

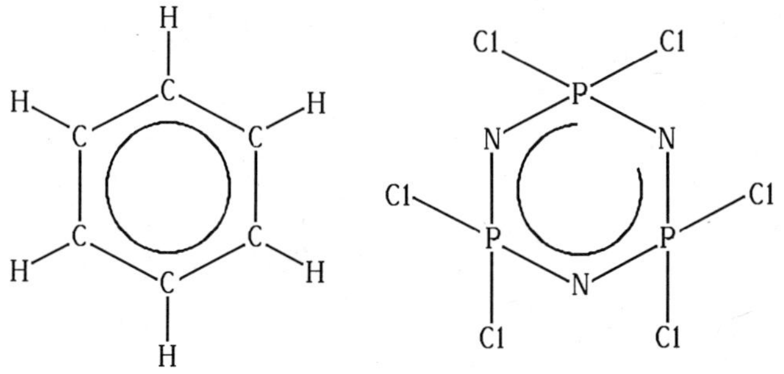

   In $P_3N_3Cl_6$ the corresponding "cloud" does not extend over the whole ring (broken circle within the ring). In $P_3N_3Cl_6$ there is less aromaticity. Show with sketches why it is impossible in $P_3N_3Cl_6$ to form a full π cloud by overlap of p and d orbitals. Assume an arbitrary assignment of the positive and negative lobes of these orbitals.

5. Consider the half-lives (in parentheses, minutes) of the following pairs of complexes:

High-spin $Cr(CN)_6^{4-}$ (0.3) $Cr^{III}(CN)_6^{3-}$ (40,000)

High-spin $Fe(H_2O)_6^{3+}$ (2)   Low-spin $Fe(CN)_6^{4-}$ (5000)

High-spin $Co(H_2O)_6^{2+}$ (1)   Low-spin $Co(NH_3)_5(H_2O)$ (2000, H_2O exch.)

     Interpret the large differences on the basis of electronic structure.

6. Based on the conclusions of problem 5, predict which of the following complexes are inert and which labile: TiF_6^{3-}, MnF_6^{3-} (high-spin), $Fe(CN)_6^{3-}$ (low-spin), $Zn(CN)_4^{2-}$.

7. Using the HOMO–LUMO formalism, predict which of the following reactions are allowed by the orbital symmetry rules and which are forbidden:

$$H_2 + D_2 \longrightarrow 2HD$$

$$H_2 + F_2 \longrightarrow 2HF$$

$$N_2H_2 + C_2H_4 \longrightarrow N_2 + C_2H_6$$

$$N_2O_2 + O_2 \longrightarrow N_2O_4$$

Assume that these reactions occur via a bimolecular 4-centered activated complex. For the forbidden reaction(s), suggest other, symmetry-allowed paths requiring activation.

6

GROUP-TRANSFER AND ATOM-TRANSFER REACTIONS

From the mechanistic point of view, two main categories of reactions can be distinguished: group-transfer and atom-transfer reactions, which do not involve changes in formal oxidation numbers, and electron-transfer reactions, which do.

It is important to keep in mind that for any reaction in solution, it is not only the reacting molecules that change, but usually also the solvent environment around them. The presence of the solvent molecules and their arrangement may not be shown explicitly in the chemical equation, but they are always there, and their role may be critical.

Group- and atom-transfer reactions include the following subcategories:

(*a*) *Substitution reactions* can in general be represented by the equation

$$RX + Y \longrightarrow RY + X \qquad\qquad 6.1$$

What this symbolism really means is that Y is transferred from the bulk of the solution to R, and X is released from R to the bulk of the solution. Initially the environment of Y is solvent molecules only; after substitution it includes R.

(*b*) *Decomposition reactions* can be represented as

$$RX \longrightarrow R + X \qquad\qquad 6.2$$

The reaction is intended to be just stoichiometric, meaning in particular that R on the right (in the bulk of the solution) has the same stoichiometry as the R on the left (bound to X). However, there can be significant differences in structure.

Again, X is transferred from an environment including R and solvent molecules to a new environment with solvent molecules only. A necessary condition for decomposition to occur is that solvated R and X be stable.

(c) *Addition (combination) reactions:*

$$XR + Y \longrightarrow XRY \qquad 6.3$$

Here, Y is transferred from the bulk of the solution to RX.

(d) *Proton transfer or neutralization reactions:*

$$BH^+ + B' \longrightarrow B + B'H^+ \qquad 6.4$$

Reaction 6.4 resembles substitution, but the transferred entity here is the mobile proton, and the interactions involved have some peculiarity (e.g., hydrogen bonding).

(e) *Metathesis.* Here the transfer is from one site in the molecule to another site in the same molecule. Schematically,

$$\begin{array}{ccc} A\!-\!B & \longrightarrow & A\!-\!B \\ | & & | \\ X & & X \end{array} \qquad 6.5$$

The mechanism may or may not involve an intermolecular step.

(f) *Nonredox reactions of coordinated ligands.* One case of a ligand decomposition reaction is

$$L_nM(X\!-\!Y) \longrightarrow L_nMX + Y \qquad 6.6$$

Such reactions can be regarded as decomposition reactions irrespective of whether or not one of the fragments includes a metal center and whether or not the reaction involves breaking a metal–ligand bond.

Similar remarks can be made for substitution and addition reactions. The important events in substitution may take place on the metal ion, but there are cases in which they take place on a ligand atom.

6.1. TYPES OF SUBSTITUTION REACTIONS

Substitution reactions can be classified in a number of different ways:

1. On the basis of coordination number and molecular geometry
2. According to the mechanism
3. On the basis of reaction type and the nature of the species involved

4. According to the periodic table and the identity of the central atom
5. According to the lability of the species involved

Here we will make a brief selective survey by classifying mainly according to coordination number and geometry, but an effort will be made to give representative examples of different labilities, central atoms, and types of reactions.

The following types of substitution reactions can be distinguished:

1. Ligand by ligand (ligand exchange, solvent exchange)
2. Solvent by ligand (complex formation, anation)
3. Ligand by solvent (solvolysis, acid hydrolysis, base hydrolysis)
4. Rearrangements (inter- and intramolecular, including isomerization and racemization)
5. Metal ion exchange

6.2. SUBSTITUTION MECHANISMS

In the usual mechanistic studies in solution, the average distance between reactant molecules is large. At large distances, only long-range interactions occur, primarily affecting the activity coefficients.

As two reactant molecules diffuse toward each other, it is useful to try to identify as many chemically distinct entities as possible. By taking species AL and BL' as our models (the charges are omitted), we can distinguish the following entities:

$$[ALL'B] \qquad [ALB] \text{ or } [AL'B] \qquad [AB]$$

$$\textbf{I} \qquad\qquad\qquad \textbf{II} \qquad\qquad\qquad \textbf{III}$$

In **I** the coordination spheres of A and B are retained. The two species are weakly held together by van der Waals forces or by ion–dipole or dipole–dipole interactions, or even by an incipient bond involving orbital overlap, if such an overlap is possible. If AL and BL are oppositely charged, there is also ion–ion attraction, and the corresponding species is often called an *ion pair*. We will use the term *outer-sphere complex*[1] for all structures that look like **I**, including cases where both A and B are metal ions or atoms and cases

---

[1]In the literature there is some confusion in terminology. Thus, in substitution reactions, a complex ML has been called inner-sphere, but in electron-transfer reactions, the inner sphere is considered the bimetallic complex MLM'. A similar-looking complex with a bridging group, MLL', has been called outer-sphere in substitution reactions. In this book we have tried to consistently use the terms defined in the text.

where one species is a metal complex and the other a solvated ligand (e.g., $ML \cdots OH_2L'$).

Species **II** is called a *bridged complex,* both in electron transfer and in substitution reactions, and it includes cases such as MLM' and MLL' (with L being regarded as the bridge).

Species **III** is called a *direct overlap complex,* and it includes cases such as MM' and ML.

Mechanisms

Three categories of substitution mechanisms can be distinguished:

Dissociative (D). The first reaction is a decomposition like that represented by equation 6.2, but R in the presence of another ligand Y is not stable; it reacts with it:

$$RX \longrightarrow R + X$$

$$R + Y \longrightarrow RY$$

Associative (A). The first reaction is an addition reaction like 6.3, and it is followed by removal of X:

$$XR + Y \longrightarrow XRY$$

$$XRY \longrightarrow X + RY$$

Interchange (I). First an outer-sphere complex is formed, and then the leaving group moves from the inner to the outer coordination sphere, and simultaneously the entering group moves from the outer to the inner coordination sphere.

As an illustration of the interchange mechanism, consider the replacement of a water molecule bound to a metal ion by an anionic ligand L^-. The reactions taking place can be represented by the following equations:

$$[L(H_2O)]^- + [(H_2O)M]^{m+} \underset{}{\overset{K_1}{\rightleftharpoons}} [L(H_2O)(H_2O)M]^{(m-1)+} \qquad 6.7$$

$$[L(H_2O)(H_2O)M]^{(m-1)+} \overset{K_2}{\rightleftharpoons} [L(H_2O)M]^{(m-1)+} + H_2O \qquad 6.8$$

$$[L(H_2O)M]^{(m-1)+} \overset{k}{\longrightarrow} [LM]^{(m-1)+} + H_2O \qquad 6.9$$

The first two equations usually correspond to fast diffusion-controlled processes, the equilibrium constants are small, and there is no independent method to measure them directly. In fact, since it is practically impossible to distinguish between the two equilibria, it is customary to use the combined equilibrium constant $K = K_1K_2$.

6.3. MOLECULAR MECHANISM

The mechanisms just described constitute analysis into elementary reactions, and they can be called stoichiometric. However, as we have seen, there is also another mechanistic aspect: what happens at the molecular level—in substitution reactions, what happens in each case at the molecular level during the critical elementary step. It is of interest whether this step is dissociative, associative, or interchange.

For a D or A mechanism, the situation looks simple. The elementary reactions reflect the molecular happenings, although there are still questions such as: What is the structure of the intermediate with the altered coordination number, and what is the role of the solvent?

For an interchange process we want to know what affects the preequilibria, but most of our attention is focused on the dynamic changes occurring during the critical concerted interchange step. This step can be visualized as being dissociative-like (I_d); the determining factor is then the release of the leaving group, while the group that will eventually enter "waits just outside" and moves in to fill the "void." The other case is an associative-like interchange (I_a), in which the incoming group helps the leaving group depart.

If the electron pair of the new bond comes from the entering ligand, the substitution is called *nucleophilic* (S_N1 if it is dissociative, S_N2 if it is associative). Less frequently, the electron pair of the new bond is contributed by M (and not by Y); the substitution is then called *electrophilic* (S_E).

More on the *A* or *I* and *D* or *I* Dichotomies

The distinction between an A and an I mechanism or between a D and an I mechanism is sometimes difficult to make. It is instructive to see why. The difficulties are not only experimental; they are also conceptual. A classification cannot be sharper than the concepts upon which it is based.

Take as an example a tetrahedral complex undergoing associative substitution. We can visualize an infinite number of different ways by which the entering group Y approaches and "sticks" to MX_4. The limiting cases can be represented as in Figure 6.1. Approach **I** seems to be a real interchange, provided that a bridged M—X—Y complex is formed. The other cases are more ambiguous; they may or may not involve direct bonding of Y to M. The distance M—Y can be smaller, equal to or larger than the distance M—X, but this is not by itself sufficient to determine whether or not Y is within the "first coordination sphere."

For a D mechanism, the intermediate with the lower coordination number is solvated. Whether or not the solvent molecules come within the first co-

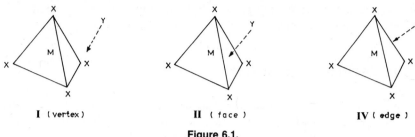

I (vertex)   II (face)   IV (edge)

Figure 6.1.

ordination sphere is somewhat a matter of conjecture. As in the case of association, it is easy to visualize situations whereby assignment of a coordination number in solution and the boundaries of the first coordination sphere are by no means clear-cut.

In conclusion, the attitude should be one of caution. It is difficult enough to assign a set of elementary reactions to a given overall reaction. It is also difficult to deduce the events occurring on a molecular scale for each of the elementary reactions of the mechanism. So we should not make it even more difficult by imagining, for example, that the coordination sphere is a perfect sphere, without fluctuations in space and time and with sharp boundaries defined by some kind of force field.

6.4. CONTOUR DIAGRAMS

The continuous gradation between the limiting $D(S_N1)$ and $A(S_N2)$ mechanisms can be illustrated by using contour diagrams (Fig. 6.2). In these diagrams, the two opposite corners (upper left and lower right) represent the reactants $(L_nMX + Y)$ and products $(L_nMY + X)$, respectively. The other two corners represent the limiting $S_N1(L_nM + X + Y)$ and $S_N2(L_nM_Y^X)$ cases.

In a reaction involving complex molecules, several bond distances and angles (simple or dihedral) change simultaneously, and this is associated with a change in the potential energy of the system, which can be expressed as a function of these "internal variables," which are usually many. Thus, the potential energy is a "surface" in an $(n + 1)$-dimensional space, where n is the number of variables. Whenever two-dimensional visual aids like those of Figure 6.2 are used, it must be kept in mind that one or two of the variables have been singled out, presumably those which give better insight into how the reaction proceeds, but it is also implied that the energy is optimized (minimized) with respect to the other variables.

In Figure 6.2 the "internal coordinates" chosen are the M—X and M—Y bond distances. The energy contours shown in this figure are arbitrary and do not result from calculations. They are given only as illustrations of various

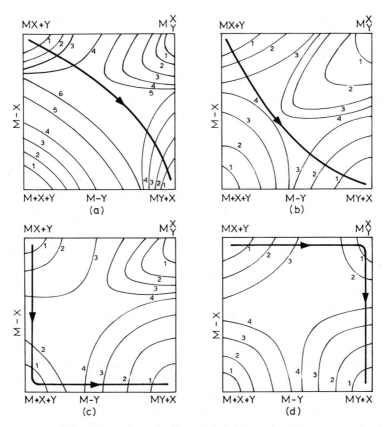

Figure 6.2. (*a*) Associative interchange mechanism (I_a); one barrier no intermediate. (*b*) Dissociative interchange mechanism (I_d); one barrier, no intermediate. (*c*) Dissociative (*D*) mechanism; two barriers, one intermediate with reduced coordination number. (*d*) Associative (*A*) mechanism; two barriers, one intermediate with an increase in coordination number. In this representation the outer-sphere complex of the interchange is not counted as an intermediate.

possibilities. The *minimum energy pathways* indicated by the arrows pass either from intermediates of reduced (*c*) or increased (*d*) coordination number (and a lifetime of at least 10^{-12} s) or they correspond to concerted mechanisms approaching one of the two limiting corners [Fig. 6.2 (*a*) and (*b*)].

6.5. EMPIRICAL CRITERIA FOR DECIDING THE MECHANISM OF SUBSTITUTION

The Observed Rate Law

Since the slow, rate-determining step in a *D* mechanism is the breaking of the M—X bond, and the entering ligand (Y) comes in later, it is reasonable to expect that the observed rate law should be first-order in [MX] and independent of [Y]. In contrast, for an A mechanism, dependence on both [MX] and [Y] is expected. Unfortunately, things are not that simple. In Chapter 2 we have already worked out several cases. An additional example will be given in Section 6.16, within the context of octahedral substitution in Co^{III} complexes.

The general conclusion is that the observed rate law by itself, without other criteria, cannot give conclusive evidence in support of one mechanism or another. The arguments should necessarily rest on more complex syllogisms.

It also must be recalled that the rate law is ambiguous regarding the concentration of the solvent (Chapter 2). A mechanism that is labeled dissociative may in fact be associative with participation of a solvent molecule, as in the substitution of square planar complexes (see below). In general, the uncertainties inherent in the "translation" of the observed rate law into a mechanism that have already been discussed in detail apply to substitution reactions as well.

Dependence on the Nature of the Entering Ligand

A small entering ligand dependence implies a dissociative (D or I_d) mechanism, whereas a large ligand dependence (comparable to the dependence on the leaving group) implies an associative mechanism (A or I_a).

In a study of the replacement of water in $Cr(NH_3)_5(OH_2)^{3+}$ by a number of anions, it was found that the values of the observed second-order rate constants vary by less than an order of magnitude. Moreover, the rate constants do not differ much from the corresponding solvent exchange rate constants. This behavior is considered typical of an I_d mechanism.

In contrast, in the anation of $Cr(H_2O)_6^{3+}$ the observed rate constants vary more markedly. For example, the ratio of the rate constants for substitution by NCS^- and Cl^- is 60. (The corresponding ratio for $Cr(NH_3)_5OH_2^{3+}$ is only ~ 6.) The large ratio is taken as strong evidence for an I_a mechanism.

Detection of Intermediates

If an intermediate with an altered coordination number is detected, it can possibly be postulated that the structure of the activated complex is similar.

The intermediate may be detected directly by some physical method, such as NMR, but it can also be inferred from the chemical behavior of the system. Since an intermediate is a genuine chemical compound, corresponding to a minimum in the potential energy profile (no matter how shallow), it can in principle react with a number of substances that have been added to the system as probes, or it can undergo intramolecular rearrangements. A special case of a "probe" is the general electrolytic environment or even the solvent itself.

The Entropy of Activation

Negative entropies of activation are in general indicative of an associative mechanism and positive entropies of a dissociative mechanism.

Figure 6.3 symbolizes different solvent exchange mechanisms of a hypothetical "flat" complex. For a D mechanism, in going from the initial state to the transition state, there is an increase in disorder, that is, the entropy has increased ($\Delta S^{\ddagger} > 0$). For an A mechanism the randomness of the transition state is smaller than that of the initial state, and $\Delta S^{\ddagger} < 0$. For interchange mechanisms, intermediate values are expected.

This brief analysis gives the salient features of the expected trends, but it is certainly idealized. It does not illustrate contributions from other changes within the first coordination sphere, nor does it include reorganization of the solvent outside this sphere. Thus, although the general trend is indeed observed, $\Delta S^{\ddagger}$ is not always a reliable criterion, even when its value is known fairly accurately.

Volume of Activation

The volume of activation $\Delta V^{\ddagger}$ is determined by high-pressure experiments, and it is frequently more accurate than $\Delta S^{\ddagger}$, because pressure can be controlled more accurately than temperature.

In a solvent exchange reaction, if there is no charge separation or cancel-

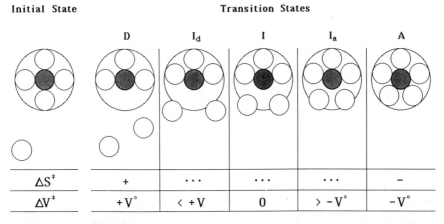

Figure 6.3. Schematic representation of solvent exchange reactions proceeding by different mechanisms. The small black circle in the middle of the idealized "flat" complex represents the central atom. The small open circles represent the bound and free ligands (solvent molecules). The large circles enclosing the small ones represent the boundaries of the first coordination sphere. For a D mechanism, the entropy and volume of activation are positive; for an A mechanism they are negative. V° is the solvent molar volume.

lation on going from the reactants to the transition state, a positive $\Delta V^{\ddagger}$ can be associated with a dissociative mechanism, and a negative $\Delta V^{\ddagger}$ with an associative mechanism. For interchange mechanisms, intermediate values are expected.

The various cases for solvent exchange reactions are depicted in Figure 6.3. For a D mechanism, in going from the initial state to the transition state, the volume increases by the volume of one solvent molecule; on a mole basis, the change roughly equals the molar volume of the solvent V^0. For an A mechanism there is a decrease by V^0. Interchange mechanisms give intermediate values. In the I_d mechanism there is only weak bonding to the entering and leaving groups. In I_a the bonding is stronger, but not quite as strong as in A. More or less bonding is symbolized in Figure 6.3 by placing the leaving and incoming groups at different distances from the central atom.

EXAMPLE

The volumes and entropies of activation for methanol exchange between solvent and complexes of the type $[M(CH_3OH)_6][ClO_4]_2$ (where M = Ni, Co, Fe and Mn) are given in Table 6.1.

The high positive value of $\Delta V^{\ddagger}$ for Ni^{II} strongly suggests a dissociative mode, and the negative value for Mn^{II} suggests an associative mode. These assignments are further supported by the values of the entropies of activation, which are positive and negative, respectively, as expected. The value of $\Delta V^{\ddagger}$ for Fe^{II} is close to zero, as expected for a "half the way" interchange mechanism (**I** in Fig. 6.3).   □

The Effect of the Nonleaving Ligands

Bulky nonleaving ligands are generally expected to inhibit the attachment of an additional (entering) group (associative mechanism) but to facilitate a reaction taking place by a dissociative path. The effects of nonleaving groups cis or trans to the leaving group will be described later.

TABLE 6.1. Entropies and Volumes of Activation for Methanol Exchange on Divalent Ions

	$\Delta S^{\ddagger}$, J K^{-1} mol^{-1}	$\Delta V^{\ddagger}$, cm^3 mol^{-1}	
Ni^{II}	+33.5	+11.4	(307 K)
Co^{II}	+30.1	+ 8.9	(279 K)
Fe^{II}	+12.6	+ 0.4	(255 K)
Mn^{II}	−50.2	− 5.0	(279 K)

6.6. THE MECHANISM OF A GIVEN SUBSTITUTION SOMETIMES CHANGES

The assignment of a mechanism to a given substitution is not of the either/ or type. A substitution can proceed by one of these mechanisms only, but examples are also available in which two such mechanisms operate simultaneously, and there are also examples in which a change in the environment causes a change in the mechanism.

EXAMPLES

Ligand exchange in the 7-coordinated uranyl complex $UO_2(DMSO)_5^{2+}$ was studied by the line-broadening NMR technique. The structure of this complex in CD_3COCD_3 solutions is a pentagonal bipyramid:

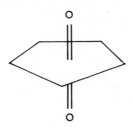

The rate of exchange in such solutions is found to have two terms:

$$k_{ex} = k_1 + k_2[DMSO]$$

(At 25°C, $k_1 = 5.5 \times 10^3 \text{ s}^{-1}$, and $k_2 = 3.2 \times 10^4 \text{ } M^{-1} \text{ s}^{-1}$.)

The reaction proceeds by two parallel paths. The first term (independent of free DMSO) has been attributed to a D mechanism and the second term to an A or I_a mechanism.

In $UO_2(acac)_2DMSO$,[2] the mechanism of the DMSO exchange can be modified; in CD_3COCD_3 it is of a D character, and the rate (at 25°C, $k = 2.4 \times 10^2 \text{ s}^{-1}$) is independent of the free DMSO concentration. However, in

---

[2] acac = acetylacetonate (the anion derived from diacetylmethane (pentane-2,4-dione):

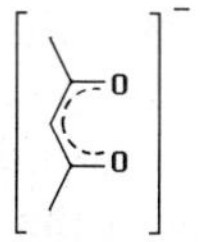

CD_2Cl_2 the observed exchange rate constant is consistent with the equation

$$k_{ex} = \frac{k_d + k_I K_{os}[DMSO]}{1 + K_{os}[DMSO]}$$

where K_{os} is the outer-sphere equilibrium constant. This means that in this system too, as in $UO_2(DMSO)_5^{2+}$, there are two parallel paths, one D and the other I, corresponding to the rate constants k_d and k_I. It also means that the same reaction proceeds by different mechanisms in different solvents (CD_3COCD_3 and CD_2Cl_2).

6.7. CHELATE RING FORMATION

Chelate ring formation and its opposite, opening and removal of the chelated ligand, are essentially substitution reactions. For example,

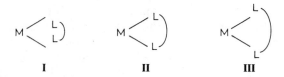

However, the fact that the *same* ligand must be substituted at the same center at least twice, from different sites, introduces a peculiarity reflected in the mechanism. Specifically, substitution now will also depend on the "flexibility" of the multidentate ligand and the distance between sites—on characteristics determined by the strong, rather "inflexible" covalent bonding within the chelate ligand. Schematically the situation can be represented as follows:

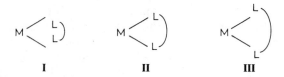

In **I** the "bite" is smaller than the optimum required by the available orbitals of M; in **II** the "bite" is just right; and in **III** it is larger. In **III** the situation can be partially remedied if the bidentate ligand moves away from M, and in **I** if it moves closer to M or if the metal or the ligand orbitals can be "bent" a little. In any case, there is a "stress" that does not exist with monodentate ligands, which will affect the rate and mechanism of substitution.

For a bidentate ligand, the rate-determining step can be either the attachment to the first site or the closing of the ring. The ring can also be formed in a concerted manner by simultaneous formation of the two bonds.

EXAMPLES

The formation of α- and β-alanine complexes with Ni^{II}, Co^{II}, and Mn^{II} was studied in alkaline solution by the temperature jump method. α- and β-alanine anions,

$$CH_3\underset{\underset{NH_2}{|}}{C}HCOO^- \quad \text{and} \quad CH_2\underset{\underset{NH_2}{|}}{C}H_2COO^-$$

respectively, form chelated complexes with these metal ions having the general formula ML_n ($n = 1,2,3$, and the charges are omitted). The data were analyzed on the basis of an interchange mechanism (I_d) involving the following stages:
  1. A fast, diffusion-controlled, ion-pair (bridged complex) formation,

$$M(aq) + \overset{\frown}{AB}(aq) \overset{K_a}{\rightleftharpoons} H_2OMOH_2 \cdots \overset{\frown}{AB} \qquad 6.10$$

In this equation only two of the water molecules of the inner coordination sphere are represented.
  2. The first substitution (second stage), which involves binding of AB to M through A:

$$H_2OMOH_2 \cdots \overset{\frown}{AB} \underset{k_{-1}}{\overset{k_1}{\rightleftharpoons}} H_2OM\overset{\frown}{AB} + H_2O \qquad 6.11$$

  3. The second substitution, involving closure of the ring,

$$H_2OM\overset{\frown}{A\ B} \underset{k_{-2}}{\overset{k_2}{\rightleftharpoons}} M\overset{\diagup A}{\underset{\diagdown B}{\Big)}} + H_2O \qquad 6.12$$

The mechanism has been written for the formation of the first complex ($n = 1$). Similar reactions can be written for $n = 2$, $n = 3$.
  The observed rate constants for the forward and reverse reactions are related to the rate constants of the mechanism as follows:

$$k_f = \frac{k_1 K_a k_2}{k_{-1} + k_2}$$

$$k_r = \frac{k_{-1} k_{-2}}{k_{-1} + k_2}$$

TABLE 6.2. Rate Constants for the Reactions of α- and β-Alanine with Co^{II}, Ni^{II}, and Mn^{II}

Metal Ion	n	α-Alanine	β-Alanine
		k_n, M^{-1} s^{-1}	
Co^{II}	1	6.0×10^5	7.5×10^4
	2	8.0×10^5	8.6×10^4
Ni^{II}	1	2.0×10^4	1.0×10^4
	2	4.0×10^4	6.9×10^3
Mn^{II}	1	—	5×10^4
		k_{-n}, M^{-1} s^{-1}	
Co^{II}	1	32	7.5
	2	280	86
Ni^{II}	1	0.022	0.23
	2	0.80	2.7

The results are summarized in Table 6.2.

For α-alanine, substitution on nickel is slower than on cobalt (for $n = 1$ and $n = 2$). For both of these ions, substitution of the second ligand is faster than substitution of the first. Substitution of the third ligand is even faster.

With Co^{II} the rate with β-alanine drops by almost an order of magnitude. Substitution on Mn^{II} is also slow. The decrease with Ni^{II} is smaller. The interpretation given is that with α-alanine the first substitution is rate-determining and with β-alanine the closure of the ring is rate-determining.

The ring with β-alanine is six-membered; with α-alanine it is five-membered. The energy barrier for ring closure is higher in β-alanine, presumably because it involves more strain and a smaller increase in entropy.

For multidentate ligands it is natural to expect more steps, which in some cases can be observed. Charge density also plays a role. Thus, because of fairly high charge density, Al^{III} is not very labile. As a result, substitution of DMSO in $Al(DMSO)_6^{3+}$ by the terdentate ligand terpy[3] can be observed, and it is found to proceed in three stages: a rapid first bond formation ($t_{1/2} <$ 20 s at 300 K) followed by two slow ring-closure steps.

[3]terpy = 2,2':6',2"-terpyridine:

6.8. COORDINATION SPHERE EXPANSION, ADDITION, AND CONDENSATION

Consider as an example the expansion of the coordination sphere of Mo^{VI} from four coordinate to six:

$$MoO_4^{2-} + H^+ \xrightleftharpoons{K_1} HMoO_4^- \qquad\qquad 6.13$$

$$HMoO_4^- + H^+ + 2H_2O \xrightleftharpoons{K_2} Mo(OH)_6 \qquad [\text{or } (H_2O)_3MoO_3] \quad 6.14$$

The values of $\Delta H°$ and $\Delta S°$ constitute a good diagnostic tool for deciding whether the expansion of the coordination sphere occurs after the addition of the second proton (as written in reactions 6.13 and 6.14) or right after the addition of the first proton.

Values of $\Delta H°$ and $\Delta S°$ for a number of similar protonation reactions that do not involve change in coordination number are collected in Table 6.3.

For reaction 6.14 there is a large negative enthalpy change (-46.5 kJ mol^{-1}) because of the formation of the two new bonds and a negative entropy change (-85 J mol^{-1} K^{-1}) because of the increased order of the 6-coordinated species. Negative values of $\Delta H°$ (-62 kJ mol^{-1}) and $\Delta S°$ (-236 J mol^{-1} K^{-1}) are also observed in the reaction:

$$IO_4^- + 2H_2O \rightleftharpoons H_4IO_6^- \qquad\qquad 6.15$$

$$\text{4-Coordinate} \qquad\qquad\qquad \text{6-Coordinate}$$

Addition

Addition is closely related to coordination sphere expansion. Consider as an example the formation of polytungstate from the tetrahedral monomeric

TABLE 6.3. Standard Enthalpies and Entropies of Some Protonation Reactions

Reaction	$\Delta H°$, kJ mol^-	$\Delta S°$, J mol^{-1} K^{-1}
$MoO_4^{2-} + H^+ \longrightarrow HMoO_4^-$	22.5	143
$CrO_4^{2-} + H^+ \longrightarrow HCrO_4^-$	5.4	142
$HPO_4^{2-} + H^+ \longrightarrow H_2PO_4^-$	-5.0	130
$HAsO_4^{2-} + H^+ \longrightarrow H_2AsO_4^-$	-3.2	119

WO_4^{2-}. The first addition can be regarded as a concerted chelate ring formation (omitting charges):

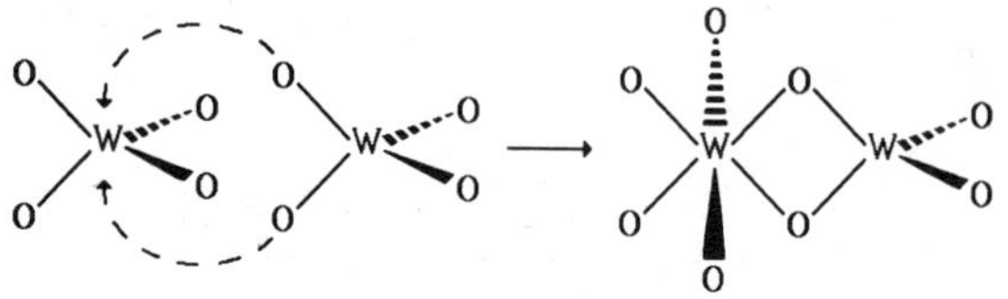

In this process the coordination number of one of the metal ions is increased by two units, while the other tetrahedron acts as a chelating agent, adding edgewise. The dimer formed can in turn also act as a chelating agent from the octahedron side, forming a trimer, and so on:

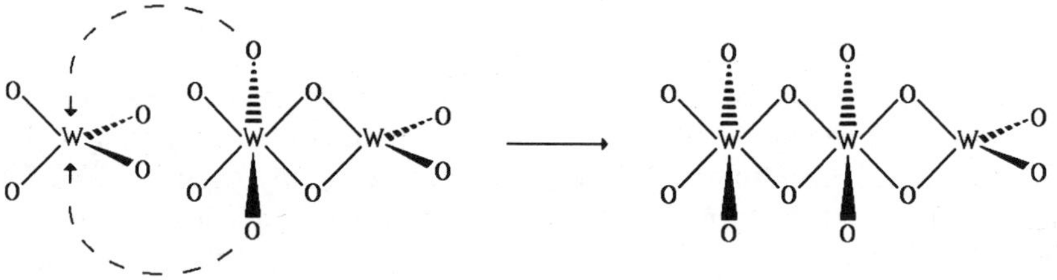

The only activation required for these processes originates in the need for changing bond lengths and angles. Accordingly, addition reactions of this kind are expected to be fast.

Condensation

In contrast to addition, condensation, as in the reaction

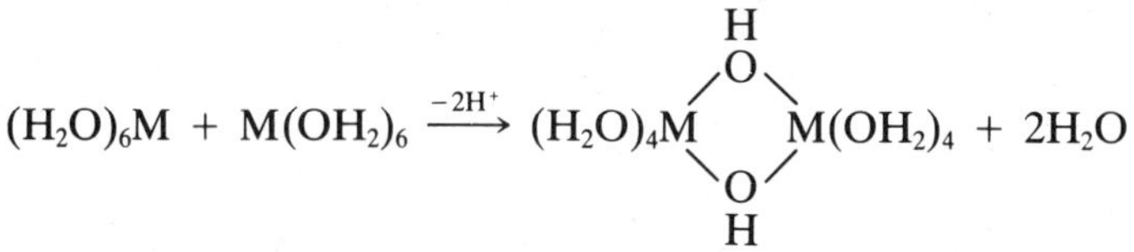

$$(H_2O)_6M + M(OH_2)_6 \xrightarrow{-2H^+} (H_2O)_4M \begin{matrix} H \\ O \\ \diagdown \\ \diagup \\ O \\ H \end{matrix} M(OH_2)_4 + 2H_2O$$

involves bond breaking or formation of an activated complex with an unusually high coordination number. In either case the process is expected to be slow. Thus, the condensation

$$2CrO_4^{2-} + 2H^+ \longrightarrow Cr_2O_7^{2-} + H_2O$$

is six orders of magnitude slower than the polymerization of tungstate.

6.9. TETRAHEDRAL SUBSTITUTION

Most d-block and p-block elements form tetrahedral molecules. Tetrahedral compounds of lithium, beryllium, and magnesium are also known. Substitution reactions on carbon are not examined here, since they are included in considerable detail in books on organic reaction mechanisms.

Boron

Substitution on tetrahedral boron is similar to that on carbon and other light elements. In the stepwise hydrolysis of boron tetrahydride, BH_4^-, the following reactions take place:

$$BH_4^- + H_2O \longrightarrow [BH_3OH]^- + H_2$$

$$[BH_3OH]^- + H_2O \longrightarrow [BH_2(OH)_2]^- + H_2$$

$$[BH_2(OH)_2]^- + H_2O \longrightarrow [BH(OH)_3]^- + H_2$$

$$[BH(OH)_3]^- + H_2O \longrightarrow [B(OH)_4]^- + H_2$$

The overall rate is determined by the first reaction. The rate law contains two terms: an acid-independent term and a term first-order in hydrogen ion concentration.

The acid-independent path is believed to proceed by a D mechanism or by a solvent-assisted dissociation. The acid-catalyzed path is believed to involve the steps

$$BH_4^- \xrightarrow{\ H^+\ } [BH_5] \xrightarrow{\ OH^-\ } [BH_3OH]^- + H_2$$

The structure proposed for BH_5 is

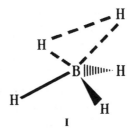

I

Three of the four sp^3 hybrid orbitals of boron are supposed to form bonds with three hydrogen atoms, whereas the fourth forms a weak "three-center" bond with the other two hydrogen atoms. It is these two hydrogens that are subsequently removed as H_2 by a dissociative mechanism and replaced by OH^-. It has been further argued that the formation of BH_5 rather than the displacement of the dihydrogen is rate-determining.

The formation of the three-center bond in BH_5 must be preceded by a stretching (and weakening) of one of the B—H bonds. Hence, it can be regarded as a special case of an interchange mechanism, since H^+ enters before H^- is removed completely, and both remain under the immediate influence of boron but at a greater distance compared to a regular bond.

If the reaction is performed in D_2O, HD is obtained almost exclusively, indicating that one of the atoms came from BH_4^- and the other from the solvent. The explanation is easy if a structure like **I** is evoked, with one of the "loose" hydrogen atoms now being a deuterium atom. However, an alternative explanation would simply involve concerted attack by the electrophilic H^+ on the HOMO of BH_4^-, according to the Scheme

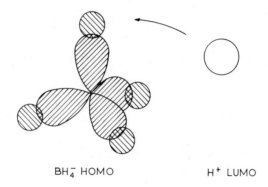

BH$_4^-$ HOMO   H$^+$ LUMO

resulting in a stripping off of one hydrogen atom.

Somewhat similar is the reaction

$$BF_3 + 3H^- \longrightarrow BH_3 + 3F^-$$

which is used in the preparation of B_2H_6. Here too, an addition compound (BF_3H^-) is formed first and then another H^- replaces the less nucleophilic F^- by an interchange mechanism.

Silicon

All substitution reactions on tetrahedral silicon that have been studied so far seem to have an associative character (A or I_a). The larger size of silicon compared to carbon and the availability of low-lying empty d orbitals are certainly factors facilitating an intermediate increase in the coordination number. Thus, the rate of the reaction

$$R_3SiH + OH^- + HS \longrightarrow R_3SiOH + H_2 + S^-$$

where HS is a protonic solvent, depends on both $[R_3SiH]$ and $[OH^-]$:

$$\text{Rate} = k[R_3SiH][OH^-]$$

The mechanism proposed is an interchange, I_a:

$$R_3SiH + OH^- \rightleftharpoons R_3SiH\cdots OH^- \qquad \text{(fast)}$$

$$R_3SiH\cdots OH^- \longrightarrow R_3SiOH + H^- \qquad \text{(slow)}$$

$$H^- + HS \longrightarrow H_2 + S^- \qquad \text{(fast)}$$

Attempts to detect intermediates of the type R_3Si^+, in which silicon has a smaller coordination number, have not been successful.

Phosphorus

Tetrahedral phosphorus exhibits all kinds of substitution mechanisms: all the way from D through I to A.

In several of its tetrahedral compounds, phosphorus is in the $+V$ oxidation state. Examples include phosphates, oxophosphates of the type R_3PO, and many organophosphorus cations.

In the hydrolysis of chloroalkylphosphinoxides, when the entering group is strongly electronegative, there is an inversion of configuration

In the intermediate trigonal bipyramid, the electronegative hydroxide ion and chloride ion ligands occupy axial positions. In other cases the configuration is retained.

It is emphasized again that attack in a substitution reaction may not always be on the central atom, in this case the phosphorus atom. Thus, the study of the reactions

$$(MeO)_3PO + {}^{18}OH^- \longrightarrow (MeO)_2PO{}^{18}O^- + MeOH$$

$$(MeO)_3PO + H_2{}^{18}O \longrightarrow (MeO)_2PO(OH) + Me{}^{18}OH$$

provides evidence supporting the view that hydroxide attacks the phosphorus atom, whereas water attacks the carbon atom

Similarly, F^- seems to attack phosphorus, but the heavier halides attack carbon, giving the corresponding alkylhalides.

Lithium

Lithium(I) species are generally very labile, but the mechanism by which they undergo substitution reactions can still be studied using techniques such as NMR line broadening.

As expected for an element of small atomic size, the mechanisms are generally dissociative. As an illustration, consider the exchange reactions of alkyl and aryl derivatives of lithium. These derivatives are usually polymeric $(LiR)_n$ ($n = 2$, 4, or 6) and undergo exchange reactions of the general type

$$x(LiR)_n + nMR'_x \longrightarrow x(LiR')_n + nMR_x$$

where R, R' are alkyls or aryls, and M can be Li, or Al, B, Zn, Cd, etc.

A particular case is the scrambling between organolithium reagents $(LiR)_n$ and $(LiR')_n$. The mechanism is believed to be dissociative. In the scrambling reaction, for example, between $(LiMe)_4$ and $(LiEt)_4$ in ether, the rate-determining step is thought to be the dissociation of the methyl tetramer into reactive dimers:

$$(LiR)_4 \xrightarrow{\text{slow}} 2(LiR)_2$$

In the dimeric structures, lithium atoms are doubly bridged by alkyl groups. In the tetramer structure, lithium atoms occupy the vertices of a tetrahedron, and the alkyl groups lie over the faces, but the system is so labile that this arrangement is not rigid. There is constant metamorphosis. The environment of the alkyl groups changes continuously, and this *intramolecular lability,* as we may call it, contributes to the NMR linewidths.

In bonding, there is no marked preference for the methyl or ethyl groups. As a result, the corresponding scrambling reaction gives a random mixture of the possible combinations. However, in the exchange between $(LiMe)_4$ and $(LiBr)_x$ the resulting distribution is not random. The NMR signals correspond to the following local environments:

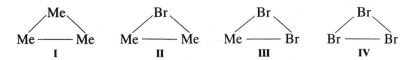

but environment **III** gives a weaker and **IV** a stronger signal than expected on the basis of random statistics.

Beryllium

Beryllium(II) is labile, but less so than other M^{II} ions. Three-coordinated species, even relatively stable ones, are well known in beryllium chemistry.

Fluoride ion transfer from solvent to BeF_4^{2-} and back to solvent can be studied by ^{19}F NMR. The equilibrium system contains three species, BeF_4^{2-}, $BeF_3(H_2O)^-$, and F^-. Two paths for the exchange are observed, corresponding to BeF_4^{2-} and $BeF_3(H_2O)^-$:

$$R_{ex}/4 = k_1[BeF_4^{2-}] + k_2[BeF_4^{2-}][BeF_3(H_2O)^-]$$

The rate-determining step for the first term has been written as

$$BeF_4^{2-} \longrightarrow BeF_3^- + F^-$$

The second term in the rate law involves formation of a fluoro-bridged species:

$$BeF_3^- + BeF_4^{2-} \rightleftharpoons [F_3BeFBeF_3]^{3-} \rightleftharpoons BeF_4^{2-} + BeF_3^-$$

which is preceded by removal of the coordinated water molecule from $BeF_3(H_2O)^-$.

6.10. TETRAHEDRAL TRANSITION METAL COMPLEXES

Tetrahedral complexes of transition metal ions in low oxidation states usually undergo substitution by mechanisms of dissociative character (D or I_d). In contrast, for higher oxidation states the mechanism is usually associative (A or I_a). Thus, in tetrahedral $Ni^0(CO)_4$, carbonyl exchange or replacement by other ligands (e.g., phosphorus ligands) proceeds by a rate that is independent of the nature of the entering group. This indicates a dissociative mechanism. In contrast, in the reaction

$$L + O_3Cr^{VI}OCr^{VI}O_3^{2-} \longrightarrow LCr^{VI}O_3 + Cr^{VI}O_4^{2-}$$

the mechanism seems to be of an associative nature.

Tetrahedral oxoanions like CrO_4^{2-}, with central metal ions at high oxidation states, belong to an important category of oxidizing agents that are inert to substitution. The mechanisms of the redox reactions will be discussed in the next chapter. Here it is only noted that these mechanisms also include substitution steps. It is therefore interesting to see how these species behave in purely substitution reactions without redox steps.

As an example, consider ^{18}O exchange between MO_4^{m-} and the solvent (water). The various rate laws observed for different oxoanions can be collected into a single expression:

$$Rate = k_1[MO_4^{m-}] + k_2[MO_4^{m-}]^2[H^+]$$
$$+ k_3[MO_4^{m-}]^2 [H^+]^2 + k_4[MO_4^{m-}][OH^-]$$

For CrO_4^{2-}, three of the four terms have been documented: k_1, k_2, and k_3; for ReO_4^-, the terms k_1, k_3, and k_4 have been observed; for MnO_4^-, the terms k_1 and k_3; and for VO_4^{3-} and FeO_4^{2-}, only the first term.

The fact that some term is not observed does not necessarily prove its absence. It is safer to say that under the conditions of the experiments, the contribution of these terms to the rate cannot be measured.

The HOMO of a d^0 tetrahedral transition metal oxoanion is a set of triply degenerate nonbonding lone pairs on the oxygens. The LUMO are the metal-centered e_g orbitals (z^2, x^2-y^2). If we consider only σ interactions, these orbitals are nonbonding; but π interaction makes them slightly antibonding (e_g^*). In a nucleophilic attack by an associative mechanism, these LUMO are populated and bonding is weakened. Strong donor solvents like water are also expected to loosen the bonds. On these grounds an interchange mechanism without direct bonding of the entering ligand to the metal, that is, a mechanism of dissociative character (I_d) and without solvent assistance, seems unlikely. A purely dissociative mechanism (D) involving breaking of a strong metal–oxygen bond and release of the unstable oxide ion seems even more unlikely.

Acid catalysis (proton-containing terms in the rate law) can be understood by using a simple Hellmann–Feynman force formalism. Before the addition of the proton, the dominant σ exchange force acting on the oxo ligand can be represented as follows:

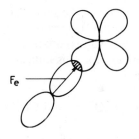

The addition of the proton creates an opposite force which causes an increase in the length of the M—O bond:

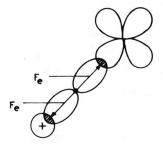

This expansion leads to considerable labilization, not only because some of the M—O bonds have been weakened, but also because the expansion facilitates the entrance of new ligands (e.g., OH^-, H_2O) and the population of the antibonding LUMO.

Thus, it seems that oxygen exchange in oxoanions is intimately related to coordination sphere expansion, addition, and condensation. In fact, some of the terms in the rate law for exchange may very well correspond to such pathways.

6.11. THE STRUCTURE CORRELATION METHOD IN TETRAHEDRAL SUBSTITUTION

The structure correlation method can be applied to any geometry, but it is more instructive to introduce it in the context of tetrahedral substitution and with a specific example.

It has been noted (Section 6.4) that, at least in principle, Y can approach a tetrahedral molecule from a face, an edge, or a vertex. However, it is reasonable to expect that one of these paths may prevail: the one corresponding to a potential energy valley (minimum energy pathway, Chapter 3). Direct mapping of this dynamic trajectory, for a given reaction taking place under ordinary conditions, is impossible. Indirect mapping, however, is feasible. Let us explain how it is carried out.

In a series of pentacoordinated organotin complexes with X and Y in trans positions,

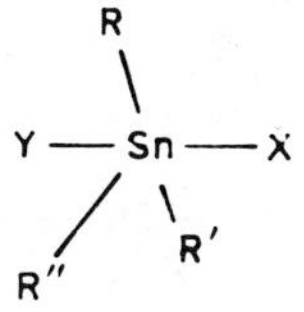

$$
\begin{array}{c}
R \\
\backslash \\
Y \,—\, Sn \,—\, X \\
/ \quad \backslash \\
R'' \qquad R'
\end{array}
$$

where R, R', R" are groups like alkyl or phenyl, the angles and bond lengths obtained from crystallographic data generally vary, but not independently. There is a correlation between the lengths d_Y and d_X of the Y—Sn and Sn—X bonds: as d_Y decreases, d_X increases. Schematically,

$$
\begin{array}{c}
R \\
\backslash \\
Y \longrightarrow Sn \longrightarrow X \\
\diagup \quad \backslash \\
R'' \qquad R'
\end{array}
$$

or in a series form (omitting the R's to simplify the presentation):

$$[Y\text{———}Sn\text{—}X] \,\cdots\, [Y\text{——}Sn\text{——}X] \,\cdots\, [Y\text{—}Sn\text{———}X]$$

This is clearly a model for the sequence of events occuring in an S_N2 mechanism, in which Y approaches the center of the face of the RR'R"SnX tetrahedron opposite the Sn—X bond, and it is bound to Sn without X being completely removed. In other words, *each of the known structures roughly represents a point on the minimum-energy pathway*. This is the "dogma" the *structure correlation method* rests on; it is essentially an analogy, but in practice it can be very useful.

The relation between bond lengths can be expressed in a statistically significant way by an equation. The differences Δd_Y and Δd_X from the bond lengths corresponding to a single bond are related by the formula

$$10^{-\Delta d_X/c} + 10^{-\Delta d_Y/c} = 1$$

where $c = 0.12$ nm (1.2 Å). This is an important aspect, and it can become the basis for testing theoretical conclusions.

There is another important aspect. The passage from the tetrahedral to the trigonal-bipyramidal geometry is accompanied by a change in the angles. There is a correlation between bond distance d_Y and the C—Sn—Y angle, with C being the carbon through which the organic ligand is bound to Sn: A large d_Y corresponds to C—Sn—Y angles smaller than 90°, and a small d_Y to angles larger than 90°. A similar correlation exists between d_X and the angle C—Sn—X. In conclusion, the known structures with trans X and Y map an S_N2 *mechanism with inversion of configuration*.

In contrast, structures with X and Y in cis positions,

$$
\begin{array}{c}
X \\
\uparrow \\
Y \longrightarrow Sn \text{——} R \\
\diagup \quad \backslash \\
R'' \qquad R'
\end{array}
$$

can be considered to lie on an S_N2 pathway with retention of configuration. However, the available data for such structures are scarce and refer to cases in which X and Y are parts of a bidentate ligand forming a chelate.

6.12. SUBSTITUTION IN SQUARE PLANAR COMPLEXES

The Rate Laws and the Mechanism

Square planar complexes have two "open" positions. The incoming ligand can approach from either side of the plane; there are no ligands perpendicular to the plane to obstruct the approach. The mechanism is associative. Yet, in spite of the absence of ligands hindering the approach, these reactions do have barriers; they are not diffusion-controlled. In the rest of this section and in the next two sections we will describe the factors determining these barriers.

Most of the available experimental data refer to d^8 low-spin square planar complexes and especially to complexes of platinum(II).

Rate

In a reaction of the type

$$T \overset{L_1}{\underset{L_2}{-\!\!\!-\!\!M-\!\!\!-}} X + Y \longrightarrow T \overset{L_1}{\underset{L_2}{-\!\!\!-\!\!M-\!\!\!-}} Y + X$$

where Y is the entering ligand, X the leaving ligand, and T the ligand trans to the leaving group, the general rate law is given by the equation

$$\text{Rate} = k_1[TL_1L_2MX] + k_2[TL_1L_2MX][Y] \qquad 6.16$$

In an associative mechanism, in which the rate-determining step is bimolecular, such as

$$TL_1L_2MX + Y \overset{k_2}{\longrightarrow} TL_1L_2MXY$$

the rate [second term in equation 6.16] depends both on the concentration of the complex and on the concentration of the entering ligand. The first term in 6.16 is also believed to be associative, but the role of the entering ligand is played here by a solvent molecule, S:

$$TL_1L_2MX + S \overset{k_1'}{\longrightarrow} TL_1L_2MXS$$

The experimentally determined rate constant is $k_1 = k_1' [S]$.

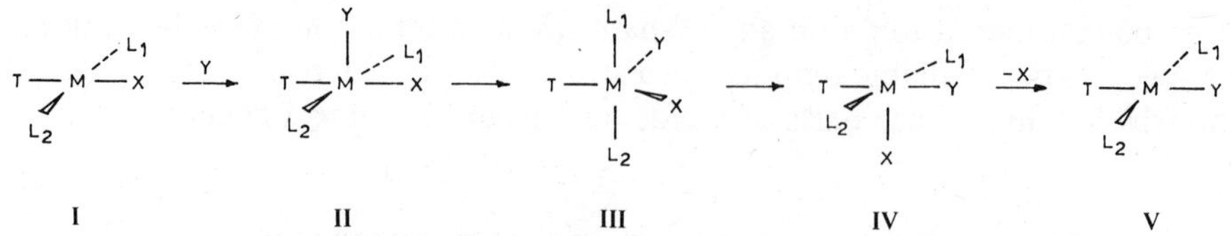

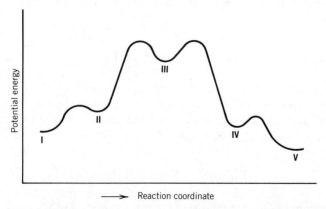

Figure 6.4. Schematic representation of the attack of Y.

For platinum(II) complexes the rate is mainly determined by the enthalpy of activation. The entropy of activation is usually negative for both paths.

It is generally accepted that the sequence of events in substitution 6.16 can be represented by the scheme shown in Figure 6.4. A square pyramid is formed first, which is transformed into a trigonal bipyramid and then back to a square pyramid and then into the final product. In this mechanism the ligand T trans to the leaving group X also becomes trans to Y, and cis ligands remain cis.

The potential energy profile of the reaction is sketched in Figure 6.5. The intermediates correspond to shallow minima.

Dependence on the Nature of the Entering Ligand

As we have seen already, in associative reactions the rate depends strongly on the nature of the entering group. For platinum(II) complexes, if $Y =$ halide ion, the rate constants follow the order

$$F^- \ll Cl^- < Br^- < I^-$$

The heavier halides react faster. A similar trend is also observed with other

Figure 6.5. Potential energy profile of a square planar substitution. The roman numerals correspond to those in Figure 6.4.

groups. For example,

$$R_3N \ll R_3As < R_3P \quad \text{and} \quad R_2O \ll R_2S$$

The reactivity of the entering ligand can be correlated with its polarizability. In the transition state, the heavier elements, under the influence of the local electronic environment, are polarized more than the lighter elements. Platinum(II) is also polarized extensively; it is one of the "soft" or "type B" metal ions.

Correlation between the rate and the proton basicity of Y does not seem to exist. For example, OH^- is a strong base, but the rate constant k_2 for $Y = OH^-$ has a very small value.

Solvent molecules S behave as entering ligands. What has been said about Y applies also to S.

Dependence on the Nature of the Leaving Ligand

The effect of the leaving ligand is often related to the strength of the M—X bond.

Substitution on $[Pt(dien)X]^+$ has been studied in detail. Diethylenetriamine, dien, is a tridentate ligand,

$$\text{HN}\begin{array}{l} \diagup \text{CH}_2\text{CH}_2\text{NH}_2 \\ \diagdown \text{CH}_2\text{CH}_2\text{NH}_2 \end{array}$$

occupying three of the four positions in the square planar complex. The dependence of the rate on the ligand X occupying the fourth position follows the trend

$$H_2O \gg Cl^- > Br^- > I^- > N_3^- > SCN^- > NO_2^- > CN^-$$

and the rate has been correlated to the strength of the Pt—X bond.

Dependence on the Nature of the Metal Center

The examples given so far have been taken from platinum chemistry. However, platinum is not unique. Nickel(II) and palladium(II) also form d^8 square planar complexes. The corresponding rates of substitution follow the trend

$$Ni^{II} \gg Pd^{II} \gg Pt^{II}$$

Pd^{II} reacts 10^5–10^6 times faster than Pt^{II}, and Ni^{II} reacts 10^7–10^8 times faster than Pt^{II}. This trend is related to the corresponding tendencies to form 5-coordinated intermediates.

Square complexes are also formed by Rh^I, Ir^I, and Au^{III}.

6.13. TRANS EFFECT IN SQUARE PLANAR SUBSTITUTION

Trends

The dependence of the rate of square planar substitution on the ligand trans to the leaving group gives important clues for understanding the details of the substitution pathway. There are also some interesting applications. This is the reason this dependence, known as the trans effect, is examined separately from the other factors that affect substitution.

By simply changing only the trans ligand and not the ligand directly involved in substitution, we may cause an impressive change in the rate—actually by several orders of magnitude.

Various ligands can be arranged according to the effect they have on the rate of substitution in the trans position. One such series is the following:

$$C_2H_4 \simeq NO \simeq CO \simeq CN^- > R_3P \simeq H^- \simeq thiourea > CH_3^- > C_6H_5^-$$

$$> SCN^- > NO_2^- > I^- > Br^- > Cl^- > amines \simeq NH_3 > OH^- > H_2O$$

For many of these ligands the trans effect is related to the weakening they cause in the bond between M and the leaving group. If the mechanism were dissociative, it could have been argued that since the M—X bond becomes weaker, the removal of X is facilitated. But the mechanism is not dissociative. It is associative; we must therefore be more careful in trying to correlate rates of substitution with M—X bond strength and bond length. Such correlations indeed exist, but not without exceptions. A notable exception is that of ethylene: its trans effect is very strong, yet its effect on the strength and length of the M—X bond is small.

A Simple Electrostatic Model

Figure 6.6a represents an early stage in the attack by Y. For simplicity, ligands cis to the leaving group are omitted. Bond M—T is drawn shorter than bond M—X to indicate that it is assumed to be stronger. This differentiation may be due to the larger polarizability of T or to π bonding. By $e^{\delta-}$ we symbolize the effective negative charge along the z axis (d^8 square planar complexes have such charges), and by $\delta+$, $\delta-$ we note the effective charges on the metal and the ligands, respectively. The main forces exerted on Y as it approaches M (Fig. 6.6a) are repulsive effective charge forces from $e^{\delta-}$, $X^{\delta-}$, and $T^{\delta-}$ (see Chapter 5, the Hellmann–Feynman theorem). The magnitude of these forces depend on distance, and Figure 6.6a clearly shows that repulsion from the right is stronger and a torque from right to left is created. As a result, Y is pushed toward X, and the equatorial triangle of a trigonal bipyramid is formed (Fig. 6.6b). Eventually X is removed, and the new square planar complex is formed (Fig. 6.6c) in which Y remains trans to T. For longer

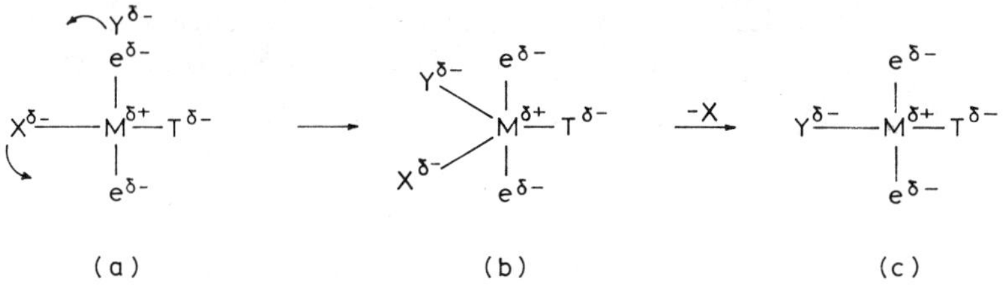

Figure 6.6. Schematic representation of the attack by Y.

M—X bond lengths, the initial repulsion is weaker, the activation needed is smaller, and the reaction is faster.

A strong trans effect is expected for short $M^{\delta+}$—$T^{\delta-}$ and long $M^{\delta+}$—$X^{\delta-}$ bond lengths. A large effective charge on T and a small charge on X will have a similar effect. Ethylene has a strong trans effect even for short M—X bond lengths, presumably because of back-donation into its π^* orbitals, which increases its effective charge while decreasing the charge on X.

The cis-Effect

Ligands in positions cis relative to the leaving group also affect the rate, but to a smaller degree than the trans ligands.

Applications to Synthesis

Trans-ligand effects are used to direct the synthesis toward the desired product. Take as an example the synthesis of $PtClBr(NH_3)(py)$. Starting with the same complex ($PtCl_4^{2-}$) and using the same reagents (NH_3, Br^-, and py), but in different order, we obtain three different isomers (the charges on the complexes are omitted):

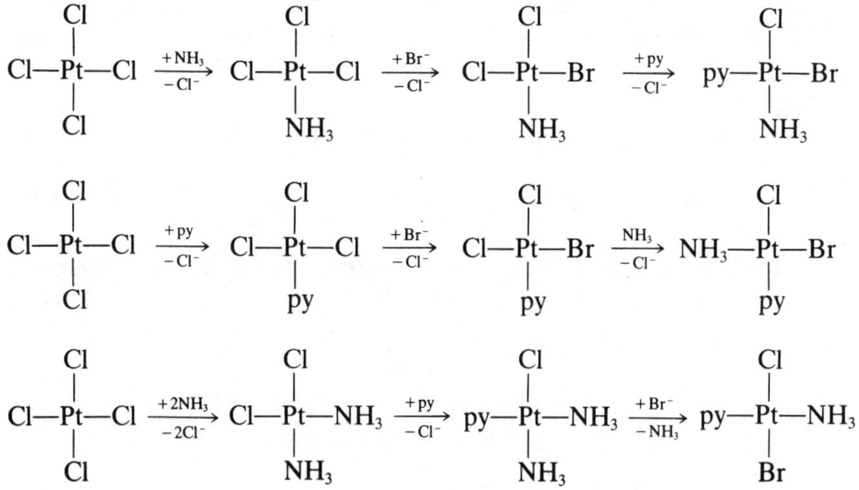

In the first two series, the leaving ligand is always the ligand trans to the ligand with the stronger trans effect. In the third series, other factors also play a role.

Antitumor Activity of Pt^{II} Complexes

The discovery that the isomer *cis*-Pt(NH$_3$)$_2$Cl$_2$ can inhibit the growth of tumors opened up new horizons in cancer chemotherapy. The isomer *trans*-Pt(NH$_3$)$_2$Cl$_2$ has no antitumor activity. The difference is partly attributed to the different behavior of these two complexes in substitution reactions. For antitumor activity to exist (which is unfortunately accompanied by some toxicity), it seems to be necessary to have two ligands that can be readily replaced in positions cis to each other.

6.14. PATHWAYS IN SQUARE PLANAR SUBSTITUTION

HOMO–LUMO interactions are not always sufficient to describe the dynamic phenomena. A more complete set of interacting orbitals is then needed. To reproduce a given structure it is sufficient to "match" the orbitals of the fragments to which this structure can be "broken" down. However, in mechanistic studies there are, in principle, a multitude of arrangements of the approaching reactant molecules and a corresponding multitude of symmetries, or asymmetries. One path may be the dominant one, but even then it is instructive to see why. Are the other paths excluded? If yes, why? What interactions are involved in each case?

Substitution at square planar complexes is as good a case as any for exploring these ideas. Here only the salient, qualitative features of the interactions will be considered.

The relevant orbitals of a low-symmetry d^8 inert square planar platinum complex TL$_1$L$_2$PtX are given in Figure 6.7. X is the leaving group and T the trans ligand, which is assumed to be a better donor than X. For this reason, the antibonding $yz(\pi^*)$ orbital, which is directed along X and T, is polarized toward T.

Even a cursory inspection of the figure reveals the following:

1. For an approaching donor Y, from any direction, the dominant interactions with the d electrons are repulsive.
2. Three of the four d-electron pairs contribute to the repulsion along each of the x and y axes. The approach is also obstructed by the ligands.
3. The xy plane is "blockaded" by the xy and z^2 electron pairs and also by the ligands.
4. Approach from the xz and yz planes is also inhibited.
5. The inhibition along z is less than the inhibition along x and y.

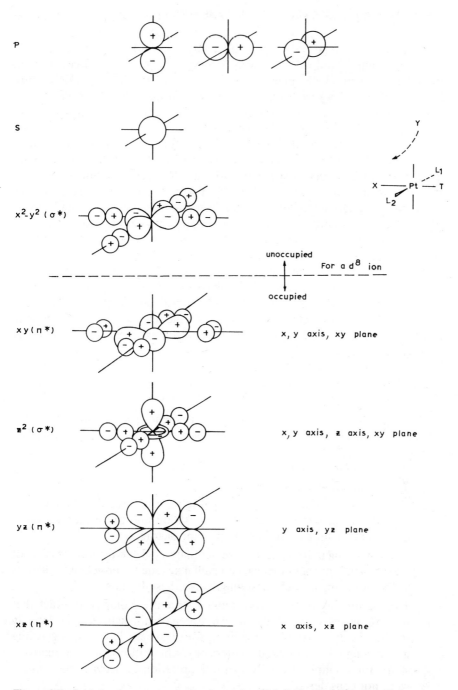

Figure 6.7. Orbitals of a square tetragonal complex TL₁L₂PtX. All *d*-derived orbitals are antibonding. Three of them are π*, two are σ*. In a *d*⁸ ion the lower four are occupied. $yz(\pi^*)$ is polarized toward T, which is assumed to be a better donor than X. Next to the occupied orbitals, the axes and planes that are "blockaded" are noted. An incoming ligand is inhibited from approaching from these directions. The ligands are assumed to use only *p*-like orbitals.

Thus, the LUMO (x^2–y^2) is quite effectively "blockaded." In order to reach it, the incoming ligand will have to overcome these repulsive forces. In fact, the following general statement can be made: The *HOMO and LUMO must have the right energy and symmetry, but they must also be accessible.*

Higher-lying s, p_x, and p_y orbitals also participate in bonding and make the "blockade" of the metal ion even tighter.

Initially, the direction of least resistance seems to be the direction along z, where the p_z orbital is available. The entering donor can then form a weak incipient bond through this high-lying orbital. The square pyramidal structure **II** in Figure 6.4 may indeed be identified with this initial species. Since p belongs to a higher energy level, it is expected to be more spread out in space (diffuse). Consequently, the overlap with it will be better if the orbitals of the entering group are also diffuse. This in turn can be correlated with the polarizability, which, as we have seen, affects the rate.

The negative charges in the electrostatic model for the trans effect (Fig. 6.6) are mainly those of the xz and yz electrons. Since yz is polarized toward T, repulsion from the right side of this occupied orbital is expected to be stronger, and Y is expected to be pushed to the left, toward X rather than toward T (Fig. 6.6) and to form the trigonal bipyramidal intermediate **III** (Fig. 6.4).

Qualifications

In describing the salient features of the interaction during square planar substitution reactions, we neglected a number of factors. More specifically,

1. Only p donors were considered.
2. Back-bonding was neglected.
3. It was assumed that the ligands themselves (in particular the entering ligand) are spherical. If not, the orientation and the rotation around the metal–ligand axis are also important factors.
4. Nonleaving ligand effects were essentially neglected. Optimization of the overlap throughout the reaction path may require considerable changes in the bond angles and bond lengths for these ligands.
5. The fluxionality of the 5-coordinated structures, that is, the fact that the differences in potential energy between such structures are relatively small, was disregarded. No comparisons were made between possible intermediate 5-coordinated isomers or configurations, and interconversions, for example, by a Berry pseudorotation or a twist mechanism, were not considered.

In a more sophisticated treatment, such factors would certainly have to be taken into account.

With the labile square planar complexes of palladium(II), nickel(II), or organometallic platinum(II), ligand displacement or removal facilitates orbital accessibility, and the reactions are faster.

6.15. SUBSTITUTION IN OCTAHEDRAL Co^{III} COMPLEXES

Octahedral geometry is most common in inorganic chemistry, and it is found throughout the periodic table. Among octahedral complexes, those of cobalt(III) are perhaps the most widely investigated coordination compounds. It is therefore natural to start the description of octahedral substitution with such complexes.

Most of the well-documented substitution reactions of cobalt(III) complexes proceed by dissociative paths (D or I_d) and are markedly affected by the nature of the leaving ligands. The effect of the entering ligand is relatively small. The rate also depends on the solvent and other factors.

Effect of the Leaving Ligand

A characteristic example is the so-called *acid hydrolysis* that occurs in aqueous media, for example,

$$Co(NH_3)_5X^{2+} + H_2O \longrightarrow Co(NH_3)_5OH_2^{3+} + X^-$$

For complexes of this general type (differing in X), there is a *linear free-energy relation* between the free energy of activation ($\Delta G^{\ddagger}$) and the free energy of the reaction (ΔG°). The correlation is 1 to 1, which suggests that the variations in $\Delta G^{\ddagger}$ are due to variations in the Co—X bond strength. Expressing this in a different way, we may say that reaching the activated complex configuration requires breaking of the Co^{III}—X bond. In the activated complex, X has essentially been removed; the mechanism is dissociative.

Stereochemical Hindrance

If there is "crowding" around the metal ion, dissociation will lead to some kind of relief. The opposite is expected for more "open" structures.

Aquation of the *meso* form of dichloro-bis(butylenediammine)cobalt(III) is 30 times faster than the corresponding reaction of a *d, l* mixture. The *meso* form is considered more sterically hindered.

Nonleaving Ligand Effects

The rate is also affected by ligands in cis and trans positions relative to the leaving ligand. In complexes of the general formula *cis-* and *trans-*

$Coen_2ACl^+$, where A is an anionic ligand which itself is not substituted but affects the lability of Cl^-, the *labilization* follows the trends:
   A in the trans position relative to Cl:

$$OH^- > NO_2^- > N_3^- > CN^- > Br^- > Cl^- > SO_4^{2-} > NCS^-$$

   A in the cis position relative to Cl:

$$OH^- > Cl^- > Br^- > NO_2^- > SO_4^{2-} > NCS^-$$

Cis complexes react faster than the corresponding trans species except for $A = NO_2^-$ or N_3^-, but generally the differences are small; there is no significant stereochemical effect.

Solvent Effects

A typical case is the solvolysis of *cis*-dibromo(tren)cobalt(III).[4] The data are summarized in Table 6.4. It is obvious from these data that there is no direct correlation with the values of the dielectric constants of the solvents.

Rate Laws

The general features of the rate laws for the different types of substitution reactions were mentioned in Section 6.6. Here some subtle aspects will be examined, using as a model a general *anation* reaction of the form

$$Co—OH_2 + Y \longrightarrow Co—Y + OH_2$$

To simplify the presentation, in this equation the five nonleaving ligands of Co^{III} and the charges have been omitted. The entering group Y is an anion;

TABLE 6.4. Rates of Solvolysis of *cis*-$Co^{III}(tren)Br_2^+$ at 25°C

Solvent	k, s^{-1}	$\Delta H^{\ddagger}$, kJ mol^{-1}	$\Delta S^{\ddagger}$, J deg^{-1} mol^{-1}
Formamide	2.7×10^{-2}	42	-134
Dimethylformamide	4.3×10^{-4}	92	0
Dimethylsulfoxide	7.7×10^{-4}	75	-50
N-Methylformamide	2.5×10^{-1}	—	—
Water	2.8×10^{-2}	63	-63

[4]tren-the quadridentate ligand $N(CH_2CH_2NH_2)_3$, 2,2',2''-triaminotriethylenamine, or tris-(2-aminoethyl)amine.

the leaving group X is a water molecule. The concentration of water is also omitted, since the medium is assumed to be aqueous.

For a D mechanism,

$$Co\text{—}OH_2 \underset{k_{-1}}{\overset{k_1}{\rightleftharpoons}} Co\text{—} + OH_2$$

$$Co\text{—} + Y \xrightarrow{k_2} Co\text{—}Y$$

Applying the steady-state approximation for the 5-coordinated intermediate, we obtain

$$\frac{d[Co\text{—}Y]}{dt} = \frac{k_1 k_2 [Co\text{—}OH_2][Y]}{k_{-1} + k_2[Y]} \tag{6.17}$$

If $k_{-1} \ll k_2[Y]$,

$$\frac{d[Co\text{—}Y]}{dt} = k_1 [Co\text{—}OH_2] \tag{6.18}$$

This is more likely to happen with excess Y. The observed first-order rate constant can be identified with k_1 and is independent of $[Y]$.

If $k_{-1} \gg k_2[Y]$,

$$\frac{d[Co\text{—}Y]}{dt} = \frac{k_1 k_2}{k_{-1}} [Co\text{—}OH_2][Y] \tag{6.19}$$

The rate depends on Y, as in an associative mechanism. Conditions favoring such a case include a low concentration of Y. The observed rate constant will be second-order.

If $k_{-1} \simeq k_2[Y]$ but Y is in excess, ($[Y] \simeq [Y]_0$), the rate will be pseudo-first-order in $[Co\text{—}OH_2]$.

$$\frac{d[Co\text{—}Y]}{dt} = k_{obs}^{(1)} [Co\text{—}OH_2] \tag{6.20}$$

but this time the observed pseudo-first-order rate constant will depend on $[Y]_0$:

$$k_{obs} = \frac{k_1 k_2 [Y]_0}{k_{-1} + k_2[Y]_0} \tag{6.21}$$

For an I_d mechanism,

$$Co—OH_2 + Y \underset{}{\overset{K}{\rightleftharpoons}} Co—OH_2 \cdots Y$$

$$Co—OH_2 \cdots Y \xrightarrow{k} Co—Y + OH_2$$

the rate is equal to

$$\frac{d[Co—Y]}{dt} = kK[Co—OH_2][Y]$$

where $[Co\text{-}OH_2]$ is the actual concentration of this species, and it does not include the concentration of the outer-sphere complex. These two concentrations are related by the expression

$$\frac{[Co—OH_2 \cdots Y]}{[Co—OH_2]} = \frac{K[Y]}{1} \qquad\qquad 6.22$$

If $K[Y] \gg 1$, equation 6.22 dictates that $[Co—OH_2 \cdots Y] \gg [Co—OH_2]$, and the outer-sphere complex should in principle be detectable. The observed rate law with respect to this dominant species will be first-order:

$$\frac{d[Co—Y]}{dt} = k[Co—OH_2 \cdots Y] \qquad\qquad 6.18'$$

Equation 6.18′ looks like equation 6.18 and is favored under similar conditions, that is, when there is excess Y.

If $K[Y] \ll 1$, the dominant species is $Co—OH_2$, and the observed rate law will have a form similar to equation 6.19:

$$\frac{d[Co—Y]}{dt} = kK[Co—OH_2][Y] \qquad\qquad 6.19'$$

Again a second-order overall rate is favored at low $[Y]$.

If $K[Y] \approx 1$, but with excess Y, $[Co—OH_2 \cdots Y]$ will have values comparable to $[Co—OH_2]$ and the rate law will be given by the equation

$$\frac{d[Co—Y]}{dt} = \frac{kK[Co—OH_2]_t[Y]_0}{1 + K[Y]_0} \qquad\qquad 6.17'$$

which looks like equation 6.17. In this expression, $[Co—OH_2]_t$ is the sum of the concentrations of $[Co—OH_2]$ and $[Co—OH_2 \cdots Y]$ at time t and equals the initial concentration minus the concentration of Co—Y formed.

Conclusions: *The observed rate law alone is not a good diagnostic criterion* *for deciding between a D and an I_d mechanism.* A combination with other criteria is needed. One such combination is with the magnitude of the rate constants and the equilibrium constants. The magnitude can be estimated theoretically or by analogy, or sometimes it may be possible to measure it independently. Thus, if in the interchange mechanism the expected value for the equilibrium constant is small and the condition $K[Y] \gg 1$ is not justified, then case 6.18′ is ruled out. Similarly, if the rate for the water-exchange process is measured independently,

$$Co—OH_2 + H_2^*O \xrightarrow{k'} Co—^*OH_2 + H_2O$$

k' must be larger than the k of the I_d mechanism.

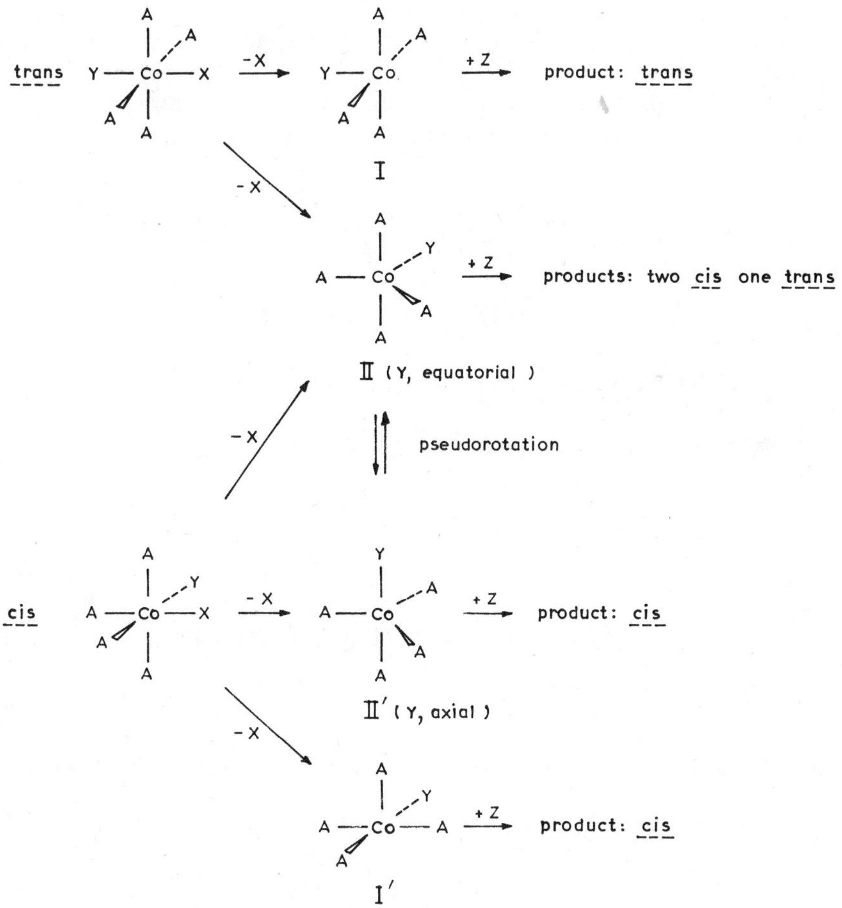

Figure 6.8. Stereochemical changes in the substitution of X by Z in trans- and cis-CoA$_4$YX. I, I′ are tetragonal pyramids; II, II′ are trigonal bipyramids.

On the basis of arguments such as these, it has been concluded that the anation reactions considered proceed by the I_d rather than the D mechanism.

Stereochemical Changes during Substitution

Typical examples are given in Figure 6.8. According to this scheme the stereochemical changes are determined by the geometry of the coordinatively unsaturated (5-coordinate) intermediate. According to another point of view, these changes also depend on the relative positions of the leaving and entering ligands in the outer-sphere complex (I_d mechanism).

6.16. ACID AND BASE HYDROLYSIS IN OCTAHEDRAL Co^{III} COMPLEXES

Acid hydrolysis is normally carried out in acidic solutions. For ligands such as I$^-$, Br$^-$, and Cl$^-$, the rate of hydrolysis is usually independent of acidity. For ligands X^{n-} ($n \geqslant 2$), which can combine with one or more protons, the rate depends on hydrogen ion concentration; for example, for $n = 2$,

$$\text{CoX(NH}_3)_5^+ + \text{H}^+ \overset{K}{\rightleftharpoons} \text{CoXH(NH}_3)_5^{2+}$$

$$\text{CoX(NH}_3)_5^+ + \text{H}_2\text{O} \overset{k_1}{\longrightarrow} \text{Co(NH}_3)_5(\text{OH}_2)^{3+} + \text{X}^{2-}$$

$$\text{CoXH(NH}_3)_5^{2+} + \text{H}_2\text{O} \overset{k_2}{\longrightarrow} \text{Co(NH}_3)_5(\text{OH}_2)^{3+} + \text{HX}^-$$

The rate is given by

$$\text{Rate} = k_{\text{obs}}[\text{complex}]_{\text{total}} = k_1[\text{CoX(NH}_3)_5^+] + k_2[\text{CoXH(NH}_3)_5^{2+}]$$

But

$$[\text{CoXH(NH}_3)_5^{2+}] = K[\text{H}^+][\text{CoX(NH}_3)_5^+]$$

and hence

$$\text{Rate} = k_{\text{obs}}(1 + K[\text{H}^+])[\text{CoX(NH}_3)_5^{2+}] = (k_1 + k_2K[\text{H}^+])[\text{CoX(NH}_3)_5^+]$$

and, with excess hydrogen ion,

$$k_{\text{obs}} = \frac{k_1 + k_2K[\text{H}^+]}{1 + K[\text{H}^+]}$$

Base hydrolysis is hydrolysis in alkaline media, for example,

$$(H_3N)_5CoX^{2+} + OH^- \longrightarrow (H_3N)_5CoOH^{2+} + X^-$$

It is generally accepted that this hydrolysis proceeds by the Garrick *conjugate base mechanism* (S_N1_{cb}):

$$OH^- + (H_3N)_5CoX^{2+} \underset{k_{-1}}{\overset{k_1}{\rightleftharpoons}} H_2O + (H_2N)(H_3N)_4CoX^+$$

$$(H_2N)(H_3N)_4CoX^+ \xrightarrow{k_2} (H_2N)(H_3N)_4Co^{2+} + X^-$$

$$(H_2N)(H_3N)_4Co^{2+} + H_2O \xrightarrow{\text{fast}} (H_3N)_5CoOH^{2+}$$

This mechanism is consistent with the finding that base hydrolysis occurs only if the complex has ligands such as H_2O or NH_3 and ethylenediamine that can donate a proton.

The second reaction is slow and of a dissociative character, like the corresponding substitutions in acid solution, except that here the presence of the conjugate base (NH_2^- in the example given) has an impressive accelerating effect.

The general rate law for base hydrolysis is given by the expression

$$\text{Rate} = k_{aq}[\text{complex}] + k_{OH}[\text{complex}][OH^-]$$

That is, it contains the term corresponding to the OH^--catalyzed path in addition to a first-order term corresponding to a noncatalyzed path.

6.17. ASSOCIATIVE AND DISSOCIATIVE MECHANISMS IN OCTAHEDRAL Cr^{III} COMPLEXES

The second octahedral substitution system to be described is that of the chromium(III) complexes. An interesting aspect in this case is the interplay between the associative and dissociative mechanisms. Evidence for associative mechanisms in some cobalt(III) complexes has also been reported, but it is rare and perhaps not conclusive. With chromium(III) the interplay is well documented. It is therefore interesting to explore. More specifically, it is interesting to ask the following questions:

1. What could be the reason for the difference between Co^{III} and Cr^{III}?
2. What factors determine the switchover from one mechanism to the other?

The available evidence indicates that an associative type of mechanism (I_a) is important in substitution reactions of chromium(III) complexes containing water as a ligand, such as substitution of X^- in $Cr(H_2O)_5X^{2+}$ (where X^- is a unidentate ligand). Similarly, an associative mechanism seems to dominate in the exchange of water between $Cr(OH_2)_6^{3+}$ and the aqueous solvent or between $Cr(DMF)_6^{3+}$ and the DMF solvent. In contrast, an I_d mechanism is favored for $Cr(NH_3)_5X^{2+}$. Also for the anation reaction,

$$Cr(NH_3)_5OH_2^{3+} + X^{n-} \xrightarrow{k_{forward}} Cr(NH_3)_5X^{3-n} + H_2O$$

the observed forward rate constants vary by less than an order of magnitude for entering groups of the same charge, and the observed rate is only slightly less than the solvent exchange rate. This behavior is typical of an I_d mechanism.

The lower-lying empty orbitals in Cr^{III}, which can conceivably be used in bond formation, are the antibonding doubly degenerate e_g^* orbitals. Ammonia is a better σ donor than water. Consequently, in ammine complexes the bonding orbitals derived from the metal d orbitals are expected to lie lower than in the aqua complexes. With the antibonding orbitals the order is reversed. The transition state for association via the antibonding e_g^* is expected to be more stable for the aqua species, and associative mechanisms are expected to be more favorable.

In arriving at this interpretation we essentially disregarded repulsions between the electrons of the entering ligand and those of the coordination compound. However, we have seen in square planar substitution that these interelectronic repulsions play a very important role in determining the path of the reaction and are generally expected to make association more difficult. This may then be the main reason association is frequent for chromium(III), with only three d electrons, but not for cobalt(III), which has six d electrons.

6.18. LABILITY OF AQUA IONS

The lability of aqua ions in water is measured by the rate of exchange with solvent water molecules. The most labile aqua ions are those of the alkali and alkaline earth metals and those of Cd^{2+}, Hg^{2+}, Cu^{2+}, and Cr^{2+}. For these ions the observed first-order rate constant exceeds 10^8 s^{-1}. The bonds between these ions and water dipoles are largely ionic. Accordingly, the lability depends on charge and size.

On the other end of the scale we have only $Cr(H_2O)_6^{3+}$ and $Rh(H_2O)_6^{3+}$, with exchange half-lives of several days. For all other ions, the exchange rate for water is fast (for the conventional use of the term "fast," see Chapter 4).

Table 6.5 contains ratios of rate constants for isoelectronic pairs. The interesting feature of this tabulation is that for M^{2+} ions exchange is several orders of magnitude faster than for the corresponding isoelectronic M^{3+} ion.

TABLE 6.5. Approximate Ratios of Water Exchange Rate Constants for Isoelectronic Pairs

Pair	Ratio of Rate Constants	Pair	Ratio of Rate Constants
V^{2+}/Cr^{3+}	10^8	Cd^{2+}/In^{3+}	10^6
Mn^{2+}/Fe^{3+}	10^5	Zn^{2+}/Ga^{2+}	10^7
Fe^{2+}/Co^{3+}	10^4	Mg^{2+}/Al^{3+}	10^5
Hg^{2+}/Tl^{3+}	10^5	Ca^{2+}/Sc^{3+}	10^6
		Sr^{2+}/Y^{3+}	10^4

6.19. AQUATION OF OCTAHEDRAL ORGANOCHROMIUM(III) COMPLEXES

Substitution reactions of organometallic compounds are often more complicated than those of ordinary complexes. Consider, for example, the following compounds:

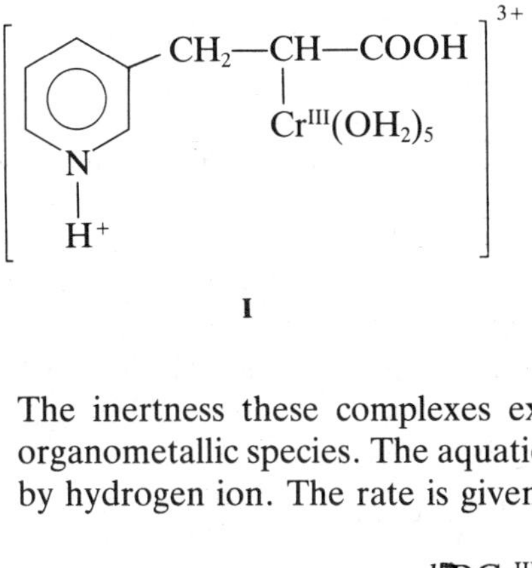

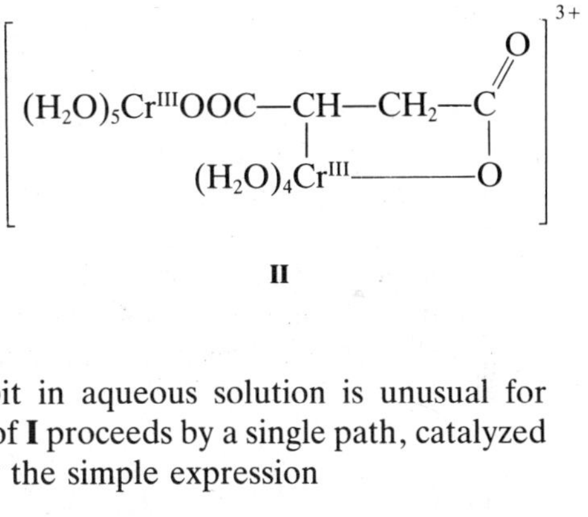

The inertness these complexes exhibit in aqueous solution is unusual for organometallic species. The aquation of **I** proceeds by a single path, catalyzed by hydrogen ion. The rate is given by the simple expression

$$-\frac{d[RCr^{III}]}{dt} = k[RCr^{III}][H^+]$$

In contrast, for complex **II**, several parallel paths are observed:

$$-\frac{d[RCr^{III}]}{dt} = \left(k_1 + k_2[H^+] + k_3[Cr^{2+}] + k_4\frac{[Cr^{2+}]}{[H^+]}\right)[RCr^{III}]$$

There is one noncatalytic path (k_1), one path catalyzed by hydrogen ions (k_2), one catalyzed by Cr^{2+} (k_3), and a fourth catalyzed by Cr^{2+} but inverse in $[H^+]$ (k_4).

The activation parameters for these paths are summarized in Table 6.6.

TABLE 6.6. Activation Parameters for the Aquation of Some Organochromium(III) Complexes

Complex/Path	E_a, kJ mol^{-1}	ln A
I.	106 ± 2.5	30.7 ± 0.9
II. Acid-catalyzed path	79 ± 3.5	17.8 ± 1.4
Noncatalyzed path	101 ± 6.5	26.9 ± 2.5
Cr^{2+}-catalyzed path, acid-independent	45 ± 2	12.7 ± 0.9
Cr^{2+}-catalyzed path inverse in [H$^+$]	63 ± 3	18.3 ± 1.3

The mechanisms proposed for **I** and **II** are quite different. The mechanism proposed for the aquation of **I** can be summarized as follows:

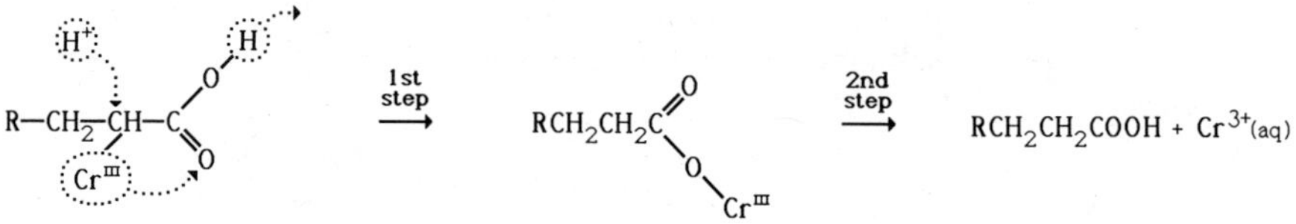

The first step is an *intramolecular transfer* of the group Cr(OH$_2$)$_5^{3+}$ from carbon to oxygen, assisted by a proton. The carbonyl oxygen is close enough to Cr^{III} for this transfer to be easy.

For complex **II** the transfer of the Cr^{III} fragment has been partially carried out from the beginning, since Cr^{III} is also bound to oxygen, not only to carbon. The proposed mechanism for aquation can be summarized as follows:

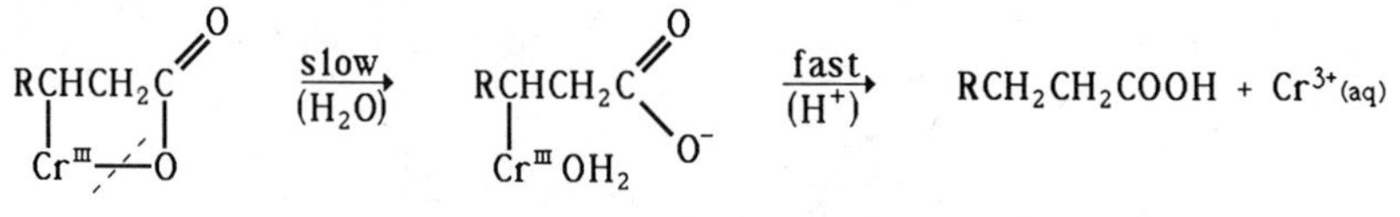

In this mechanism the rate is determined by the breaking of the Cr—O bond, which is facilitated by acid, probably by the attachment of H$^+$ to the carboxylic oxygen bound to Cr^{III}:

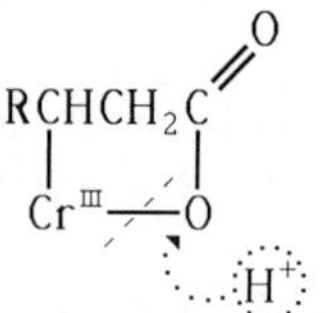

For the Cr^{2+}-catalyzed path, the following mechanism has been proposed:

$$Cr^{III}OOCCHCH_2C \xrightarrow[\text{(H}_2\text{O)}]{\text{slow}} \left[Cr^{III}OOCCHCH_2C \right] \xrightarrow[\text{(H}^+)]{\text{fast}} Cr^{III}OOCCH_2CH_2COOCr^{III} + Cr^{2+}$$

The electron is transferred from Cr^{2+} to the Cr^{III} bound to carbon. This Cr^{III} becomes labile Cr^{II}, and the Cr—C bond breaks. In this scheme, Cr^{2+} is shown to attack the carboxylic group. This attack corresponds to the third term in the rate law (k_3). At lower acid concentrations, hydrolyzed forms participate in the reaction, and the corresponding term in the rate law has an inverse dependence on $[H^+]$ (the k_4 term).

Acid Catalysis

In the class of organometallic compounds examined here, three cases are generally observed:

a. The rate law contains only acid-independent terms.
b. The rate law contains both acid-dependent and acid-independent terms.
c. The rate law contains only acid-dependent terms.

Case (a). Complexes are of the general formula $Cr^{III}:CHXX'$, where

$$\left| \begin{array}{c} X \\ X' \end{array} \right| = \left| \begin{array}{c} Cl \\ H \end{array} \right| \text{ or } \left| \begin{array}{c} Cl \\ Cl \end{array} \right| \text{ or } \left| \begin{array}{c} Br \\ H \end{array} \right| \text{ or } \left| \begin{array}{c} Br \\ Br \end{array} \right| \text{ or } \left| \begin{array}{c} I \\ H \end{array} \right| \text{ or } \left| \begin{array}{c} COOH \\ H \end{array} \right|$$

The formal charge on X (or X') is small. Hydrogen bonding of the form

$$-X\text{----}H-O \qquad \text{or} \qquad -X\text{----}H-O \begin{array}{c} H^+ \\ \diagdown \\ H \end{array}$$

is expected to be weak, and it is not expected to affect the rates of aquation appreciably. In such cases the mechanism may involve a concerted dipole–dipole, two-center attack:

$$\begin{array}{c} H \\ | \\ O-H \\ \downarrow \quad \downarrow \\ \boxed{\delta+ \quad \delta-} \end{array}$$

Attack on the oxygen side can be regarded as nucleophilic, whereas from the hydrogen side it is electrophilic. Two subcategories can be distinguished:

$$
\begin{array}{cc}
\overset{\delta+\ \ \delta+\ \ \delta-}{\text{(aa)}\quad \text{Cr}-\text{C}-\text{X}} & \overset{\delta+\ \ \delta-\ \ \delta-}{\text{(ab)}\quad \text{Cr}-\text{C}-\text{X}} \\
\qquad\quad \uparrow\ \ \uparrow & \qquad\quad \uparrow\ \ \uparrow \\
\qquad\quad \text{O}-\text{H} & \qquad\quad \text{O}-\text{H} \\
\qquad\quad | & \qquad\quad | \\
\qquad\quad \text{H} & \qquad\quad \text{H} \\
\text{for X=Cl, Br} & \text{for X=I, COOH}
\end{array}
$$

Different bonds break in the two cases, and different products are formed.

If, instead of H_2O, the attacking species is H_3O^+, its proton is pushed out by the formal positive charge:

It is again attack by a water molecule. There is no acid catalysis.

Case (*b*). Examples are $CrCF_3$, $CrCHXX'$ (X = HO, CHO). The mechanism is similar to (aa), above; this time, however, bond breaking is facilitated by hydrogen bonding. For example,

(ba) doesn't differ much from (bb), and for this reason the activation energies do not differ much either (they differ by less than 10 kJ mol^{-1}).

Case (*c*). Examples are

$$
\begin{array}{ccc}
\text{RCHCH}_2\text{COOH} & \text{and} & \text{CH}_3. \\
\quad | & & \quad | \\
\quad \text{Cr}^{\text{III}} & & \quad \text{Cr}^{\text{III}}
\end{array}
$$

It has been proposed that the mechanism involves direct proton attack on carbon and simultaneous removal of Cr^{III}.

$$\text{H}^+$$
$$\downarrow$$
$$-\text{CH}-$$
$$\backslash$$
$$\text{Cr}^{III}$$

The reaction is not affected by hydrogen bonding, and there is no dipole–dipole two-center attack. Only one path is observed, and it depends on the hydrogen ion concentration.

In platinum(II), cobalt(III), and chromium(III) substitution reactions, the factors (e.g., nature of leaving group or nature of entering group) that affect associative and dissociative substitution were examined. Here the emphasis was shifted: various catalytic paths for substitution were examined, and it was shown that the interesting events may be quite different from dissociation or association. It was further demonstrated that the same catalyst, for example, hydrogen ion, may act differently depending on the substrate and the conditions.

6.20. INTRAMOLECULAR METATHESIS

The transfer of the $Cr^{III}(OH_2)_5$ fragment from carbon to oxygen discussed in the previous section is an example of an intramolecular metathesis mechanism. The fragment migrates from one site to a neighboring site of the same molecule, without leaving it. A mechanism of this kind can be characterized as a concerted dissociative–associative mechanism: $Cr^{III}(OH_2)_5$ "dissociates" from carbon and simultaneously "associates" with oxygen:

$$\text{C} - \text{C}$$
$$| \quad |$$
$$\text{Cr}^{III} \quad \text{O}$$

Two cases of intramolecular methathesis ("slip") mechanisms will be discussed in more detail: the metatheses involved in the nucleophilic attack on transition metal π-complexed olefins and olefin insertion reactions.

Nucleophilic Attack on Transition Metal π-Complexed Olefins

According to the Dewar–Chatt–Duncanson model, the olefin donates π electrons to an empty metal-like orbital of the ML_n fragment, and this fragment back-donates from a filled orbital to the π^* orbital of the olefin (Fig. 6.9).

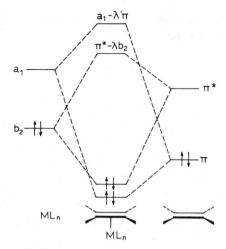

Figure 6.9. Energy levels for a metal–olefin η^2 complex.

The LUMO ($\pi^*-\lambda b_2$ in Fig. 6.9) of the olefinic complex is localized mostly on the olefin, but not as much as in the free olefin; moreover, its energy is higher. One would thus expect the addition of an electron donor (nucleophile) to the free olefin to be easier than addition to the coordinated olefin. Yet the opposite is observed: the metal ion facilitates the reaction.

It must be concluded that in the transition state the olefinic complex is not in its equilibrium η^2 structure. A deformation should precede the addition of the nucleophile, for example, a *slipping of the ML_n fragment:*

This makes the interaction of the HOMO of the nucleophile with the LUMO of the coordinated olefin stronger; the level of the $\pi^*-\lambda b_2$ orbital is lowered, but not all the way back to the original π^* level. If only this interaction is considered, free olefin is still expected to be more reactive. However, slipping of the ML_n fragment does not affect only the interaction between the filled orbital of the donor and the empty orbital of the acceptor. It also affects the interaction between the filled donor and acceptor orbitals, which in the free olefin and in the symmetrically coordinated olefin is repulsive. In simplified terms it may be said that as ML_n slips, it sweeps π electron density away from the far carbon atoms. As a result, repulsion on the approaching nucleophile diminishes. In fact, if this kind of electron withdrawal is extensive enough, it can turn the repulsive interaction into an attractive one and facilitate dramatically the addition of the nucleophile.

6.21. INSERTION REACTIONS

The Olefin Insertion Reaction

Olefin insertion in metal–hydrogen or metal–alkyl bonds is important in the catalytic hydrogenation of multiple bonds, in hydroformylation, in the Wacker process, in isomerizations, in polymerizations, and in other processes. Here it will be exemplified by the reaction (charges are omitted)

$$
\underset{\textbf{I}}{\overset{\displaystyle H}{P\!-\!\underset{\displaystyle P}{\overset{|}{Pt^{II}}}\!-\!X}}
\quad\xrightarrow[-X]{(+\,\text{olefin})}\quad
\underset{\textbf{II}}{\overset{\displaystyle H}{P\!-\!\underset{\displaystyle P}{\overset{|}{Pt^{II}}}\!-\!\|}}
\quad\xrightarrow{+X'}\quad
\underset{\textbf{III}}{\overset{\displaystyle P}{X'\!-\!\underset{\displaystyle P}{\overset{|}{Pt^{II}}}\!-\!Et}}
$$

where P = phosphine ligands. The formation of the 4-coordinated species **II** with the hydride ion and the olefin in *cis* positions is well documented. The transformation **I** → **II** should necessarily involve 5-coordinated (associative) or 3-coordinated (dissociative) intermediates, but it is not clear which. Nevertheless, formation of such intermediates usually leads to complex **II**; a 4-coordinated intermediate with the olefin and the hydride ion in trans positions is unlikely. Also, there is no evidence for a free radical mechanism.

If the olefin is in the coordination plane, the movement that leads to insertion can be pictured as follows:

$$
\overset{\displaystyle H}{P\!-\!\underset{\displaystyle P}{\overset{|}{Pt^{II}}}\!-\!\overset{CH_2}{\underset{CH_2}{\|}}}
\qquad\longrightarrow\qquad
P\!-\!\overset{\displaystyle H}{\underset{\displaystyle P}{Pt^{II}}}\!\!\overset{\cdots CH_2}{\underset{CH_2}{<}}
$$

In this picture olefin insertion can hardly be distinguished from hydride ion transfer from platinum to the olefin. After this transfer, the new ligand X' may take the place of the hydride ion, but it can also enter between the phosphine ligands.

Diversion

Caution! Intermediates and stable structures are not transition states. It was stressed in Chapter 3 that the isolation and characterization of stable intermediates is usually a good indicator of how the reaction proceeds. Intermediates were characterized as "stations" in the "itinerary" of the reaction that may *resemble* some activated states. In Chapter 4, however, some general reservations were expressed.

The specific examples presented here justify this caution. In the nucleophilic attack of a coordinated olefin, the stable η^2 structure *is not* the structure of the activated complex. Among the 5-coordinated species with composition $Pt(H)(Cl)(PH_3)_2(olefin)$, the most energetically favored one is

but for olefin insertion by way of a 5-coordinated intermediate this structure is avoided.

Other Insertion Reactions

A mechanism similar to that for the insertion of an olefin into a Pt—H bond can also be proposed for the insertion of other unsaturated molecules into Pt—H or platinum–alkyl bonds. The most important among these unsaturated molecules are the acetylenes, allene, CO_2, CS_2, C≡N—R and CO.

Replacing Pt^{II} (d^8) by another metal ion may introduce new features. Thus, in the insertion of carbon monoxide into a Mn—CH_3 bond:

$$(OC)_5Mn—CH_3 + CO \longrightarrow (OC)_5Mn—\overset{\displaystyle O}{\overset{\displaystyle \|}{C}}—CH_3$$

the mechanism proposed is of a dissociative nature:

$$(OC)_5Mn—CH_3 \longrightarrow (OC)_4Mn—\overset{\displaystyle O}{\overset{\displaystyle \|}{C}}—CH_3$$

(6-Coordinated)   (5-Coordinated)

$$(OC)_4Mn—\overset{\displaystyle O}{\overset{\displaystyle \|}{C}}—CH_3 \underset{-L}{\overset{+L}{\rightleftharpoons}} L(OC)_4Mn—\overset{\displaystyle O}{\overset{\displaystyle \|}{C}}—CH_3$$

The actual insertion takes place in the first step, which can be regarded as an intramolecular methyl transfer from the metal to a coordinated carbon monoxide.

α-Hydride Ion Transfer

In the olefin insertion discussed above, the hydride ion transferred was initially bound directly to the metal, but there are cases where it comes from a ligand. One such case is the so-called α-hydride ion transfer, for example,

$$
\begin{array}{c}
\text{CH}_3 \\
\mid \\
\text{H}_3\text{C}\cdots\text{W}\cdots\text{CH}_3 \\
\text{H}_3\text{C} \quad \mid \quad \text{CH}_3 \\
\text{CH}_3
\end{array}
\quad \longrightarrow \quad 3\text{CH}_4 + \text{W}[\text{CH}_2]_3
$$

(Permethyl compound)

This is clearly a multistep process, since it involves breaking of three carbon–hydrogen bonds, formation of three new ones, breaking of three metal–carbon bonds, and transformation of the remaining three bonds into double bonds. These steps may be intermolecular, involving homolytic cleavage,

$$(\text{H}_3\text{C})_{n-1}\text{M}:\text{CH}_3 \longrightarrow (\text{H}_3\text{C})_{n-1}\dot{\text{M}} + \dot{\text{C}}\text{H}_3$$

(the dots represent the electrons of the bond) followed by hydrogen abstraction,

$$(\text{H}_3\text{C})_{n-2}\dot{\text{M}}{-}\text{CH}_3 + \dot{\text{C}}\text{H}_3 \longrightarrow (\text{H}_3\text{C})_{n-2}\text{M}{=}\text{CH}_2 + \text{CH}_4$$

but there are also well-documented cases where the transfer is intramolecular:

$$
\begin{array}{c}
\text{H}_2\text{C}(\colon\text{H}) \\
\mid \\
\text{M}\odot\text{CH}_3
\end{array}
\quad \longrightarrow \quad
\begin{array}{c}
\text{H}_2\text{C} \\
\parallel \\
\text{M}
\end{array}
+ \text{CH}_4
$$

6.22. TOPOLOGICAL MECHANISMS

The term "topological mechanism" refers to intramolecular changes of bond lengths and/or angles without breaking of bonds or formation of new ones or the joining of fragments and generally not involving discontinuities.

The associative–dissociative dichotomy is not applicable here, but the atom or group transfer formalism is still valid, provided that the topological transformation is regarded as an intramolecular transfer.

Strictly speaking, all vibrations of a molecule are topological transformations, but here we consider only those with large enough amplitude, which lead to permutations of nuclear positions or change the symmetry of the molecule.

Most tetrahedral or octahedral complexes have very high barriers for such rearrangements. In contrast, for coordination number 5 or higher than 6, this kind of nonrigidity is quite common.

The Berry Mechanism

This is perhaps one of the best-known topological rearrangements. If a bending mode has sufficiently large amplitude, the following rearrangement becomes possible:

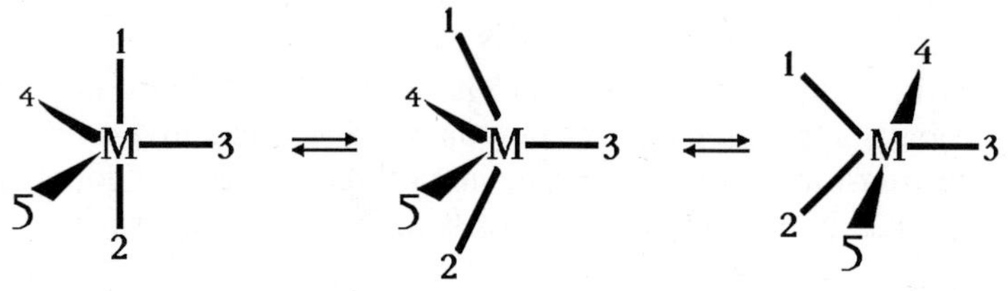

In the original trigonal bipyramidal structure, ligands 1 and 2 are axial; after rearrangement they end up in equatorial positions. Ligands 4 and 5 originally in equatorial positions become axial. This is accomplished without dissociation or association, through a square pyramidal configuration.

The ^{19}F NMR spectrum (from -197 to $60°C$) of the trigonal bipyramidal molecule PF_5 shows that all fluorines are equivalent. There is no differentiation between equatorial and axial positions. The barrier for this Berry rearrangement is very low.

Generally, 5-coordinated complexes with identical ligands tend to have low Berry rearrangement barriers. The barriers become higher if the ligands are not identical, because then there are preferences. In trigonal bipyramidal complexes, the more electronegative ligands tend to occupy the axial positions, whereas an ethylene will strongly prefer an equatorial planar orientation, for example,

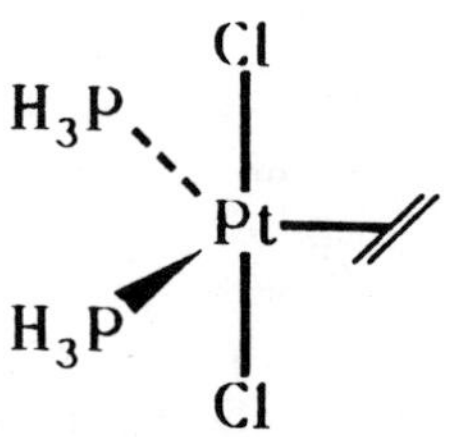

In a square pyramidal structure, the ethylene will prefer a basal position.

Inversion of Configuration of Pyramidal Molecules

Many cases of inversion of configuration of pyramidal molecules (**I**) have been studied. In **I** the central atom can be nitrogen, sulfur, etc., and R_1, R_2, R_3 are other atoms or group of atoms, including groups of complexed metals (as in **II**).

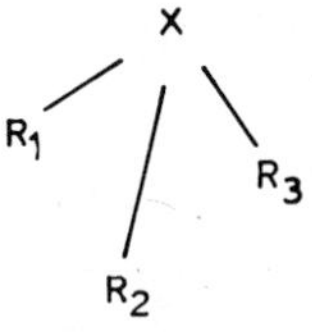

I 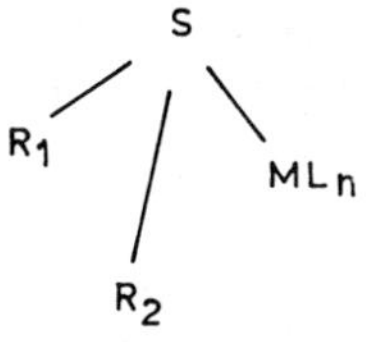   II

An example of a simple molecule undergoing inversion is ammonia. An example of a molecule containing a metal atom is *mer*-Ti(Me)Cl$_3$(MeSCH$_2$CH$_2$SMe); usually molecules like this are presented as octahedral, with the metal at the center—and of course they are—but by focusing on the sulfur atoms, they can also be presented as in **II**

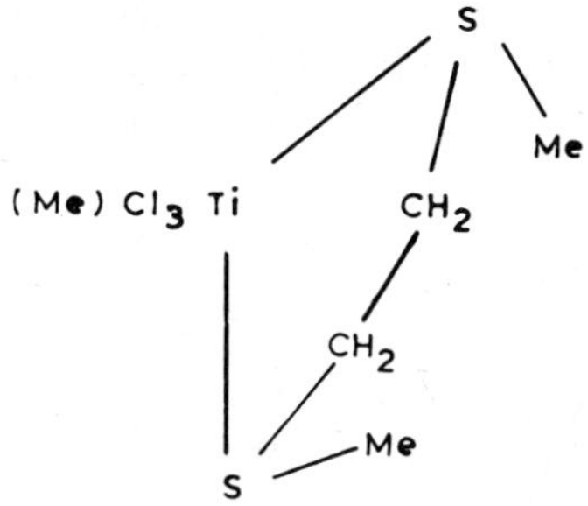

(Me) Cl$_3$ Ti ... CH$_2$ ... Me

syn

(Me) Cl$_3$ Ti ... CH$_2$ ... Me

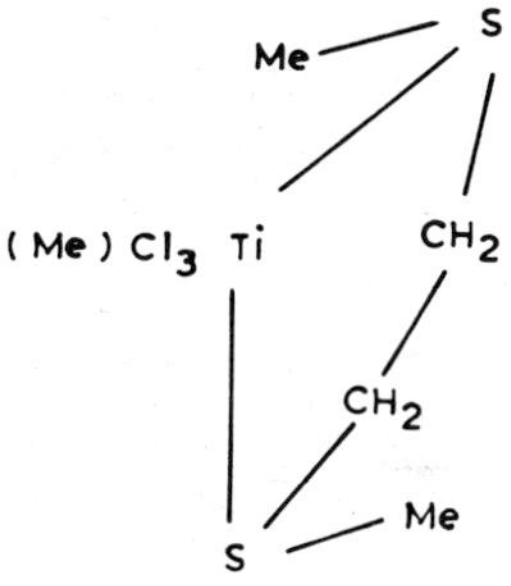

anti

It is then clearly seen that the rapid *syn* ⇌ *anti* transformation that is observed by NMR at ambient temperatures is a typical inversion.

The usual mechanism for the inversion of pyramidal molecules is intramolecular. A vibrational mode leads to inversion through a planar arrangement:

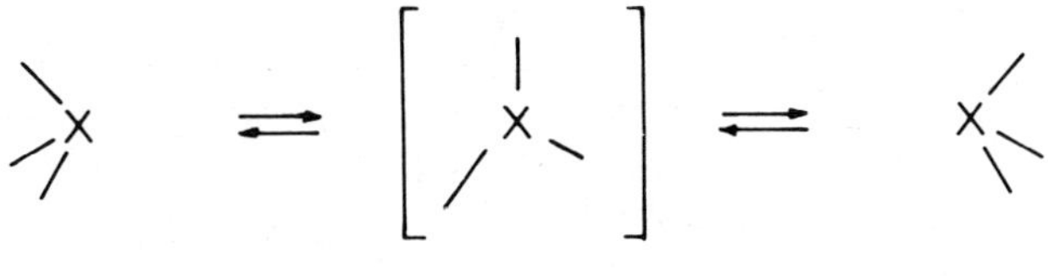

$$$$

The rate of the process depends on the activation required to reach the intermediate planar configuration. However, it is noted that the system may go over the potential barrier, but there are cases where quantum-mechanical tunneling through the barrier should be considered.

Trigonal Twist

If we look at an octahedron (trigonal antiprism, symmetry O_h or D_{3d}) from one of its faces down, the picture looks like this:

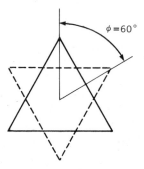

The "twist angle" ϕ is equal to 60.° The solid and dashed lines represent two opposite trigonal faces. Suppose now that the bottom face is twisted (dashed line), while the other is kept in its place. The resulting picture is

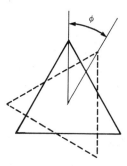

The angle ϕ has decreased, and in the limit where $\phi = 0$, a trigonal prism is obtained (D_{3h} symmetry):

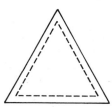

This "trigonal twist" from the octahedron to the trigonal prism is a topological transformation. Is it possible? As already stated, such transformations in 6-coordinate complexes in general have high barriers. Indirect evidence that the trigonal twist is indeed difficult is obtained by applying the structure correlation method to acetylacetonate and tropolonate complexes. Data are collected in Table 6.7.

The twist goes from 60° to about 34°, namely only up to about the midpoint of the D_{3d} to D_{3h} pathway. However, a complete twist in octahedral complexes is not impossible. In fact, the octahedron can sometimes be transformed into a trigonal prism and then back to an octahedron again, for example,

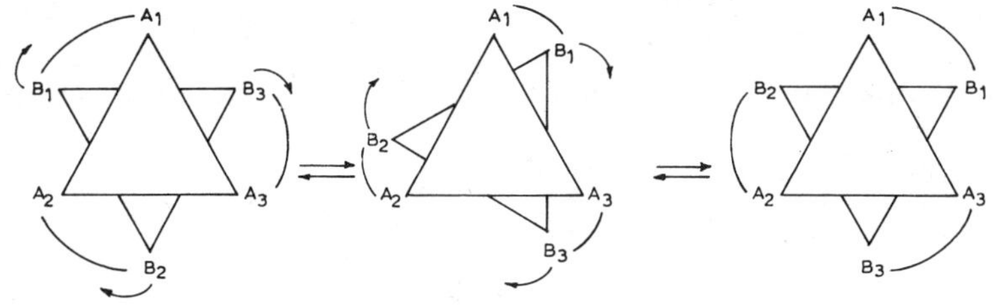

cisλ   The transformation passes   cisΔ
                        through a trigonal prism

The net result is that one enantiomer is transformed into the other:

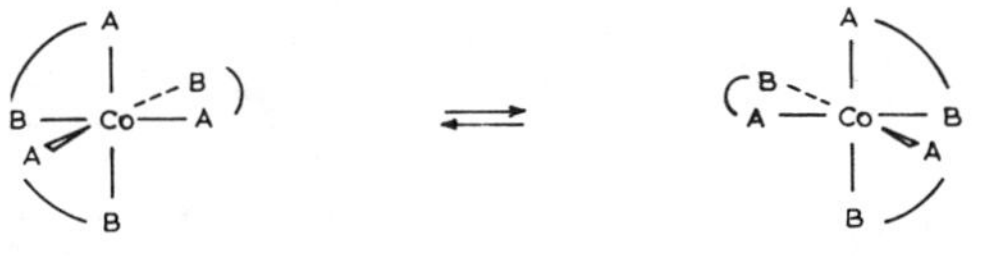

cisλ   cisΔ

A specific example, where such a transformation has been observed is

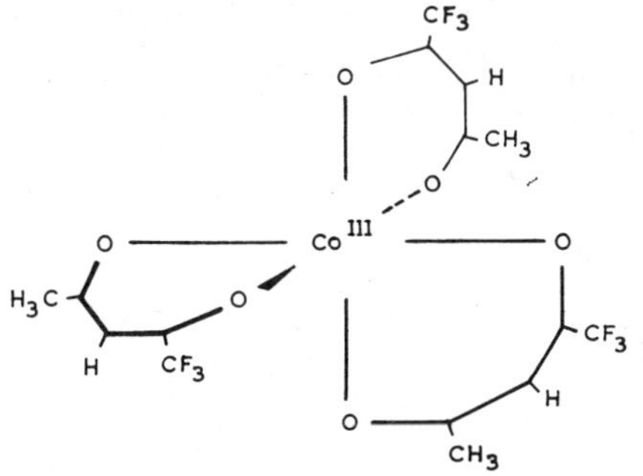

TABLE 6.7. Twist Angles of Acetylacetonate and Tropolonate Complexes[a]

Complex	ϕ
Sc(acac)$_3$	47.4°
V(acac)$_3$	56.0°
Cr(acac)$_3$	61.6°
Mn(acac)$_3$	60.2°
Fe(acac)$_3$	54.0°
Co(acac)$_3$	67.9°
Al(acac)$_3$	61.6°
Sc(trop)$_3$	33.9°
Mn(trop)$_3$	49.4°
Fe(trop)$_3$	38.5°
Al(trop)$_3$	48.1°

[a] acac$^-$ = acetylacetonate (the anion derived from diacetylmethane(pentane-2,4,-dione)):

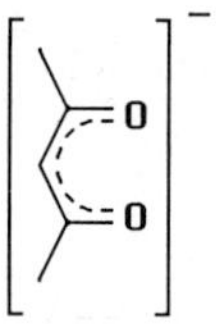

trop$^-$ = tropolonate (the anion derived from 1-hydroxycyclohepta-3,5,7-trien-2-one):

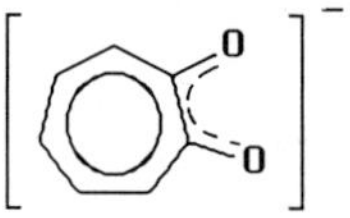

It is also believed that the racemizations of Cr(phen)$_3^{3+}$ and Cr(ox)(phen)$_2^+$, which have $\Delta V^\ddagger$ values close to zero, proceed by a twist mechanism involving, to a first approximation, only deformation, without a change in the metal–ligand bond length.

Other Topological Mechanisms

Simple bending and twisting are not the only ways a topological change can occur. Other motions can in principle also be effective. Three more examples are listed.

1. The following type of transformation has been observed in $(Ph_2EtP)_2NiBr_2$:

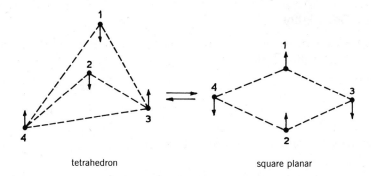

tetrahedron   square planar

2. In the transformation of the pentagonal bipyramid to the capped trigonal prism at the left of the diagram, two axial positions (1 and 7) of the pentagonal bipyramid and two equatorial (3 and 5) form the tetragonal face which is capped by 4. In its transformation to the prism at the right, three edges (not bonds) break (edges 1–4, 7–3, and 7–5) and a new one is formed (3–5), while simultaneously vertex 4 moves below the equatorial plane. For example, $K_4Mo(CN)_7$ undergoes this type of transformation.

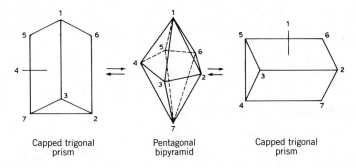

Capped trigonal   Pentagonal   Capped trigonal
prism   bipyramid   prism

3. In the bicapped tetrahedron, the hydride moves from one face to another. An example would be $H_2Fe[P(OC_2H_5)_2C_6H_5]_4$.

Bicapped tetrahedron

6.23. INTER- AND INTRAMOLECULAR PROTON TRANSFER

General Description

Proton-transfer reactions are among the faster reactions in solution. They are studied by relaxation techniques.

Neutralization,

$$H^+ + OH^- \underset{k_{-1}}{\overset{k_1}{\rightleftharpoons}} H_2O$$

is essentially a proton transfer from $H_9O_4^+$ to $H_7O_4^-$, i.e.,

$$H_9O_4^+ + H_7O_4^- \longrightarrow 8H_2O$$

The values of k_1 and k_{-1} (at 25°C) are $1.4 \times 10^{11} M^{-1} s^{-1}$ and $2.5 \times 10^{-5} s^{-1}$, respectively. $H_9O_4^+$ is H_3O^+ hydrogen-bonded to three water molecules, and $H_7O_4^-$ is OH^- hydrogen-bonded to three water molecules.

The rate of neutralization is controlled by the diffusion together of $H_9O_4^+$ and $H_7O_4^-$. The actual transfer of the proton is fast.

The obvious generalization of the neutralization reaction is the transfer of the proton from $H_9O_4^+$ to bases other than $H_7O_4^-$, for example, to SO_4^{2-}, CH_3COO^-, F^-, HS^-, and $Cu(OH)^+$. Proton transfer from other than H_3O^+ or $H_9O_4^+$ (e.g., from NH_4^+) or intramolecularly is also possible. All these reactions are diffusion-controlled. The differences in rates among them are generally small.

Reactions of OH^- transfer are considerably slower than proton-transfer processes, because of the greater mobility of the latter.

HOMO and LUMO in Proton Transfer

In neutralization, as a first approximation, it can be considered that a pair of electrons is transferred from OH^- to a proton. Hence the LUMO is $1s$-like, and the HOMO is one of the p-like orbitals of OH^-. The overlap is easy and effective:

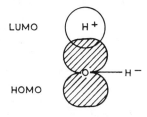

For other bases the HOMO might have different symmetry, but the overlap is always effective.

If the proton from the acid AH is not transferred completely, but instead an adduct AH $\cdots$ B is formed, the approximation that the LUMO is that of the proton is no longer valid. Instead, the LUMO of the whole AH molecule must be used. Moreover, repulsive HOMO–HOMO interactions,

may also be important. The net effect is attractive, and again there is no barrier for the reaction, apart, of course, from that needed to bring the reactant molecules together.

Covalent Hydrate and Pseudobase Formation

Some reactions of coordinated compounds have already been discussed—for example, olefin insertion to M—H bonds and nucleophilic attack on coordinated olefins. Other examples will be given in Chapter 7. Here one more case is mentioned, the so-called covalent hydrate and pseudobase formation, such as the one illustrated.

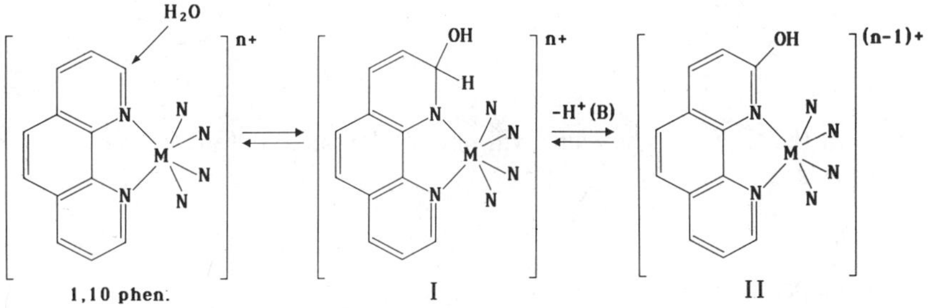

Proton abstraction from the covalent hydrate I results from the action of a base B. Pseudobase II is generally more reactive than I. Ligands like pyridine, 2,2'-bipyridine, and 2,2':6',2"-terpyridine give similar reactions.

Lewis Acids and Bases

If proton transfer is not the only possibility, new mechanistic aspects arise. An illustration is given here, based on the chemistry of substituted amines.

The exchange of BMe_3 in the addition compound Me_3NBMe_3 with BMe_3

in the bulk of the solution is an ordinary substitution reaction proceeding by a dissociative mechanism:

$$Me_3NBMe_3 \underset{fast}{\overset{slow}{\rightleftharpoons}} Me_3N + BMe_3$$

$$Me_3N + B^*Me_3 \underset{slow}{\overset{fast}{\rightleftharpoons}} Me_3NB^*Me_3$$

The recombination of Me_3N with BMe_3 or B^*Me_3 can be regarded as a neutralization reaction or as an electrophilic attack by BMe_3.

Substituted amines like H_2NX generally react by the following three mechanisms:

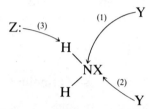

In cases 1 and 2, there is nucleophilic attack on nitrogen and on X, respectively. The mechanism is I_d. In case 3, one proton is stripped out by a base Z: and HNX^- is formed, which is the conjugate base of H_2NX (conjugate base mechanism). Subsequently HNX^- reacts further with the nucleophile Y:

$$HNX^- + Y \longrightarrow HNY + X^-$$

Which of the three mechanisms will prevail depends on both X and Y. Some examples are:

Mechanism 1: $X = OSO_2^-$, $Y = Ph_3P$
Mechanism 2: $X = OH^-$, $Y = Ph_3P$
Mechanism 3:   Hydrolysis of nitramine (H_2NNO_2)

GENERAL REFERENCES

E. W. Abel, S. K. Bhargava, and K. G. Orrell, The Stereodynamics of Metal Complexes of Sulfur-, Selenium-, and Tellurium-Containing Ligands, in S. J. Lippard, ed. *Progr. Inorg. Chem.*, Vol. 32, Wiley, New York, 1984.

H. B. Burgi and J. D. Dunitz, From Crystal Statics to Chemical Dynamics, *Acct. Chem. Res.* **16**, 153 (1983).

O. Eisenstein and R. Hoffmann, Transition-Metal Complexed Olefins: How Their Reactivity toward a Nucleophile Relates to their Electronic Structure, *J. Am. Chem. Soc.* **103**, 4308 (1981).

J. H. Espenson, Reactions and Reaction Mechanisms of Organochromium(III) Complexes, A. G. Sykes, ed., *Adv. Inorg. Bioinorg. Mech.* **1**, 1–63 (1982).

H. Gamsjäger and R. K. Murmann, Oxygen-18 Exchange Studies of Aqua and Oxo Ions, A. G. Sykes, ed., *Adv. Inorg. Bioinorg. Mechanisms* **2**, 317–380 (1983).

D. A. House, Stereochemistry Reaction Rates of Anionopentaamine Complexes of Cobalt(III) and Chromium(III), *Coord. Chem. Rev.* **23**, 223 (1977).

D. L. Kepert, Isopolytungstates, F. A. Cotton, ed., *Prog. Inorg. Chem.* **4**, 199 (1962).

C. H. Langford and J. P. K. Tong, Preferential Solvation and the Role of Solvent in Kinetics. Examples from Ligand Substitution Reactions, *Acct. Chem. Res.* **10**, 258 (1977).

C. H. Langford and H. B. Gray, *Ligand Substitution Processes*, W. A. Benjamin, New York, 1965.

D. A. Palmer and H. Kelm, The Use of High Pressure Kinetics During the Elucidation of Reaction Mechanisms of Coordination Compounds in Solution, J. Osugi, ed., 4th ed., *Int. Conf. High Pressure,* (1974), *Phys. Chem. Soc. Japan, Kyoto,* 657 (1975).

D. R. Stranks, The Elucidation of Inorganic Reaction Mechanisms by High Pressure Studies, *Pure Appl. Chem.* **38**, 303 (1974).

T. W. Swaddle, Substitution Reactions of Divalent and Trivalent Metal Ions, A. G. Sykes, ed., *Adv. Inorg. Bioinorg. Mechanisms* **2**, 95–138 (1983).

T. W. Swaddle, Activation Parameters and Reaction Mechanism in Octahedral Substitution, *Coord. Chem. Rev.* **14**, 217 (1974).

D. L. Thorn and R. Hoffmann, The Olefin Insertion Reaction, *J. Am. Chem. Soc.* **100**, 2079 (1978).

M. Tobe, in T. C. Waddington, ed., *Inorganic Reaction Mechanisms,* Thomas Nelson, London, 1972.

PROBLEMS

1. If the reaction

$$Co(NH_3)_4CO_3^+ \xrightarrow{2H^+, H_2O} Co(NH_3)_4(H_2O)_2^{3+} + CO_2(g)$$

is performed in $H_2^{18}O$, it is found that one of the two oxygens in the new complex originates from the solvent, and the other from the coordinated CO_3^{2-}. Propose a mechanism consistent with these observations.

2. Water exchange for *trans*-$Co(en)_2(OH_2)_2^{3+}$ has $\Delta V^\ddagger = 5.9 \pm 0.2 \text{ cm}^3 \text{mol}^{-1}$, and this value is independent of pressure. In contrast, the trans to cis isomerization of this compound has $\Delta V^\ddagger = 13.7 \pm 0.7 \text{ cm}^3 \text{ mol}^{-1}$ (in 0.05 M HClO$_4$) and depends on pressure. Also, anation of *cis*-$Co(en)_2(OH_2)_2^{3+}$ by oxalic acid (H$_2$ox) and hydrogen oxalate (Hox$^-$) has $\Delta V^\ddagger = 4.8 \pm 0.2 \text{ cm}^3 \text{ mol}^{-1}$, which is independent of pressure. Interpret these data, and comment on the differences.

3. In the temperature range 14–95°C the equilibrium

$$Cr(OH_2)_6^{3+} + NCS^- \rightleftharpoons Cr(OH_2)_5NCS^{2+} + H_2O$$

is established slowly, and the value of the equilibrium constant equals $\sim 10^3$. The rate of the reaction from left to right is given by the relation

$$R_1 = (k_1 + k_2[H^+]^{-1} + k_3[H^+]^{-2})[Cr(OH_2)_6^{3+}][NCS^-]$$

The rate of the reaction from right to left is given by the relation

$$R_{-1} = (k_{-1} + k_{-2}[H^+]^{-1} + k_3[H^+]^{-2})[Cr(OH_2)_5NCS^{2+}]$$

The three terms in each of these rate laws correspond to three parallel paths:

(a) What are the compositions of the activated complexes?

(b) Propose a mechanism for each path.

(c) At 25°C, $k_1 = 1.9 \times 10^{-6} \, M^{-1} \, s^{-1}$ and $k_2 = 7.2 \times 10^{-9} \, s^{-1}$. At what hydrogen ion concentration do the rates corresponding to these two paths become equal?

4. Derive the rate equation for the substitution mechanism:

$$Co(CN)_5OH_2^{2-} \longrightarrow Co(CN)_5^{2-} + H_2O \quad \text{(slow)}$$

$$Co(CN)_5^{2-} + I^- \longrightarrow Co(CN)_5I^{3-} \quad \text{(fast)}$$

5. Under ordinary conditions, carbon tetrachloride is not hydrolyzed. It is, however, hydrolyzed when exposed to intense sunlight. Propose a mechanism.

6. The rates by which R is substituted in sulfur compounds of the formula R_2SO_2 are generally smaller than the rates in compounds of the type R_2SO. The same is true of the corresponding selenium compounds. Also, at the same hydrogen ion concentration, NO_2^- reacts faster than NO_3^-. Finally, the rates by which oxoanions of chlorine react follow in general the trend

$$ClO^- > ClO_2^- > ClO_3^- > ClO_4^-$$

Correlate and interpret these observations.

7. With ^{35}S and ^{18}O as tracers the following observations were made:

Oxygen exchange:

$$S^*O_2 + SO_3 \rightleftharpoons SO_2 + S^*O_3 \qquad \text{(1) fast}$$

Sulfur exchange:

$$^*SO_2 + SO_3 \rightleftharpoons SO_2 + {}^*SO_3 \qquad \text{(2) slow}$$

Sulfur exchange:

$$^*SO_2 + SOCl_2 \rightleftharpoons SO_2 + {}^*SOCl_2 \qquad \text{(3) slow}$$

Oxygen exchange:

$$S^*O_2 + SOCl_2 \rightleftharpoons SO_2 + S^*OCl_2 \qquad \text{(4) slow}$$

Sulfur exchange:

$$^*SO_2 + (NMe_4)_2S_2O_5 \rightleftharpoons SO_2 + (NMe_4)_2{}^*S_2O_5 \quad \text{(5) fast}$$

It is also known that a solution of SO_3 in liquid SO_2 is a good conductor of electricity.

(a) What are the ions in the solution of SO_3 in liquid SO_2? Write the chemical equation for the formation of these ions. Hint: Take into account that SO_3 is one of the strongest (Lewis) acids.

(b) Is the transfer of a neutral oxygen atom from SO_3 to SO_2 fast or slow? (*Caution*: It is a neutral atom, not an ion.)

(c) Is self-ionization of SO_2 (in the absence of SO_3) fast and extensive?

(d) In liquid SO_2, $S_2O_5^{2-}$ ions can be considered as solvated SO_3^{2-} ions, $SO_3^{2-} \cdot SO_2$. Can this provide an explanation for the fast exchange (5)?

(e) How can the fast exchange in (1) be explained?

8. The general rate law for substitution in square tetragonal platinum(II) complexes is also valid for the reaction

$$Pt(NH_3)_4^{2+} + Cl^- \longrightarrow Pt(NH_3)_3Cl^+ + NH_3$$

Design the experiments needed to verify this and to determine the rate constants. What kind of experimental data will be obtained, and how are the data to be treated?

9. The isoelectronic anions $H_2SiO_4^{2-}$, HPO_4^{2-}, SO_4^{2-}, and ClO_4^- exchange oxygen with the solvent (water) with rates having the following trend:

$$H_2SiO_4^{2-} > HPO_4^{2-} > SO_4^{2-} > ClO_4^-$$

Correlate and interpret these results.

10. Would the reaction

$$Co(solv)_6^{3+} + 6NH_3 \longrightarrow Co(NH_3)_6^{3+} + 6(solv)$$

(solv = solvent molecules) be faster in water or in ethanol? Explain.

7

ELECTRON-TRANSFER REACTIONS

In this chapter we examine the second major mechanistic category: *electron-transfer* or *oxidation–reduction* or *redox reactions*—reactions involving changes in formal oxidation states. The ultimate objective is to see how one or more electrons are transferred (directly or indirectly) from one molecule to another, or, as Henry Taube popularized it, to see how the "glue" that holds the atoms together is redistributed.

7.1. BACKGROUND

Historical Note

The most interesting chemical theory during the 18th century was the phlogiston theory, which was intended to explain combustion. One of the better-known supporters of this theory, George Ernest Stahl (1666–1734), believed that

All substances contain a component called phlogiston.

The nature of phlogiston is the same in all substances; only the way it is bound with each substance varies.

During their combustion, metals lose phlogiston and are transformed into "ashes."

Substances that have lost phlogiston may regain it if they are heated with other substances rich in phlogiston (e.g., charcoal).

However, the objections to this theory were very strong. The opponents argued that

Phlogiston is impossible to isolate, weigh, and characterize. Its existence is hypothetical.

The ashes obtained in the combustion of metals are heavier than the metals they were obtained from. Loss of "phlogiston matter" should cause a decrease in weight, not an increase.

In the end, the "battle" was won by Lavoisier and his supporters. The phlogiston theory was abandoned. The 19th and early 20th century chemists were very firm in its condemnation.

Today, this condemnation is not justified. In fact, we know that many of Stahl's ideas were right. If we simply interchange the words "phlogiston" and "electron," it is not difficult to see that substances indeed contain electrons, lose them, and regain them and that the distribution of electrons determines the properties in each case. The "ashes" weigh more than the metal because of the addition of oxygen, but the metals are nevertheless oxidized (have lost "phlogiston")—and free electrons *have* been weighed and characterized.

Lavoisier's ideas about combustion with air were, of course, correct, but phlogiston theory was not completely wrong.

Definition

Redox reactions involve changes in the formal oxidation numbers of two or more chemical elements. Most redox reactions involve an increase in the oxidation number of only one element and an equivalent decrease in the oxidation number of another or of the same element in the same or in a different compound and with different oxidation state. Reactions involving change of the oxidation number of more than two elements, such as

$$14ClO_3^- + 3As_2S_3 + 24OH^- \longrightarrow 14Cl^- + 6H_2AsO_4^- + 9SO_4^{2-} + 6H_2O$$

are rather rare.

The definition of a redox reaction implies electron transfer or displacement, from the reducing to the oxidizing agents. For example,

$$\overset{\overset{\displaystyle e^-}{\frown}}{Cr^{2+}(aq) + Fe^{3+}(aq)} \longrightarrow Cr^{3+}(aq) + Fe^{2+}(aq)$$

$$\overset{\overset{\displaystyle 2e^-}{\frown}}{H^- + Cl_2} \longrightarrow H^+ + 2Cl^-$$

$$\overset{\overset{\displaystyle 10e^-}{\frown}}{16H^+ + 5S^{2-} + 2MnO_4^-} \longrightarrow 5S + 2Mn^{2+} + 8H_2O$$

In these examples there is a change in the charges of the ions.

In the reaction

$$LiH + ClCl \longrightarrow LiCl + HCl$$

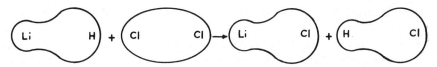

Figure 7.1. Schematic presentation of the change in relative electronegativities.

there is a change in the partners and in the relative electronegativity of the combined atoms (Fig. 7.1).

It should also be noted that electron transfer is often accompanied by atom or group transfer, bond breaking, bond formation, and so on. For example,

$$\underset{H}{ClO} + NO_2^- \longrightarrow [\underset{H}{ClONO_2^-}] \longrightarrow \underset{H}{Cl} + ONO_2^-$$

$$NO_2^+ + C_6H_6 \xrightarrow{\text{Base B}} C_6H_5NO_2 + HB^+$$

Redox behavior refers to the giving and taking of electrons. The same is true, however, for the generalized acid–base behavior. If we compare the redox reaction

$$2Ag^+ + Zn \longrightarrow 2Ag + Zn^{2+}$$

with the acid–base reaction

$$Ag^+ + 2NH_3 \longrightarrow Ag(NH_3)_2^+$$

we note that both depend on the ability of Ag^+ to accept electrons. In fact, in some cases the distinction is not sharp. For example, if the bound NO is considered to remain neutral, the reaction

$$Fe(OH_2)_6^{3+} + NO \longrightarrow Fe(OH_2)_5NO^{3+} + H_2O$$

must be considered a nucleophilic substitution. If, however, NO is thought to have become NO^+, the reaction is a redox process.

Relation Between Oxidation Number, Molecular Geometry, and the Composition of the First Coordination Sphere

It is evident that a change in composition is associated with a change in molecular geometry. It is perhaps trivial to say, for example, that NO_2 (where the oxidation number of nitrogen is IV) has a different geometry than NO (where the oxidation number of nitrogen is II). It is less evident that even the solvation geometry of metal ions in solution changes when the oxidation number changes. Geometry is generally a very sensitive indicator of chemical change.

Even the same element bonded to the same number and kind of other atoms gives molecules with different geometries upon change of its oxidation state. The stable oxidation states of chromium in aqueous solutions are II, III, and VI. The geometry of $Cr(H_2O)_6^{2+}$ is tetragonally distorted octahedral. The complex $Cr(H_2O)_6^{3+}$ is also octahedral, but without distortion. Removal of three more electrons from $Cr(H_2O)_6^{3+}$ leads to more change; at this stage, the composition changes also, and the geometry becomes tetrahedral.

$$Cr(H_2O)_6^{3+} \xrightarrow{\ 3e^-\ } HCrO_4^- + 7H^+ + 2H_2O$$

$$2HCrO_4^- \rightleftharpoons Cr_2O_7^{2-} + H_2O$$

It is also noted that even a simple electronic excitation, which in a sense is an internal electron transfer, leads to a change in geometry. Excitation of water, for example, may lead to an opening of the bond angle,

Useful Generalizations

The accumulation of six positive charges in the space around chromium is electrostatically unfavorable. The "tension is relieved" by the removal of protons. This is a general phenomenon. It is rare to have small cations with $4+$ charge in aqueous solution. The only known stable simple ions $M^{4+}(aq)$ in $1\ M$ $HClO_4$ belong to the $5f$ series. Ions like $Ce^{4+}(aq)$ and $Sn^{4+}(aq)$ lose charge by hydrolyzing or by forming complexes with anionic ligands. Stable small cations with charge $\geqslant 5$ are unknown.

The positive charge of a central metal ion M^{n+} attracts electrons from the coordinated water molecules and weakens the O—H bonds. As a result, coordinated water is more acidic than water in the bulk of the solution. With increasing n and increasing strength of the M—O bond (Fig. 7.2), the equilibria

are shifted to the right.

It is seen in Figure 7.2 that for high oxidation numbers (high n) the dom-

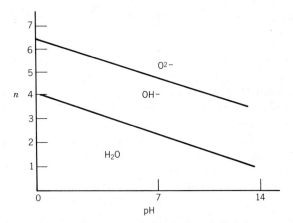

Figure 7.2. pH ranges where H_2O, OH^-, or O^{2-} ligands to an M^{n+} metal ion tend to dominate.

inant ligand is O^{2-}, even in strongly acidic solutions. Common examples are CrO_4^{2-} [chromium(VI)] and MnO_4^- [manganese(VII)]. For low oxidation numbers, OH^- or H_2O dominate.

Monomeric hydroxo compounds are seldom stable in aqueous solution. Ions like $Al(OH)(H_2O)_5^{2+}$ tend to dimerize by forming OH bridges

$$2[Al(OH)(H_2O)_5]^{2+} \rightleftharpoons \left[(H_2O)_4Al \begin{array}{c} H \\ O \\ \diagdown \diagup \\ \diagup \diagdown \\ O \\ H \end{array} Al(OH_2)_4 \right]^{4+} + 2H_2O$$

This aluminum dimer has a higher positive overall charge than the monomer, and in turn it hydrolyzes and new hydroxides are formed along with new bridges and there is further condensation. In short, it polymerizes. In this example, this process leads to the formation[1] of the ion

$$[Al_{13}O_4(OH)_{24}(H_2O)_{12}]^{7+}$$

which contains not just H_2O or OH^- or O^{2-}, but all three of them. The charge is high, but now it is spread out over a large molecule.

In this kind of inorganic polymer it is also possible to have linear bridges of the form M—O—M or bent bridges of the form $M \diagup^{O}\diagdown M$.

In nonpolar solvents, charged species are not favored at all.

---

[1] In inorganic chemistry, one or a few of the polymeric species may dominate; this can be contrasted with the usual organic polymers, where a mixture of molecules with a wide distribution of molecular weights is obtained.

Tendency toward Electroneutrality: Formal Oxidation Numbers and Real Charges

Chemical systems contain electric charges—ions, dipoles, transient dipoles, and so on—but under ordinary conditions the overall charge is zero. In fact, there is a tendency for the neutralization of charges at the smallest possible scale (the *Pauling electroneutrality principle*). For example, we may obtain an overall neutral system from an equal number of positive and negative charges by placing them in layers. In solution, however, the ions are mobile. Opposite charges can be mutually neutralized at a smaller scale, and this is indeed what happens. The ions mix.

Even microscopic volume elements tend to be neutral (Fig. 7.3).

The real charge on beryllium in $Be(H_2O)_4^{2+}$ is ~ -0.1, and in $Be(H_2O)_6^{2+}$ ~ -1.1. Here the smallest possible volume for the neutralization of the charges is essentially the first coordination sphere of beryllium. The numbers quoted indicate that neutralization is more effective in $Be(H_2O)_4^{2+}$. As a result this ion is more stable than $Be(H_2O)_6^{2+}$. For the same reason, $Al(H_2O)_6^{3+}$ (the real charge on aluminum is ~ -1.0) is more stable than $Al(NH_3)_6^{3+}$ (real charge ~ -1.1).

It must be concluded that what we customarily call one-electron or two-electron changes do not refer to real changes in the charges around the atoms oxidized or reduced. As a rule, real charges change by much less. What actually change (e.g., by one or two units) are the formal oxidation numbers.

The examples given above also imply that formal oxidation numbers are far from being equal to the *real* (*effective*) *charges*. The oxidation number, for example, of beryllium in the above compounds is II and that of aluminum is III, but the real charges are quite different! In fact, these real charges—that is, the excess or deficiency of charge compared to the free atom—are even negative!

Effective charges on cations are almost always smaller than typical charges, and sometimes it even happens that an increase in typical charge leads to a decrease rather than an increase of the real charge. The effective charge on

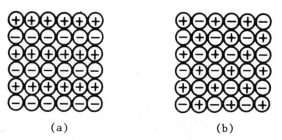

(a)   (b)

Figure 7.3. If a system can "choose" between a layered (*a*) and an alternative (*b*) arrangement (if the other laws of chemistry and physics allow) it will "choose" the latter.

iron(II) in high-spin complexes is a little less than $+2$. In the corresponding complexes of iron(III), it is only $+1.2$.

The effective charge on monatomic anions is usually less (absolutely) than 1. Coulomb repulsions prevent accumulation of more than one negative charge (more than one electron) on each atom. Higher negative charges tend to spread out over the molecule. Even in oxides like MgO we do not really have oxide ions, O^{2-}. The representation Mg^+—O^- is more realistic than the representation $Mg^{2+}O^{2-}$; the bond is partially covalent, and in Mg^+—O^- it is symbolized by the dash.

7.2. MECHANISMS AND RATE LAWS

The following mechanistic classes can be distinguished in oxidation–reduction reactions.

1. *In the simplest possible case, the mechanism consists of only one elementary reaction*:

$$Red + Ox \xrightarrow{k} Products$$

where Red is the reducing agent and Ox the oxidizing agent. The rate law is also simple:

$$Rate = k[Red][Ox]$$

Among the examples in this class are all of the simple outer-sphere electron-transfer reactions (see Chapter 3).

2. *The mechanism may again contain only one elementary redox reaction but also include atom- or group-transfer step(s) not involving change of the formal oxidation number* (e.g., substitution, proton transfer, or bond breaking). The form of the rate law will then depend on which of these reactions controls the rate.

In the mechanism

$$Co(NH_3)_4(C_2O_4)^+ + V^{2+}(aq) \longrightarrow [Co(NH_3)_4(C_2O_4)V]^{3+} \qquad (slow)$$

$$[Co(NH_3)_4(C_2O_4)V]^{3+} \xrightarrow{H^+} Co^{II} + V^{III} + H_2C_2O_4 + 4NH_4^+ \qquad (fast)$$

it is believed that the overall rate is determined by the rate of formation of the bridged binulcear intermediate (first reaction), which is essentially the substitution of a coordinated water molecule in $V^{2+}(aq)$ by the complex moiety. In fact, many reductions by $V^{2+}(aq)$ are believed to be controlled by this substitution and to have comparable rates and activation parameters. In the specific example, the rate is

$$- \frac{d[\text{Co}^{\text{III}}]}{dt} = k[\text{V}^{2+}][\text{Co}^{\text{III}}]$$

where $[\text{Co}^{\text{III}}]$ represents the concentration of the Co^{III} complex.

Reductions by $\text{Cr}^{2+}(\text{aq})$ often proceed by inner-sphere mechanisms, too, but their rate is determined by the electron transfer step rather than the substitution process. In the reduction of $(\text{H}_3\text{N})_5\text{Co}^{\text{III}}\text{Cl}^{2+}$ by $\text{Cr}^{2+}(\text{aq})$, the mechanism includes, in addition to the electron-transfer step, a number of substitution and bond-forming or bond-breaking steps:

$$\text{Cr}(\text{H}_2\text{O})_6^{2+} \longrightarrow \text{Cr}(\text{H}_2\text{O})_5^{2+} + \text{H}_2\text{O}$$

$$(\text{H}_3\text{N})_5\text{CoCl}^{2+} + \text{Cr}(\text{H}_2\text{O})_5^{2+} \longrightarrow (\text{H}_3\text{N})_5\text{CoClCr}(\text{H}_2\text{O})_5^{4+}$$

$$(\text{H}_3\text{N})_5\text{CoClCr}(\text{H}_2\text{O})_5^{4+} \longrightarrow (\text{H}_3\text{N})_5\text{Co}^{2+} + \text{ClCr}(\text{H}_2\text{O})_5^{2+} \quad \text{(redox)}$$

$$(\text{H}_3\text{N})_5\text{Co}^{2+} + 5\text{H}^+ + 6\text{H}_2\text{O} \longrightarrow \text{Co}(\text{H}_2\text{O})_6^{2+} + 5\text{NH}_4^+$$

The dimer $(\text{H}_3\text{N})_5\text{CoClCr}(\text{H}_2\text{O})_5^{4+}$ is a real intermediate and lives long enough to participate in fast acid–base reactions. The activated complex of the redox step, which determines the overall rate, resembles this intermediate. The observed rate is given by

$$\text{Rate} = k_{\text{obs}}[\text{Cr}^{2+}][\text{Co}^{\text{III}}]$$

3. Finally, in some oxidation–reduction reactions *the mechanism includes more than one redox step.* Such is the case, for example, with Fenton's reagent, which is mentioned in Chapter 2.

Electron transfer by way of a bridged activated complex is intimately related to substitution. In fact, as has already been stated, the overall rate may be determined by a substitution rather than an electron-transfer step. The general classification for substitution reactions into dissociative and associative processes is also expected to be applicable to substitutions accompanying electron transfer.

7.3. CLASSIFICATIONS

The essential similarities or differences between oxidation–reduction reactions are revealed only if one examines the elementary electron-transfer steps themselves after having carefully identified them.

Elementary electron-transfer reactions between metal complexes are class-ified into two distinct categories: those proceeding by way of *an inner-sphere* (*bridged*) activated complex and those proceeding by way of an *outer-sphere* activated complex. In outer-sphere processes, the compositions, of the co-ordination spheres of the reductant and the oxidant do not change. In inner-sphere reactions, the two reactants share a common ligand, "the bridge."

Direct and Indirect Electron Transfer

The distinction between inner- and outer-sphere activated complexes was made initially for reactions between coordination compounds, but it was subsequently somewhat generalized. A reaction between a coordination com-pound and a free ligand (e.g., an organic molecule) is characterized as inner-sphere if the organic molecule enters the first coordination sphere of the complex, and outer-sphere if it stays out. For the organic molecule itself, the concept of inner- and outer-sphere reactions is not considered to be useful. This generalization, however, can create some misunderstandings, as can be seen from the following illustrations.

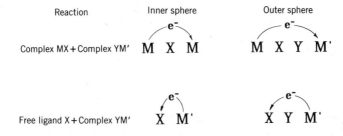

In the second case, what we call the outer sphere (XYM') is actually a bridged structure (via Y), and for what we call inner sphere there is really no counterpart in reactions between complexes. Electron transfer directly from metal ion to metal ion, involving metal–metal interaction, has not been studied very much.

Misunderstandings can be avoided if electron-transfer reactions are class-ified as *direct* and *indirect*. In the first case, the electron is transferred directly from the original to the final orbital; in the second case, a third orbital is implicated in between. The transfer, for example, from one transition metal ion to another is direct if there is direct overlap between the e_g^* or t_{2g} orbitals. It is indirect if the electron first goes to a ligand-centered orbital before ending up in the e_g^* or t_{2g} orbital of the acceptor.

The distinction between direct and indirect transfer can be made only if the corresponding initial and final orbitals can be correlated unambiguously.

EXAMPLES

In the following cases, electron transfer is direct.

$$Br^- + OH \longrightarrow Br + OH^-$$

$$NO^- + O_2 \longrightarrow ONO_2^-$$

$$Pt^{II}Cl_4^{2-} + Cl_2 \longrightarrow Pt^{IV}Cl_6^{2-}$$

$$(HOO^-)Ce^{IV} \longrightarrow HOO + Ce^{III}$$

The first is a free radical reaction, the second and third are oxidative additions, and the last is a reductive elimination. □
   In the examples cited there is intimate contact between the atom that is oxidized and the one that is reduced. There is no other atom in between. However, direct transfer can be conceived, even if other atoms or groups are interposed, provided there is sufficient overlap between the donor and the acceptor orbitals. Thus, the following reactions can be considered to proceed by direct electron transfer:

$$Ru(NH_3)_6^{3+} + Ru(ND_3)_6^{2+} \longrightarrow Ru(NH_3)_6^{2+} + Ru(ND_3)_6^{3+}$$

$$(H_2O)_5Cr^{II} + FCr^{III}(OH_2)_5^{2+} \longrightarrow [(H_2O)_5Cr^{II}FCr^{III}(OH_2)_5]^{4+} \longrightarrow$$

$$(H_2O)_5Cr^{III}F^{2+} + Cr^{II}(OH_2)_5^{2+}$$

The first of these reactions proceeds by an outer-sphere and the second by an inner-sphere activated complex. In both cases the transfer is direct, because it is energetically unfavorable to have an electron transferred initially to the ligand. The empty orbitals available on these particular ligands have high energy.
   The examples of indirect transfer through a monatomic bridge are rather ambiguous. Thus, if the reaction

$$I^- + ICl \longrightarrow I_2 + Cl^-$$

proceeds by an attack on the iodine side, it can be argued that the transfer of the electron to chlorine is indirect. Similarly, the reactions

$$Cl_5Pt^{IV}Cl^{2-} + I^- \longrightarrow [Cl_5Pt^{IV}ClI]^{3-} \longrightarrow Cl^- + Pt^{II}Cl_4^{2-} + ICl$$

$$I^- + \overset{O}{\underset{O}{ICrOH}} \longrightarrow [\overset{O}{\underset{O}{IICrOH}}]^- \longrightarrow I_2 + HCrO_3^-$$

can be considered to involve indirect electron transfer provided they proceed as indicated.

However, most reactions between complexes with monatomic ligands seem to proceed by direct transfer. Monatomic ligands usually have high ionization energies and low electron affinities. They give or take electrons with difficulty, and this makes them poor mediators for indirect transfer. The same is true of many polyatomic ligands, such as NH_3 and H_2O, but there are other ligands having low-lying empty orbitals or loosely bound electrons that can act as good mediators. Among these are organic ligands with a conjugated double-bond system as in the reaction

$$\longrightarrow \quad Co^{II} \;+\; Cr^{III} \;+\; isonicotinamide \;+\; NH_4^+$$

In this reaction the electron is first transferred from Cr^{2+}(aq) to the isonicotinamide ligand and then to Co^{III}.

Another example of indirect electron transfer is the reaction

$$\longrightarrow \quad Products \;(Co^{II}, NH_4^+, Cr^{III}, oxalate\ ion)$$

which has been postulated to proceed by way of a double oxalato bridge.

In outer-sphere reactions, indirect electron transfer cannot be excluded. Electron exchange between $Co(phen)_3^{3+}$ and $Co(phen)_3^{2+}$ is more than 10^{12} times faster than the exchange between $Co(NH_3)_6^{3+}$ and $Co(NH_3)_6^{2+}$. The difference may very well reflect the ability of the phen ligand (1,10-phenanthroline) to accept electrons and play the role of a mediator. With ammonia there is no such possibility.

7.4. SOLVENT MEDIATION AND SOLVATED ELECTRONS

Transfer of an electron from the reductant to the solvent and then from the solvent to the oxidant can take place, but only under special circumstances and not in the usual inorganic oxidation–reduction reactions. Thus, active metals like sodium in liquid ammonia give solvated electrons, which react with other dissolved substances, for example,

$$M \xrightarrow{NH_3} M^+ + e(am)^-$$

$$RX + e(am)^- \longrightarrow RX^-$$

$$RX^- \longrightarrow R + X^-$$

$$R + e(am)^- \longrightarrow R^-$$

$$R^- + NH_3 \longrightarrow RH + NH_2^-$$

$$M^+ + X^- \longrightarrow MX$$

where $e(am)^-$ symbolizes the solvated (ammoniated) electron.

Solvated electrons can also be generated in aqueous solution by the ionization of the solvent (water) molecules themselves by γ rays or other types of high-energy radiation.

7.5. OXIDATION-REDUCTION REACTIONS OF OXO AND HYDROXO COMPOUNDS

The Role of Lability

In oxidation–reduction reactions of oxo and hydroxo compounds involving substitution, two aspects of lability are of interest: the lability that oxo and hydroxo groups induce on other ligands, and their own lability.

A terminal hydroxide ion is generally much more labile than water. Convincing evidence is provided by the conjugate-base path in the base hydrolysis of aquo ions. In contrast, a multibonded terminal oxide ion is less labile than water. Thus,

$$O^{2-} < H_2O < OH^-$$

Other trends that seem to have been established are

$$\text{trans to } O^{2-} > \text{cis to } O^{2-} > M\overset{\displaystyle O}{\underset{\displaystyle O}{\diagup\!\!\!\backslash}}M > O^{2-} \text{ itself}$$

and

$$\text{Terminal OH}^- > \text{bridging OH}^- > \text{bridging O}^{2-}$$

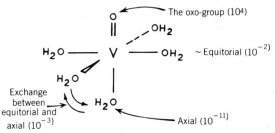

Figure 7.4. Half-lives (in seconds) of exchange between the oxo group or coordinated water molecules and water in the solvent, and between axially and equatorially coordinated water in $VO(H_2O)_5^{2+}$.

In comparing labilities, reference should be made to the same metal ion, the same oxidation state, and the same other ligands. Unfortunately, the data available are rather limited.

If the rate of an oxidation–reduction reaction depends on a preceding substitution process, it is expected that the path involving the fastest substitution will dominate. It is therefore useful to know the substitution rates at all sites.

The order-of-magnitude lifetimes for exchange with the solvent at the various sites of $VO(OH_2)_5^{2+}$ are shown in Figure 7.4.

It is reasonable to expect that in redox reactions involving bridging of V^{IV}, it is the most labile trans water that is removed.

Acid and Base Catalysis

Replacement of an oxide ion by hydroxide ion results in dramatic changes in both self-lability and the lability of the other ligands. If unfavorable lability is a serious obstacle to a bridging path, this obstacle must be removed if the electron is to be transferred. An effective way is the attachment of a proton on the oxide. In fact, there are cases, like the tetraoxo anions

$$\left[\begin{array}{c} O \\ | \\ O^{\diagdown} \underset{|}{\overset{M}{\diagup}} O \\ O \end{array} \right]^{n-}$$

where there are no labile groups at all, and where fast protonation of the oxide groups is a prerequisite for a bridged pathway.

Three cases of proton transfer are considered:

1. From the solvent to the oxide group or to an oxygen bridge
2. From coordinated water to the solvent

3. Intramolecularly, for example, from a coordinated water molecule to a terminal oxide or a bridging oxygen

In the first case, the rate will be first- or second-order in hydrogen ion depending on how many protons are needed. Indeed, many oxidations by oxo-anions proceed with a rate given by the general expression

$$\text{Rate} = k[\text{XO}_m^{n-}][\text{Y}^-][\text{H}^+]^l \quad (l = 1,2)$$

where Y^- is the reducing agent.

Lewis acids in general, not just protons, are also expected to have an accelerating effect.

Protons are also directly involved in electrochemical redox reactions. For example, the dominant Mo^{VI} species in strongly acidic solutions at low concentrations ($[\text{Mo}^{\text{VI}}] < 0.1$ mM) are the trioxo and the cis-dioxo species in equilibrium with each other:

$$(\text{H}_2\text{O})_3 \overset{\displaystyle \text{O}}{\underset{\displaystyle \text{O}}{\text{MoO}}} + \text{H}^+ \rightleftharpoons (\text{H}_2\text{O})_3 \overset{\displaystyle \text{O}}{\underset{\displaystyle \underset{|}{\overset{|}{\text{O}}}}{\text{MoO}^+}} $$

Electrochemical reduction of the cis-dioxo form, coupled with transfer of two protons to the same oxo group, seems to lead to the formation of the

transient species $\left[(\text{H}_2\text{O})_4 \overset{\displaystyle \text{O}}{\underset{\displaystyle \underset{|}{\overset{|}{\text{O}}}}{\text{Mo}^{\text{V}}}} \right]^{2+}$:

$$\left[(\text{H}_2\text{O})_3 \overset{\displaystyle \text{O}}{\underset{\displaystyle \underset{|}{\overset{|}{\text{O}}}}{\text{Mo}^{\text{VI}}\text{O}^+}} \right] + 2\text{H}^+ + e^- \rightleftharpoons \left[(\text{H}_2\text{O})_4 \overset{\displaystyle \text{O}}{\underset{\displaystyle \underset{|}{\overset{|}{\text{O}}}}{\text{Mo}^{\text{V}}}} \right]^{2+}$$

which quickly dimerizes:

$$2 \left[(\text{H}_2\text{O})_4 \overset{\displaystyle \text{O}}{\underset{\displaystyle \underset{|}{\overset{|}{\text{O}}}}{\text{Mo}^{\text{V}}}} \right]^{2+} \rightleftharpoons \left[(\text{H}_2\text{O})_3 \text{Mo}^{\text{V}} \overset{\text{O}}{\underset{\text{O}}{\diagup \diagdown}} \text{Mo}^{\text{V}}(\text{OH}_2)_3 \right]^{2+} + 2\text{H}^+ + 2\text{H}_2\text{O}$$

Enhancement of lability caused by proton transfer to the solvent (case 2) can also be viewed as base catalysis. The corresponding terms in the rate laws will be inverse in the concentration of hydrogen ion. This seems to be the case, for example, in the oxidation of $Mo_2O_4^{2+}$ by $IrCl_6^{2-}$ or $Fe(phen)_3^{3+}$. With both oxidants, the rate law contains two terms inverse in hydrogen ion:

$$Rate = k_1[Mo_2O_4^{2+}][H^+]^{-1} + (k_2 + k_3[H^+]^{-1})[Mo_2O_4^{2+}][Oxid]$$

Strictly speaking, the third case of proton transfer is not an acid catalysis. The rate is expected to be independent of hydrogen ion concentration. An intramolecular proton transfer is kinetically indistinguishable from a case without proton redistribution. An acid-independent term in the rate law can be attributed to either one.

Coordination Sphere Expansion

Another path for bridge formation is by way of coordination sphere expansion. From the point of view of chelate ring formation, this expansion has already been discussed in Chapter 6, where it was shown that it can be induced by the addition of one or two protons to the oxo ligands. The significance for redox reactions is that expansion can be achieved by direct addition of the reductant to the oxidant.

In oxidation reactions by an oxoanion MO_4^{2-}, the following two paths can be considered:

1. Formation of an oxidant–reductant complex *after* coordination sphere expansion. The formation of this complex is facilitated by the increased lability of the 6-coordinate species (only the usual 4 to 6 expansion is discussed):

$$MO_4^{2-} + H^+ \overset{K_1}{\rightleftharpoons} HMO_4^-$$

$$HMO_4^- + 2H_2O + H^+ \overset{K_2}{\rightleftharpoons} (H_2O)_3MO_3$$

$$(H_2O)_3MO_3 + red \longrightarrow [ox\text{-}red]$$

$$(ox)$$

2. Coordination sphere expansion *due* to an oxidant–reductant adduct formation, such as

$$MO_4^{2-} + H^+ \overset{K_1}{\rightleftharpoons} HOMO_3^-$$

$$HOMO_3^- + H^+ \overset{K_2}{\rightleftharpoons} (HO)_2MO_2$$

$$(HO)_2MO_2 + H_2O + red \longrightarrow (red)(H_2O)_2MO_3$$

Electron Transfer without Significant Structural Change

Electron exchange reactions like

$$\text{MnO}_4^{2-} + \text{MnO}_4^{-} \rightleftharpoons \text{MnO}_4^{-} + \text{MnO}_4^{2-}$$

proceed by an outer-sphere mechanism without significant modification of the inertness of the oxide group or significant structural change.

More complex structures like isopoly- and heteropolymolybdates undergo redox reactions coupled with proton transfer, but again without significant structural change. For example,

$$\text{SiMo}_{12}\text{O}_{40}{}^{4-} + ne^- + nH^+ \longrightarrow H_n\text{SiMo}_{12}\text{O}_{40}{}^{4-}$$

where n can be as high as 8. The reduced products are deep blue ("molybdenum blues") and are often readily reoxidized back to the original state.

Qualitative Consideration of Electronic Factors

The electron distribution of an oxo compound can be affected either by placing one or two protons on an oxygen atom or by placing one or more electrons on the metal ion. In either case the bonding between M and O is weakened, and substitution or substitution-dependent electron transfer is facilitated. Protons on oxygens will attract donor (bonding) electrons, and electrons on the metal will push them away. This proton- or electron-induced redistribution will also be associated with more electron density on the oxygen atom, increased basicity, and preference for the aqua or hydroxo groups over the oxo groups.

Schematically,

$$M\!-\!O \quad H^+ \qquad\qquad M\!-\!O$$

It is obvious that a positive metal ion will act in a manner similar to the proton:

$$M\!-\!O \quad M^{n+}$$

Some of the practical consequences can be summarized as follows:

> An oxygen bridge is likely to be more basic than a terminal oxide group.
> M—O bonds in bridged species are likely to be weaker than terminal M—O bonds; bond strength is expected to decrease as the number of oxygen atoms bound to the same metal ion increases.
> Protons and metal ions are expected to catalyze substitution and substitution-dependent electron transfer.
> With lower oxidation states, hydroxo and aquo groups are expected to be favored over oxo groups.
> Less polymerization is expected with lower oxidation states.
> Hydroxide ion is probably more basic than a terminal oxide ion bound to the same metal ion. A second proton is likely to be attached to the same oxygen (to the hydroxide ion) rather than to another oxide ion.

These trends will, of course, be affected by other factors, such as the number and kind of the other ligands. The simple qualitative model, however, is quite adequate for obtaining insight into some aspects of the known chemistry of the oxo and hydroxo compounds.

7.6. OXIDATIVE ADDITION

Definition

There is some confusion as to what should be considered oxidative addition—and the reverse, reductive elimination—and what should not. It is therefore necessary to define the terms carefully.

In this section we restrict our attention to what we might call "dissociative oxidative addition," which includes reactions like the following:

Complex		Addenda		Adducts	
L_mM^n	+	X—Y	$\longrightarrow$	$L_mM^{n+2} \big\langle \begin{smallmatrix} X \\ Y \end{smallmatrix}$	7.1
$L_mM^nM'^+$	+	X—Y	$\longrightarrow$	L_mM^{n+2}—Y + M^+X^-	7.2
$2L_mM^n$	+	X—Y	$\longrightarrow$	L_mM^{n+1}—X + L_mM^{n+1}—Y	7.3
$L_{2m}M_2^{2n}$	+	X—Y	$\longrightarrow$	L_mM^{n+1}—X + L_mM^{n+1}—Y	7.4

These reactions have the following characteristics:

1. They involve an unambiguous increase in the formal oxidation number by one or two units.
2. The "added" molecules X—Y (the *addenda*) can exist independently of the *adduct*.
3. The X—Y bond breaks completely upon addition, and in the adduct one of the fragments or both remain (at least transiently) attached.

By adopting this rather restrictive view of oxidative addition we can concentrate on the dissociative activation of the X—Y bond, a subject of the utmost importance in catalysis.

The examples will be taken from the field of organometallic chemistry of transition metals. Compounds of representative elements also undergo reactions designated as oxidative additions, such as

$$S^{IV}F_4 + F_2 \longrightarrow S^{VI}F_6$$

but as a rule such reactions find no application in catalysis.

Dihydrogen, dihalogens, and the alkyl halides are among the molecules that can add oxidatively. Usually, the complexes to which these molecules add have electron-rich central ions (d^7, d^8, d^{10}).

In many cases, oxidative addition is accompanied by coordination sphere expansion (equations 7.1–7.4). In other cases it may also be accompanied by the expulsion of one or two of the original ligands. The oxidative transformation, for example, of a 5-coordinated complex into a 6-coordinated one requires removal of one of the original ligands.

Clarification

In a grammatical sense, one may consider as oxidative addition any addition that is associated with an increase in the formal oxidation number of the element that accepts the new substituents.

This "grammatical" outlook encompasses, for example, reactions between oxo anions, such as

$$S^{IV}O_3^{2-} + Cl^{V}O_3^{-} \longrightarrow S^{VI}O_4^{2-} + Cl^{III}O_2^{-}$$

and reactions such as (only the directly relevant skeletons are shown):

$$Co^{III}\!-\!Cl + Cr^{II} \longrightarrow [Co^{III}\!-\!Cl\!-\!Cr^{II}] \longrightarrow Co^{II} + ClCr^{III}$$

which are characteristic of the inner-sphere electron transfer path. It also encompasses atom-abstraction free-radical reactions such as

$$HO + HR \longrightarrow HOH + R$$

However, it should be clear that such a generality is not helpful; lumping together so many diverse cases does not serve any practical purpose.

The Multiple Bond Addition Controversy

The question also arises whether the addition of unsaturated compounds such as

$$L_m M + \begin{matrix} X \\ \| \\ Y \end{matrix} \longrightarrow L_m M\!\!\begin{matrix} X \\ | \\ Y \end{matrix} \quad \text{or} \quad L_m M\!\!-\!\!\begin{matrix} X \\ \| \\ Y \end{matrix}$$

$$\textbf{I} \qquad\qquad \textbf{II}$$

are oxidative or not. Presentation of the product as **I** implies breaking of one of the two bonds of the unsaturated system, and in this sense classification of the reaction as oxidative addition is indeed justified. The justification is less obvious if the product is represented by formula **II**.

If the mixing between the metal d orbitals and the double- (or triple-) bond π^* orbitals is strong, X—Y approaches a single bond, and it can be argued that one of the two bonds has been replaced by an $X \diagdown \quad Y$ bridge. If the $\diagdown M \diagup$

mixing is weak, X=Y retains its double bond character, and the reaction cannot be regarded as oxidative addition.

In practice, there is a continuum of interactions depending on M, on the double bond, and on the other ligands and ranging all the way from very weak to very strong. The distinction is difficult to draw. Moreover, in specific cases we may not even have experimental evidence or criteria to judge whether the interaction is strong or weak.

Accordingly, we will avoid characterizing the addition to double or triple bonds as oxidative. Some of these additions may indeed involve an appreciable activation, comparable to that of breaking a single bond, but it is felt that they should be discussed separately elsewhere (Section 7.8).

(a)   (b)

Figure 7.5.

The Intimate Mechanism

Generally, metal ions act as Lewis acids. However, in oxidative addition reactions it can be considered that they also act as bases or nucleophiles. This is essentially the reason why only electron-rich ions undergo such reactions. Correspondingly, the added molecule acts in part as an electrophile (acid).

From this perspective, it is not surprising that the most common of all acids, the proton, should also act as an addendum, for example,

$$V^{II} + H^+ \rightleftharpoons V^{III}\!-\!H$$

The ligands in the original complex are also among the factors determining the tendency for oxidative addition.

Good electron acceptors such as CO and some olefins decrease the basicity of the metal ion and have a negative influence on the tendency for oxidative addition.

The details of oxidative addition on a molecular scale are not yet quite clear and may vary in different systems. What seems, however, rather safe to say is that in dissociative oxidative addition the crucial role is played by the σ^* antibonding orbitals of the addendum, insofar as these are the *last* ones to be filled before breaking the X—Y bond.

There are two ways these orbitals can be populated; either by a side-on or by linear overlap with occupied metal d orbitals (Fig. 7.5). The first overlap corresponds to a transition state with a three-center interaction (a'), the other to a transition state with two-center interactions (b').

$$\begin{bmatrix} X \cdots Y \\ \diagdown \diagup \\ M \end{bmatrix}^{\ddagger} \qquad\qquad \overset{\delta+}{[M} \cdots \overset{\delta-}{X \cdots Y]^{\ddagger}}$$

(a')   (b')

Two cases can be distinguished:

1. The overlaps (a) or (b) themselves (Fig. 7.5) determine the mode of coordination. If (a) is more effective, side-on coordination prevails. If (b) is more effective, the coordination is linear. It is also conceivable to have both types of coordination, as parallel pathways.
2. The mode of coordination is determined by the overlap of other orbitals, and *then* σ^* *is forced* to overlap in a given way. In this case, parallel occurrence of (a') and (b') is impossible.

The first case is that of dihydrogen, simply because there are no other low-lying empty orbitals. In H_2 the mode of coordination is expected to be determined solely by which of the overlaps of its σ^* antibonding orbitals will be better and will best match the energy levels of L_nM. Given the large assortment of available complexes, it is natural to expect examples of both modes of coordination and of the corresponding transition states, or even their parallel occurrence.

In contrast, with first-row homonuclear diatomic molecules and heteronuclear molecules like CO, CN^-, and NO, linear coordination seems to be forced by the overlap of other orbitals.

Hoffmann and coworkers examined how the coordination of diatomics change as the sum of d and π^* electrons varies from 4 to 12. The conclusions are summarized in Table 7.1.

For $[d + \pi^*]^4$, the coordination is η^2, for $[d + \pi^*]^6$ it becomes linear, for $[d + \pi^*]^8$ it is bent or kinked, for $[d + \pi]^{10}$ it becomes η^2 again, and for $[d + \pi^*]^{12}$ it is again linear. At this point, with 12 electrons, the number of antibonding electrons becomes equal to the number of bonding electrons and X—Y will depart from M unless there is an overlap decreasing the metal–diatomic antibonding character by a flow of d electrons into the X—Y $\sigma^*(3\sigma_u)$ orbital. But then the decrease of the antibonding character of M—X is accompanied by an increase in the antibonding character of X—Y. In the case of dihalogens, this leads to a breaking of the X—Y bond.

TABLE 7.1.   Coordination Modes of Diatomic Ligands with Transition Metals

$d + \pi^*$ Electrons	4	6	8	10	12
Coordination[a]	X—Y \/ M Side-on (η^2)	Y | X | M Linear (η^1)	Y / X | M Y X \ M Bent or kinked	X—Y \/ M Side-on	Y | X | M Linear

[a]M is also coordinated with four or five other ligands. Coordination of X—Y is determined by the overlap of the metal d orbitals with the π^* and n orbitals of the diatomic molecule.

Thus, *just before* oxidative addition, a linear geometry is imposed by the overlap of orbitals other than the X—Y σ^* orbital. X—Y is not coordinated in an η^2 manner, and Y is not already coordinated to M. Bonding of Y to M *follows* the breaking of the X—Y bond or happens in a concerted fashion:

$$\text{Y...}\text{—X—}M^n \longrightarrow M^{n+2} \overset{X}{\underset{Y}{\diagup\!\!\!\diagdown}} \qquad\qquad 7.5$$

Alternatively, Y may escape into the solution or react with another molecule, either as Y^-,

$$\text{Y—X—}M^n \longrightarrow M^{n+2} - X \; + \; Y^- \qquad\qquad 7.6$$

or as Y (free-radical mechanism):

$$\text{Y—X—}M^n \longrightarrow M^{n+1} - X \; + \; Y \qquad\qquad 7.7$$

Reaction 7.7 is followed by addition of Y to another ML_m. It is noted that in all three reactions 7.5–7.7, either part or the whole X—Y remains bound to the metal; otherwise, it is not an addition.

The strengths of the M^{n+2}—X, M^{n+1}—X, M^{n+2}—Y bonds, the stability of the $n + 2$ and $n + 1$ oxidations states of M, and the solvation energies are among the factors that determine the course of the reaction.

Alkyl Halides

Methyl fluoride is isoelectronic with difluorine, and by analogy it is expected to have a similar behavior. Generally, for alkyl halides a "polar" (linear) activated complex is expected. Their antibonding σ^* orbital is polarized toward the alkyl end, and the (b)-type overlap is large.

Stereochemical evidence bears this out. An activated complex of type (b') is expected to lead to an inversion of the configuration at the carbon atom or to scrambling. In contrast, an activated complex of type (a') should result in retention of configuration. The experimental data indicate inversion or scrambling.

Stereochemical observations at the metal center are less instructive. In equations 7.5–7.7, Y may add in different ways, giving either cis or trans products;

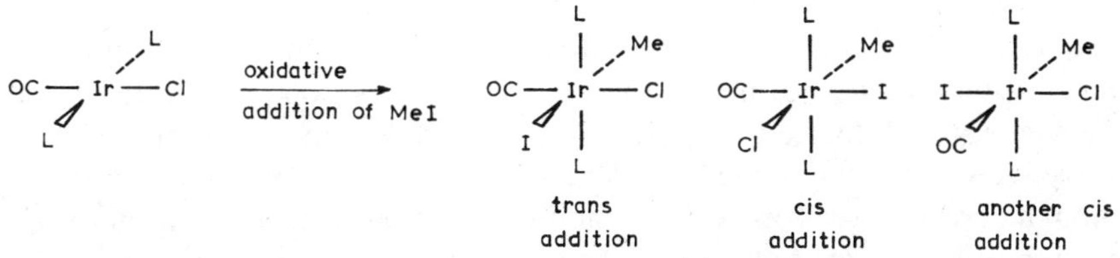

| trans addition | cis addition | another cis addition |

Additions in which original ligands trans to each other became cis have not been reported.

With alkyl halides, evidence can be cited for all three paths 7.5, 7.6, and 7.7. A concerted S_N2 mechanism (equation 7.5) has been proposed, for example, for the addition of alkyl halides to d^8 Co^I chelate complexes, but a free-radical mechanism is postulated for the one-electron additions of alkyl halides to $Co^{II}(CN)_5^{3-}$ and to $Cr^{II}(en)_3^{2+}$. The reaction

$$(R_2Cu^ILi)_n + nR'X \longrightarrow nLiX + [R_2Cu^{III}R']_n$$

can be cited as an example of a reaction occurring according to equation 7.6.

Cyclometalation

Cyclometalation is an intramolecular oxidative addition of a C—H group leading to cyclic (chelate) structures. The hydrocarbon group is part of a ligand already bound to the metal ion.

The first step in the process can be represented by the following scheme:

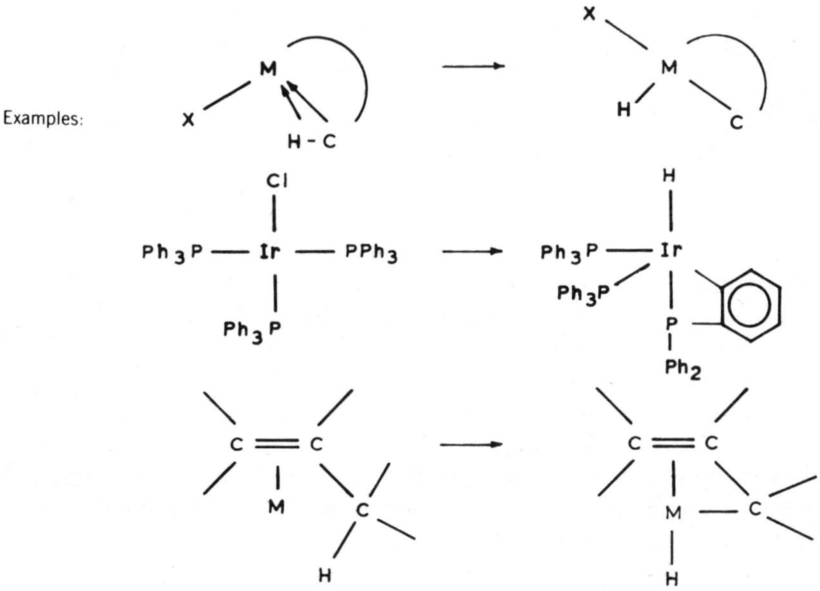

7.7. REDUCTIVE ELIMINATION

Classification and Clarification

Reductive elimination is the opposite of oxidative addition—the opposite in a conceptual sense, not necessarily the reverse reaction. As a rule, it is not the same groups that add oxidatively to a complex and are then eliminated reductively from the same complex. Apart from anything else, if there is enough driving force for oxidative addition, the chances are that there is not enough driving force for the corresponding reductive elimination.

Reductive elimination is the opposite of oxidative addition in another sense too. In oxidative addition the "destination" of the "added" fragment(s) may vary. The "added" fragment(s) may end up bound to the same metal ion or partly escape into solution. By analogy, we may consider as reductive eliminations all processes leading to the formation of a molecule from fragments bound to the same metal ion or to different metal ions or to a nonmetallic atom or coming partly from the solution. However, the designation "reductive" makes it necessary that in the process the formal oxidation number of the metal ion center must decrease.

Thus, the following general reactions can be classified as reductive eliminations:

$$L_mM^{n+2}\diagup_{\diagdown Y}^X \longrightarrow L_mM^n + X\!-\!Y \tag{7.8}$$

$$L_mM^{n+1}\!-\!X + Y\!-\!M^{n+1}L_m \longrightarrow 2L_mM^n + X\!-\!Y \tag{7.9}$$

$$2L_mM^{n+1}\!-\!\overset{|}{\underset{|}{C}}\!-\!\overset{|}{\underset{|}{C}}H \longrightarrow 2L_mM^n + H\overset{|}{\underset{|}{C}}\!-\!\overset{|}{\underset{|}{C}}H + C\!=\!C \tag{7.10}$$

$$L_mM^{n+2}(\!-\!\overset{|}{\underset{|}{C}}\!-\!\overset{|}{\underset{|}{C}}H)_2 \longrightarrow L_mM^n + H\overset{|}{\underset{|}{C}}\!-\!\overset{|}{\underset{|}{C}}H + C\!=\!C \tag{7.11}$$

In equations 7.8 and 7.9, if X = H and Y = C (e.g., alkyls or aryls bound through carbon), reductive elimination results in the formation of a carbon–hydrogen bond. If X = C and Y = C, reductive elimination results in the formation of a carbon–carbon bond. If X = H and Y = H, reductive elimination results in the formation of dihydrogen. In all these cases the fragments are initially bound to a metal.

In equations 7.10 and 7.11 there is again formation of C—H and C—C bonds, but now it is not only the bonds to the metals that break, but also some of the original C—H bonds.

Reductive elimination is not limited to carbon–hydrogen compounds, but it is these compounds that have the greatest practical interest.

Mechanisms

The mechanism can be either intra- or intermolecular.

Intramolecular

These are the reverse of the paths described in oxidative addition. The following examples can be cited (L = phosphine, R = CH_3, CH_2CH_3, $CH_2CH_2CH_3$)

The groups that combine intramolecularly and are eliminated must be cis to each other, either from the beginning or after fast position exchange.

It is noted that the reactions just cited can also be considered as insertions into metal–hydrogen or metal–carbon bonds.

Intermolecular

The following subcategories of intermolecular mechanisms have been observed.

(*a*) *Free radical (intermolecular).*

$$L_mM\text{—}X \longrightarrow L_mM + \dot{X}$$

The free radical produced reacts further with another complex,

$$\dot{X} + Y\text{—}ML_m \longrightarrow XY + ML_m$$

or with the solvent or with some scavenger, and/or it recombines with other radicals, for example,

$$\dot{X} + HR \longrightarrow XH + \dot{R}$$

$$\dot{R} + \dot{R} \longrightarrow RR$$

$$\dot{X} + \dot{R} \longrightarrow XR$$

EXAMPLE I

$$RMn(CO)_4L \longrightarrow \dot{R} + Mn(CO)_4L$$

$$\dot{R} + HMn(CO)_4L \longrightarrow RH + Mn(CO)_4L$$

$$2Mn(CO)_4L \longrightarrow Mn_2(CO)_8L_2$$

where $L = (p\text{-}CH_3OC_6H_4)_3P$, $R = p\text{-}CH_3OC_6H_4CH_2$.   □

EXAMPLE II

$$Ph_2Hg^{II} \longrightarrow 2\dot{P}h + Hg^0$$

$$\dot{P}h + HOCH_2R \longrightarrow PhH + \dot{O}CH_2R$$

$$\dot{P}h + \dot{O}CH_2R \longrightarrow PhH + RCHO$$ □

(b) *Concerted* (*intermolecular*). The following subcases can be distinguished, differing on how the reacting complexes approach each other.

(b1) Outer-sphere:

$$M—X + Y—M \longrightarrow [M \cdots X \cdots Y \cdots M] \longrightarrow XY + 2M$$

For example,

$$^{3-}[(NC)_5Co—H] + [H—Co(CN)_5]^{3-} \longrightarrow H_2 + 2Co(CN)_5^{3-}$$

(b2) Four-center interaction:

$$\begin{matrix} M & M \\ | & | \\ X & Y \end{matrix} + \longrightarrow \begin{bmatrix} M \cdots M \\ \vdots \quad \vdots \\ X \cdots Y \end{bmatrix} \longrightarrow XY + 2M$$

(b3) X (or Y) transfer involving metal–metal interaction:

$$\begin{matrix} M & + & M—Y \\ | & & \\ X & & \end{matrix} \longrightarrow \begin{bmatrix} M \cdots M \cdots Y \\ \vdots \\ X \end{bmatrix} \longrightarrow \begin{bmatrix} M \cdots M \cdots Y \\ \vdots \\ X \end{bmatrix} \longrightarrow M_2 + XY$$

For example, in nonpolar solvents,

$$\begin{matrix} (CO)_5 & & (CO)_5 \\ | & & | \\ Mn & + & Mn—R \\ | & & | \\ H & & H \end{matrix} \xrightarrow{-CO} \begin{matrix} (CO)_5 & (CO)_4 \\ | & | \\ Mn \cdots Mn—R \\ | \\ H \end{matrix} \longrightarrow$$

$$\begin{matrix} & (CO)_4 \\ & | \\ (OC)_5Mn—Mn—R \\ & | \\ & H \end{matrix} \xrightarrow{+CO} RH + Mn_2(CO)_{10}$$

where R = p-CH$_3$OC$_6$H$_4$CH$_2$.

Reductive Disproportionation

The overall reductive elimination reaction 7.10 is also known as reductive disproportionation. It is believed that the first step in this reaction is an intramolecular transfer of hydrogen to the metal (β-*elimination*):

$$\begin{matrix} M^{n+1}—\overset{|}{\underset{|}{C}}— \\ \quad \underset{\cdots H}{\overset{|}{—\underset{|}{C}}—} \end{matrix} \longrightarrow M^{n+1}—H + \overset{}{\underset{}{>}}C=C\overset{}{\underset{}{<}}$$

followed by further reaction of MH with another M^{n+1}—$\overset{|}{C}$—$\overset{|}{C}H$, by the intermolecular mechanisms described above. A prerequisite for such a mechanism is, of course, the availability of a β-hydrogen (relative to the metal).

Transfer of α-hydrogen (α-*elimination*), giving a carbene, has also been reported:

$$M\overset{|}{C}H \longrightarrow M\overset{H}{\underset{C}{\diagup}}$$

Reaction 7.11 can also be characterized as reductive disproportionation and can be thought to proceed by a similar mechanism:

$$M\underset{C-CH}{\overset{C-CH}{\diagup\diagdown}} \longrightarrow M\underset{C-CH}{\overset{H}{\diagup\diagdown}} + {\diagup\overset{C=C}{\diagdown}}$$

except that now the reductive elimination from the alkyl metal hydride takes place intramolecularly.

7.8. REDUCTION OF MULTIPLE BONDS BY METAL IONS OF LOW VALENCE

The commercial importance of the reduction of unsaturated compounds such as olefins, alkynes, carbon monoxide, dinitrogen, carbon dioxide, and carbon disulfide is well known, and there is no need to discuss it further here.

Before starting the discussion of the reduction of multiple bonds, and in order to avoid confusion, let us first impose the limits: Only stoichiometric reduction of carbon–carbon and carbon–oxygen multiple bonds by metal ions undergoing a net one-electron change will be discussed in this section. Catalytic reductions will be considered in Chapter 8.

The reduction of the double bond is an overall two-electron change. The two electrons may end up on the same molecule:

$$2M^n + {\diagup\overset{C=C}{\diagdown}} \xrightarrow{2H^+} 2M^{n+1} + H\overset{|\,|}{\underset{|\,|}{CCH}} \qquad 7.12$$

or a dimer may be formed:

$$2\overset{\cdot}{M^n} \; + \; 2\overset{/}{\underset{/}{C}}{=}\overset{\backslash}{\underset{\backslash}{C}} \xrightarrow{\;2H^+\;} 2M^{n+1} \; + \; H\overset{||||}{\underset{||||}{C C C C}}H \qquad 7.13$$

In the systems considered, the attack by the metal ion proceeds in two successive steps; the probability for simultaneous attack by two metal ions is very low.

First Step

The following two modes of attack have been documented:

$$\begin{array}{cc} C{=}X & C{=}X \\ \uparrow & \nearrow \\ M & M \\ (a) & (b) \end{array}$$

where X = O or C.

In aqueous organometallic chemistry, V^{2+}(aq) (d^3) and Ti^{3+}(aq) (d^1) attack the unsaturated molecule in an (a) manner, Cr^{2+}(aq) (d^4) and Eu^{2+}(aq) (f^7) in a (b) manner.

It is noted that these two modes correspond to the two modes of oxidative addition of alkyl halides (Section 7.6), except that configuration (b') has been depicted as linear, whereas geometry (b) for the metal–olefin interaction is nonlinear.

Structures (a) and (b) can also be considered as *ion radicals* (abbreviated M^{n+1}–òl or simply M-òl). They are not yet alkyl metals. One electron is still missing. They could be regarded as alkyl metals, M^{n+2}—R, only if the oxidation state of the metal is considered to have been changed by two units.

Second Step

The intermediate ion radical formed in the first step can undergo a number of reactions.

(*a*) *Back-reaction:*

$$M^{n+1}\text{—òl} \longrightarrow M^n + \text{ol}$$

The effect is to decrease the overall rate.

(*b*) *Intra- or intermolecular atom or radical transfer* for example,

$$\begin{array}{c} \text{H} \text{-} \text{-} \text{\textbackslash} \\ | \quad \vdots \\ \text{M} \text{—} \dot{\text{o}}\text{l} \end{array} \longrightarrow \quad \text{M} \text{—} \text{R}$$

$$\begin{array}{c} \text{R} \text{-} \text{-} \text{\textbackslash} \\ | \quad \vdots \\ \text{M} \text{—} \dot{\text{o}}\text{l} \end{array} \longrightarrow \quad \text{M} \text{—} \text{R}'$$

where R, R' are alkyls. Here the resulting complexes are coordinatively unsaturated and react readily with other ligands. The reactions can also be regarded as intramolecular radical recombination reactions.

We can also write analogous intermolecular (bimolecular) reactions.

(*c*) *Further attack by another M,* proceeding by way of formation of a transient carbon–carbon or carbon–oxygen bridged species:

$$\text{M}^{n+1}\text{—}\dot{\text{o}}\text{l} + \text{M}^{n} \longrightarrow [\text{M}^{n+1}\dot{\text{o}}\text{l}\text{M}^{n}] \xrightarrow{\text{2H}^{+}} 2\text{M}^{n+1} + \text{olH}_2$$

where olH_2 is the hydrogenated olefin.

In mode (b), the second M should attack X:

$$\underset{\text{M}}{\overset{\text{M}}{\diagdown}}\text{C}\text{—}\dot{\text{X}}\overset{\text{M}}{\diagup}$$

In mode (a), attack by a second M requires a previous slipping of the first M toward C.

(*d*) *Free radical formation.* Since M—$\dot{\text{o}}$l itself is an ion radical, it can be split into an ion and an organic free radical:

$$\text{M}\text{—}\dot{\text{o}}\text{l} \longrightarrow \text{M} + \dot{\text{o}}\text{l}$$

Both species on the right are unstable: $\dot{\text{o}}$l because it is a free radical, and M because it is coordinatively unsaturated. However, in the presence of a strong donor (like a water molecule in aqueous solutions), M can be stabilized, and the reaction is driven to the right.

(e) *Dimerization.*

$$M^{n+1}-\dot{o}l + \dot{o}l-M^{n+1} \longrightarrow 2M^{n+1} + RR$$

where RR is the dimer. Dimerization can also be regarded as an intermolecular free radical recombination.

(f) *Disproportionation.*

$$M^{n+1}-\dot{o}l + \dot{o}l-M^{n+1} \xrightarrow{2H+} 2M^{n+1} + ol + olH_2$$

where olH_2 is again the hydrogenated olefin.

(g) *Rearrangement or breaking of the ligand.*

(h) *Attack by another molecule,* for example, attack by another olefin molecule, by a nucleophile, or by an electrophile. With more than one double bond or with triple bonds, the number of theoretically possible combinations increases.

7.9. INTRAMOLECULAR ELECTRON TRANSFER: MIXED VALENCE COMPOUNDS

The basic mechanism of an inner-sphere electron transfer can be written as follows:

First stage
$$M^n-X + Y-M^m \xrightarrow{-Y^-} M^n-X-M^m$$

Second stage
$$M^n-X-M^m \longrightarrow M^{n+1}-X-M^{m-1} \longrightarrow products$$

7.14

The second stage is an *intramolecular electron transfer* and can be studied separately without the complications of the first by simply starting with a dimeric compound

$$M^n-X-M^m \qquad 7.15$$

which contains the element M in two different oxidation states (mixed valence) or two different elements M and M'.

Three well-known examples of mixed valence compounds are the following:

$$[(H_3N)_5Ru^{II}N \bigcirc - \bigcirc NRu^{III}(NH_3)_5]^{5+}$$

(a)

$$[(H_3N)_5RuN \bigcirc NRu(NH_3)_5]^{5+}$$

(b)

$$[(H_3N)_5RuN \equiv C - C \equiv NRu(NH_3)_5]^{5+}$$

(c)

In complex (a) the interaction between Ru^{II} and Ru^{III} is weak. The orbitals of the two metal ions are largely localized, and we may say that one ruthenium is in oxidation state II and the other in oxidation state III. In contrast, the interaction in complex (c) is strong, and the orbitals are nonlocalized. The extra electron is shared by the two ruthenium ions, which are equivalent; the oxidation state of each one of them may be considered to be 2.5. The case of complex (b) is intermediate.

Mixed valence skeletons can be found in distinct dimers like those just described or in oligomers or polymers or even in crystals as the famous Prussian blue crystals. In all cases, as in the examples just cited, one may recall the general aphorism, "a system is not simply the sum of its parts, but it is this sum *plus* the interaction between them." This interaction may be weak or strong or intermediate, but it is always there; otherwise there is no system.

In mixed valence systems the most *visible* (in a real sense) effect of this interaction is the color. For very weak interactions, the spectrum is nearly the sum of the spectra of the components. For intermediate interactions (type B in the Day and Robin classification), an additional band appears—the so-called *intervalence* band. In strong interactions, the spectrum changes radically.

Other properties are also affected accordingly. In the $[Ru^{II}, Ru^{III}]$ complex (b), for example, the rocking vibration of ammonia is observed at 790 cm^{-1}, whereas in the analogous $[Ru^{II}, Ru^{II}]$ complexes it is observed at 750 cm^{-1} and in $[Ru^{III}, Ru^{III}]$ complexes at 820 cm^{-1}. In complex (c), the two triple bonds are equivalent. If we had distinct Ru^{II} and Ru^{III} ions, these two triple bonds would have different stretching frequencies, which is not the case.

Some Mechanistic Implications

In complex (c) the energy expenditure for transferring the electron from one site to the other is negligible, as in metals. In complex (a) an appreciable amount of energy is required, as in insulators. Complex (b) is intermediate, as in semiconductors.

In a mechanism like that given by equations 7.14, electron transfer via a complex of type (c) would then have a small contribution to the overall energy of activation; via a complex of type (b), larger; and via a complex of type (a), even larger.

It is also instructive to recall that the factors affecting thermal electron transfer (without absorption of light) and photoinduced electron transfer are essentially the same. A correlation is therefore expected between the rate of electron transfer and the position of the intervalence band. But more about this in Section 7.12.

7.10. "TWO-ELECTRON" TRANSFER

Complementary and Noncomplementary Reactions

Oxidation–reduction reactions involving two-electron changes can be *complementary* or *noncomplementary*. In the former, the changes in the oxidation numbers of the oxidant and of the reductant are the same, for example,

$$Sn^{II} + Tl^{III} \longrightarrow Sn^{IV} + Tl^{I}$$

in the latter they differ, for example,

$$2Fe^{II} + Tl^{III} \longrightarrow 2Fe^{III} + Tl^{I}$$

The reduction of double bonds by one-electron reagents (Section 7.8) is also a noncomplementary process.

In cases like these, the question naturally comes to mind whether the transfer of the two electrons takes place in successive one-electron steps or simultaneously in a single two-electron step.

In noncomplementary reactions, simultaneous transfer of two electrons seems unlikely, because it must involve a three-body collision or formation of an unstable intermediate with an unusual oxidation number. In either case the probability is expected to be low.

In complementary reactions, the chances for two-electron transfer are better in spite of the low quantum-mechanical probability. In fact, there are several cases where such a simultaneous transfer seems to be well-documented, for example,

$$Tl^I + {}^*Tl^{III} \longrightarrow Tl^{III} + {}^*Tl^I$$

$$Hg^0 + Tl^{III} \longrightarrow Hg^{II} + Tl^I$$

Two-Electron, Atom, and Positive Ion Transfer

The atom-transfer reaction:

$$RX + Cr^{2+} \longrightarrow \dot{R} + XCr^{III}$$

involves a change of the oxidation number of chromium by one unit. There are similar reactions involving changes by two units, for example,

$$Cl^IO^- + S^{IV}O_3^{2-} \longrightarrow Cl^- + OS^{VI}O_3^{2-}$$

However, it should be noted again that in spite of the fact that we often use terms like "one-electron transfer" or "two-electron transfer," the changes in the real charges around each atom are much smaller.

Tracer experiments have shown that in the first step of the reaction between ClO_3^- and SO_2 there is a net transfer of one oxygen atom from the oxidant to the reductant, which can be understood by postulating the following path:

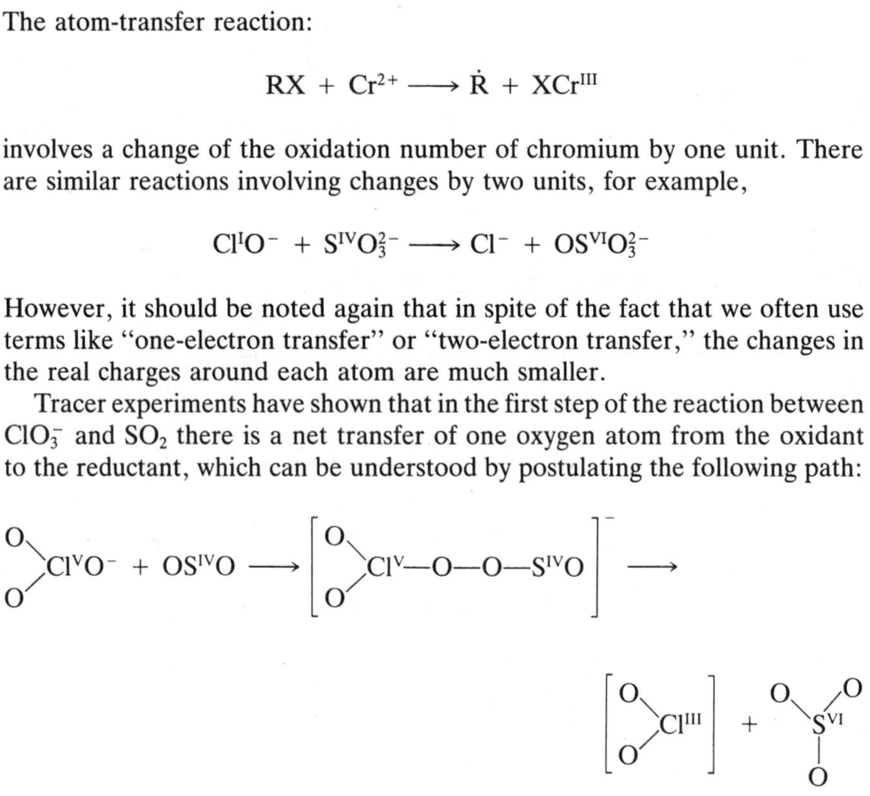

$$\begin{array}{c}O\\ \diagdown\\ \diagup\\ O\end{array}Cl^VO^- + OS^{IV}O \longrightarrow \left[\begin{array}{c}O\\ \diagdown\\ \diagup\\ O\end{array}Cl^V{-}O{-}O{-}S^{IV}O\right]^- \longrightarrow$$

$$\left[\begin{array}{c}O\\ \diagdown\\ \diagup\\ O\end{array}Cl^{III}\right]^- + \begin{array}{c}O\\ \diagdown\ \diagup\ O\\ S^{VI}\\ |\\ O\end{array}$$

Since in assigning formal oxidation numbers the oxygens are counted as oxide ions (O^{2-}), transfer of the oxygen atoms as illustrated above is formally equivalent to a two-electron change.

This simple scheme for oxygen atom transfer is by no means the only way to interpret the results. More complicated schemes may be proposed by implicating water (the solvent) and proton or hydroxide ion transfer steps. In any case, the net effect is an atom transfer and a concomitant "two-electron" change.

An example involving the net transfer of a positive ion is provided by the reaction of $Pt^{II}(NH_3)_4^{2+}$ with trans-$Pt^{IV}(NH_3)_4Cl_2^{2+}$, which is catalyzed by chloride ion and can be thought to occur in the following concerted manner:

$$\text{Cl}^- \dashrightarrow \text{Pt} \dashleftarrow \text{Cl}^+ \quad \text{Pt} \quad \text{Cl}^- \longrightarrow$$

Again, there are other reasonable mechanisms that can be postulated, but in any case the apparent result is the transfer of a positive ion, independently of whether or not this ion is really transferred as such.

7.11. MODELING AND SIMULATION

The need often arises to model or simulate complex biological systems with simpler inorganic systems or to simulate difficult-to-study molecules undergoing complex reactions by others that look and act in a somewhat similar manner. The idea is to bring out and illuminate certain aspects that are buried under the complexity of the real system. There is always, however, the danger of making the wrong analogies and forgetting that this modeling involves simplifications and omissions of parts of the system or some of its interactions that may really be vital to its function. Thus, modeling and simulation are by no means substitutes for a study of the real system, but they are extremely valuable in gaining insight and inspiration.

As an example, consider systems that can be regarded as models for the biologically important reduction of pyruvic acid, $CH_3COCOOH$, but also as simulations of carbon dioxide reduction.

Pyruvic acid plays an important role in the metabolism of carbohydrates. Its reduction to lactic acid, $CH_3CHOHCOOH$, is the last step in the glycolytic path:

$$
\begin{array}{ccc}
\text{COOH} & & \text{COOH} \\
| & & | \\
\text{C}{=}\text{O} + \text{NADH}_2 & \xrightarrow{\text{catalyst}} & \text{HOCH} + \text{NAD} \\
| & & | \\
\text{CH}_3 & & \text{CH}_3
\end{array}
\qquad 7.16
$$

where $NADH_2$ = dihydronicotinamideadeninedinucleotide. This reaction is necessary for the continuation of the metabolism of glucose under oxygen-deficient conditions (e.g., under conditions of intense muscular effort). In some microorganisms, pyruvic acid undergoes decarboxylation and is eventually transformed into alcohol rather than lactic acid. In other cases it is transformed into L-alanine, CH_3CHNH_2COOH, or oxaloacetic acid, $HO_2CCH_2COCOOH$, or acetyl-coenzyme A. Also, there are several other reactions resembling reaction 7.16. Some examples are

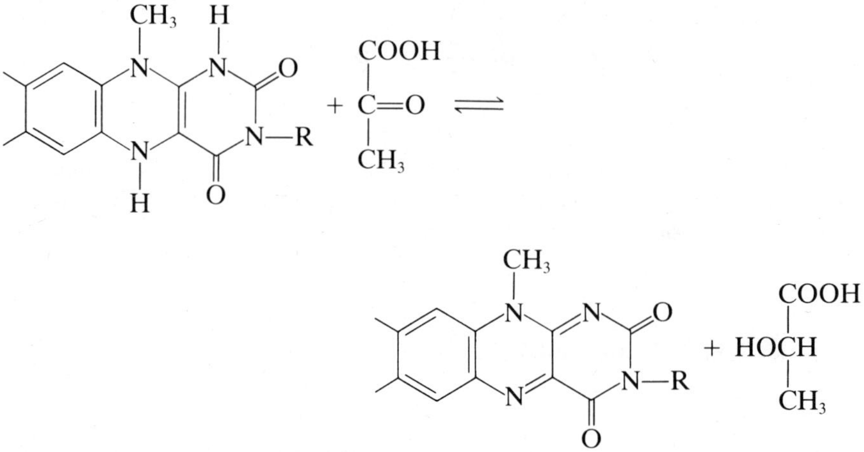

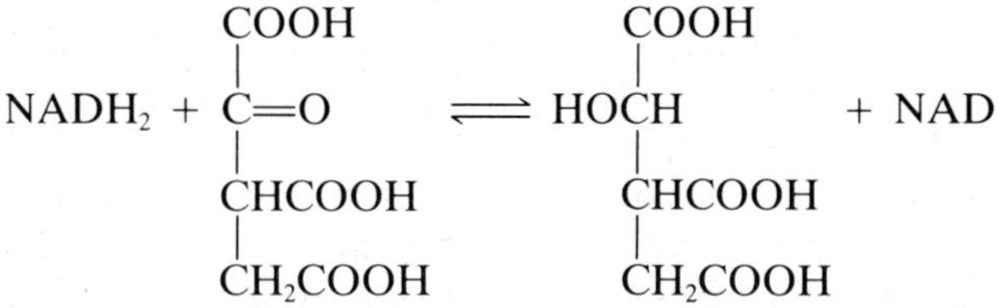

$$NADH_2 + \begin{array}{c} COOH \\ | \\ C=O \\ | \\ CHCOOH \\ | \\ CH_2COOH \end{array} \rightleftharpoons \begin{array}{c} COOH \\ | \\ HOCH \\ | \\ CHCOOH \\ | \\ CH_2COOH \end{array} + NAD$$

Some of the aspects of this complex behavior can be "modeled" by the reduction of pyruvic acid by low-valence metal ions, such as $Cr^{2+}(aq)$, $V^{2+}(aq)$, $Ti^{3+}(aq)$, and $Eu^{2+}(aq)$.

As stated above, these same systems can also be regarded as simulating the final stages in the reduction ("fixation") of carbon dioxide. Along the simulation line:

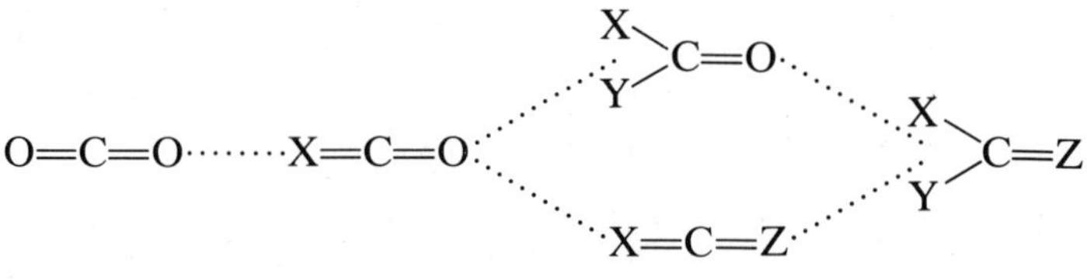

(X, Y, Z ≠ oxygen atom), ketones (pyruvic acid is a keto acid, X = CH_3, Y = COOH) lie closer to the end. After reducing the carbonyl group, the molecule is no longer CO_2-like. Before this reduction, the molecules retain some CO_2 characteristics, and it is reasonable to expect that their study may reveal features in the mechanism that are difficult to discover from the study of carbon dioxide itself. A summary of the information obtained from these model-simulation systems follows.

With chromium(II) (a $t_{2g}^3 e_g^1$ ion, $E_{reduction}^0 = -0.40\ V$), the stoichiometry of the overall reaction is two Cr^{2+} ions for each pyruvic acid molecule, and the final product is lactic acid. In other words, two electrons are transferred from two Cr^{2+} ions to each pyruvic acid molecule.

Pyruvic acid exists in aqueous solutions in two forms: the keto form, $CH_3COCOOH$, and the hydrated form, $CH_3C(OH)_2COOH$. The keto form reacts very fast with Cr^{2+}. The hydro form reacts with Cr^{2+} after being transformed into the keto form. As a result, the rate of the overall reaction depends on this hydro-to-keto transformation, which is first-order in the concentration of H^+ and is independent of the concentration of Cr^{2+}.

With vanadium(II) (a $t_{2g}^3 e_g^0$ ion, $E_{reduction}^0 = -0.255\ V$), the picture changes radically. The stoichiometry becomes 1:1, and the product is dimethyltartaric acid. One electron is transferred to each pyruvic acid molecule. The rate law is

$$\text{Rate} = (k[V^{2+}] + k'[V^{2+}]^2)[P]^2$$

where [P] is the concentration of pyruvic acid. The rate is independent of hydrogen ion concentration. Keto-to-hydro transformation is fast compared to the redox reaction.

With titanium(III) (a $t_{2g}^1 e_g^0$ ion, $E_{reduction}^0 = 0.04\ V$), the stoichiometry is again 1:1, and the product is a polymerized complex of dimethyltartaric acid. The rate is

$$\text{Rate} = \frac{k[Ti^{3+}]^2[P]}{[H^+]^2}$$

Note the inverse dependence on the square of the hydrogen ion concentration.

Finally, for europium(II) (an f^7 ion, with $E_{reduction}^0 = -0.43\ V$), the product is lactic acid, and the rate is given by the expression

$$\text{Rate} = \frac{k[Eu^{2+}][H^+]}{k'[Eu^{2+}] + k''[Eu^{3+}]}[P]$$

The dependence on $[H^+]$ is linear.

Two general conclusions emerging from these observations are that the reaction depends strongly on the nature of the reducing agent and that pyruvic acid becomes inactive when hydrated.

7.12. MAPPING THE COURSE OF A BIMOLECULAR REDOX REACTION

Let us try to trace the course of a redox reaction from the time the reactants are separated by a long distance with no interaction between them, as they diffuse toward each other, interact, and react, until the products are again separated. Some aspects of the interaction itself have already been examined in Chapter 5, but they will be partly repeated here within the context of a redox reaction. Additional concepts will also be introduced, in an attempt to draw a qualitative, but as complete as possible, picture of the physical processes involved.

The Encounter

The factors determining the rate of encounter of the reactants in the case of a redox reaction are not different from those affecting the rate of encounter in any other reaction. In the first place, the reactants must diffuse toward each other. The contribution of diffusion to the energy of activation amounts to a few kilocalories per mole ($\sim$10 kJ/mole). In the limit, a (diffusion-controlled) reaction takes place at each encounter. Usually, however, the donor (D, reductant) and the acceptor (A, oxidant) must stay in contact long enough for the necessary rearrangements to take place. This corresponds to what is called the formation of a *precursor* complex.

The net interaction between D and A at the precursor state must be attractive. An incipient or even a strong bond must form if D and A (or modifications of D and A) are to stay together for a while. This attractive interaction may be coulombic if the charges are opposite, or it may be the result of a positive overlap between the HOMO of the donor and the LUMO of the acceptor. In any case, the precursor corresponds to a minimum in the potential energy curve, which is not reached, however, without activation, an activation that must be clearly distinguished from the activation of the subsequent electron-transfer process itself.

One contribution to the activation energy for the formation of the precursor complex is diffusion, which was already mentioned. Another contribution is from coulombic repulsion when the charges are of the same sign. Still another is the energy required for rearrangement and possible substitution.

The Franck–Condon Principle

The Franck–Condon principle, which is of fundamental importance for understanding electron transfer, may be stated as follows: *The probability for an allowed transition from a vibrational level v of the initial electronic state i to a vibrational level v' of the final electronic state f is proportional to the square of the electronic transition moment $\int \psi_i \bar{\mu} \psi_f d\tau_e$ and to the square of the overlap integral $\int \psi_v \psi_{v'} d\tau_n$ of the vibrational levels.*

This formulation applies to the so-called dipole moment transitions. The angular momentum operator μ equals $\Sigma q_k \bar{r}_k$, where $\bar{r}_k$ is the vector from the origin to the point charge q_k.

The electronic transition is obviously a redistribution of the electron "cloud," which, however, will not occur unless the nuclei are in a position and state of movement (phase) that corresponds not only to the original electronic state

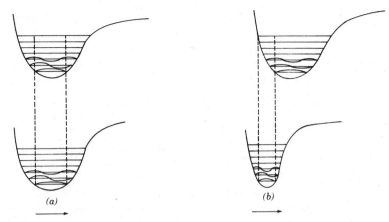

Figure 7.6. Illustration of "common" positions of the nuclei in diatomic molecules. Each of the electronic states corresponds to a manifold of vibrational levels.

but also to the new one. If they are not, there will probably be a fluctuation, but the "cloud" will not be stabilized into a new stationary state. This is essentially the physical significance of the dependence on the overlap integral.

In Figure 7.6a the range of common positions is broader than in Figure 7.6b and the overlap between vibrational levels better, in the sense that there are more values of internuclear distances for which the two electronic states have nonvanishing vibrational functions. Accordingly, in (a) the probability for transition is higher. Moreover, the transitions in (b) are to higher vibrational levels only, which also contributes to making the probability smaller because of a poorer match between the phases of the vibrational motions.

In descriptive language it is said that in (b) the *Franck–Condon barrier* is higher.

Strictly speaking, the formulation of the Franck–Condon principle given above applies only to allowed transitions (a nonvanishing transition moment). However, even for nonallowed transitions it is still required that the nuclei be in positions and phases corresponding to both electronic states—and the broader the range the better.

Optical Electron Transfer

Consider a "symmetric" reaction, in which the products are chemically identical with the reactants, for example, the exchange reaction

$$\overset{e^-}{\overgroup{Fe(OH_2)_6^{2+}}} + Fe(OH_2)_6^{3+} \longrightarrow Fe(OH_2)_6^{3+} + Fe(OH_2)_6^{2+}$$

A schematic presentation of a reaction of this kind is given in Figure 7.7a. The circles in this figure symbolize the transferring electron, which is pre-

sumed to move independently of the other electrons. This assumption constitutes a rather serious sacrifice of rigor, but it serves instructional purposes.

Note that the acceptor and donor are drawn with different initial equilibrium distances. Also note that after the formation of the precursor complex
the electron is transferred without change in the relative positions of the
nuclei (symbolized by full circles and x's) and without a preceding rearrangement. All this is in accordance with the Franck–Condon principle.

Immediately after electron transfer—that is, after the donor has just given
the electron and the acceptor has taken it—the internuclear distances are not
changed; they were "frozen," in the sense that they remain as in the original
species. As a result of this, the successor complex is at first vibrationally
excited. However, vibrational relaxation to the final equilibrium is fast.

Figure 7.7a includes the potential energy diagram corresponding to the

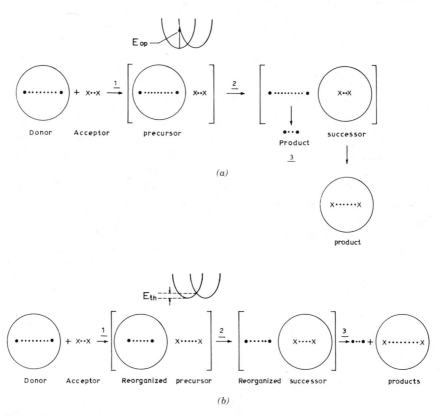

Figure 7.7. Schematic representation of electron-transfer modes satisfying the Franck–Condon
principle: (a) without prior reorganization, (b) after reorganization. **1.** Formation of the precursor. **2.** Precursor-to-successor transformation, i.e., electron transfer. **3.** Decomposition of the
successor, relaxation. The potential energy profiles refer to the electron-transfer step.

precursor-to-successor transformation just described. The energy, E_{op}, required for the "vertical" transition depicted in Figure 7.7a can in principle be supplied thermally, for example, from the translational kinetic energy of the approaching reactant molecules or from large vibrational fluctuations. In fact, early investigators believed that this was indeed the prevailing path in electron transfer reactions. Later, it was recognized that E_{op} is too large for this path to be effective since it is typically a hundred times larger than the average (under ordinary conditions) kinetic energy ($\mathbf{k}T$) of the molecules.

The necessary energy for the kind of transition depicted in Figure 7.7a can also be supplied by a photon, $h\nu = E_{op}$, which is absorbed by the precursor complex, but this requires a long lifetime for this complex, relatively high concentrations, and sufficiently large absorptivities. In short, it requires conditions that are met in the light absorption by mixed valence compounds but not in most of the ordinary bimolecular reactions.

Thermal Electron Transfer

For bimolecular reactions and thermal electron transfer in mixed valence compounds, the appropriate model is illustrated in Figure 7.7b, which differs from Figure 7.7a in that before the transfer of the electron the donor and acceptor nuclei have assumed an intermediate configuration corresponding to the intersection of the two potential energy profiles. This is a kind of compromise, and obviously the activation energy required in now much less, $E_{th} < E_{op}$.

The necessity of a nuclear rearrangement prior to the thermal transfer of the electron is the basic reason electron-transfer reactions are generally slower than proton-transfer reactions in spite of the fact that the electron is much lighter. It is not really the displacement of the electron itself that determines the rate, but the movement and reorganization of groups of heavy nuclei, as dictated by the Franck–Condon principle.

Two points must be emphasized here:

1. The compromise is not reached without a price, because the preexponential factor becomes smaller. However, the gain in lowering the activation enters into the equation exponentially, whereas the losses are in the preexponential factor. As a result, the compromise finally pays off.

2. There is not just one distance over which the electron is transferred (as depicted in Figure 7.7b for instructional purposes), but an entire range of distances, each with its own probability, in accordance with the Franck–Condon principle. Even the "optical" transition (Fig. 7.7a) has a non-vanishing, albeit small, probability of occurring thermally. The "fast paths" will, of course, dominate, but the net result will be determined by the integrated probability over all distances.

Solvent Reorganization

In the above discussion, solvent reorganization was neglected. Yet this reorganization is by no means negligible; it is not only the precursor that should be reorganized and achieve optimum configuration, but also the surrounding solvent. And there is a contribution to the energy of activation originating in this reorganization, which in fact may even be rate-determining.

Tunneling

The configuration at the intersection point (Fig. 7.7*b*) is energetically the most favored, and the system will have to go at least over the barrier, as in any other reaction. However, at low temperatures another path becomes important: quantum-mechanical nuclear tunneling *through* the barrier:

Precursor   Successor

The transmission coefficient for penetrating a barrier is generally smaller if this barrier is high, if the width at the point of crossing is large, and if the mass of the "tunneled" nuclei is large.

7.13. A DONOR–ACCEPTOR MODEL

Let us now have a closer look at what happens after the donor and acceptor have reached an optimum position of the nuclei for electron transfer.

Consider again a self-exchange reaction of the type:

$$M^{II} + M^{III} \longrightarrow M^{III} + M^{II}$$

$$D \qquad A$$

As stated, the formation of the precursor complex is the result of an attractive coulomb interaction (ion pair or ion cluster), a net positive overlap of the frontier orbitals (HOMO, LUMO), or both. In any case, what this means is that the acceptor finds itself within the potential of the donor and vice versa. It is obvious that if there is no interaction, that is, if the donor and acceptor don't "know" about each other, there is no electron transfer either. In the discussion so far this interaction was not considered explicitly, and in the potential energy diagrams of Figure 7.7 it is not indicated. In reality, if the two reactants are close enough to each other, there is mutual polarization and the potential energy curves at the intersection are perturbed (Fig. 7.8).

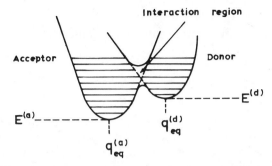

Figure 7.8. Potential energy curves indicating the interaction between the acceptor and the donor. Note that $q_{eq}^{(a)} < q_{eq}^{(d)}$, $E^{(a)} < E^{(d)}$.

For the sake of simplicity, we continue to consider a binuclear donor and a binuclear acceptor. The equilibrium internuclear distance of the acceptor here is assumed to be shorter than that of the donor.

At the intersection region, the donor is on the "repulsive" side of its own potential energy curve in the sense that in relation to its equilibrium position there is more interelectronic repulsion. This is equivalent to an activation toward removing the electron, achieved by way of thermal population of the corresponding vibrational levels.

On the other hand, the acceptor at the intersection is on the "attractive" side of its own potential energy curve. This means there is a decrease in interelectronic repulsion relative to the equilibrium position. The electron cloud is more "rarefied." In this sense, at the intersection the acceptor can better attract the electronic density of the "excited" donor. The efficiency of the electron transfer will depend on how effectively the two systems will stay at the intersection, not simply in contact (interacting) but also at the appropriate vibrational levels and having the right phases. This contact time is clearly shorter than the time (period) for a complete vibration, since only part of the phase is effective, and it may be of the order of magnitude of the time needed for the "hopping" of the electron, that is, the period of the electronic movement (of the order of 10^{-15}–10^{-16} s). If it is shorter, there is simply not enough time for this hopping, and it is more likely that the electron will stay on the donor, although there is still some probability for transfer; the system comes to the intersection many times before it goes to products. Reactions of this kind are called *nonadiabatic*.

If, on the other hand, the time spent by the donor and the acceptor at the configuration of effective interaction is sufficient and there is no other barrier, such as a spin-change barrier, the probability for the transfer approaches unity and the reaction is termed *adiabatic*.

Three factors affecting the time of effective contact will now be considered:

1. The difference between bond lengths of the donor and the acceptor [$q_{eq}^{(d)}$ and $q_{eq}^{(a)}$ in Fig. 7.8].

2. The strength of the interaction.
3. The difference in electronegativity (the difference in the depths of the wells).

Differences in Bond Lengths

If the donor and acceptor have the same equilibrium distances, there is no need for reorganization. Such reactions are expected to be fast even if the other two factors are not favorable. Electron exchange between $Fe(phen)_3^{2+}$ and $Fe(phen)_3^{3+}$ is very fast (the rate constant at 25°C is greater than $3 \times 10^7 \ M^{-1} \ s^{-1}$). This can be attributed mainly to the fact that Fe—N bond lengths in the two oxidation states are very similar (0.1971 and 0.1973 nm, respectively).

A word of caution! In this book we examine solution reactions. Bond lengths are determined accurately only in crystals. If there is no direct x-ray determination in solution and the solid state values are used, there is always some uncertainty. There is, in fact, evidence that the environment can sometimes play an important role. Take as an example the exchange

$$Fe(CN)_6^{3-} + Fe(CN)_6^{4-} \rightleftharpoons Fe(CN)_6^{4-} + Fe(CN)_6^{3-}$$

The rate constant for this reaction depends strongly on the concentration of the cations present. For example, with $0.1 \ M \ K^+$, the rate constant k is $9.6 \times 10^3 \ M^{-1} \ s^{-1}$, but as the concentration of K^+ approaches zero, k tends to the value $25 \ M^{-1} \ s^{-1}$. The Fe^{II}—C and Fe^{III}—C bond lengths are 0.1900 and 0.1926 nm, respectively. This is a large difference, and the exchange is expected to be slow, as indeed it is at the limit of zero ionic strength. A reasonable explanation for the increase in the value of the rate constant at higher positive ion concentrations is that these ions tend to equalize the charges by being placed in bridging positions, and as a result bond distances tend also to equalize. In prussian blue, the mixed valence solid compound $K_4Fe(CN)_6$—$K_3Fe(CN)_6$, the bridging is repeated over the entire crystal. In solution the bridging is only local and probably extends only over small clusters, but the effect is similar: a tendency to make the sites equivalent, which is, of course, in accordance with the electroneutrality principle.

If cations can act in such a manner, it is reasonable to expect that dipoles may have a similar effect, especially strong dipoles like water.

Finally, it must be emphasized that the interaction between the donor and the acceptor may lead to a decrease of the difference between bond distances.

The Strength of Interaction

If equilibrium distances differ, the effective contact time and therefore the probability for electron transfer will depend on the strength of interaction, that is, the strength of the (incipient) bond, and on how this strength varies during the vibration.

Factors determining the strength of interaction include:

Coulomb attraction between opposite charges
HOMO–LUMO interaction
Surrounding solvent

It should be stressed that orbital overlap at the intersection is not the same as the average orbital overlap, that is, the overlap at the equilibrium positions. A similar remark can be made for coulomb attractions and for solvent effects. The electron cloud of the molecule follows the vibrating nuclei, and during a vibration it expands and contracts. The empty orbitals undergo analogous fluctuations. The overlap is different during the different phases of the vibration.

Differences in Electronegativity

It was shown in Chapter 5 that attractive interaction is a necessary condition for the reaction between a donor and an acceptor, but it is not sufficient. There must also be a difference in electronegativity. In "symmetric" reactions, like the ones discussed so far in this section, the electron will go from the donor to the more electronegative acceptor—more electronegative in the sense that the metal is in a higher oxidation state and hence it is more electron-attracting. After the transfer, the situation is reversed. The acceptor becomes donor, and the donor becomes the acceptor. The picture of the potential energy diagram will be the same, except that now the two sides will correspond to different molecules. This interchange of potential energy curves does not correspond to a change in internuclear distances, since at the intersection the internuclear distances of donor and acceptor are equal.

At some instant in this swapping process, the electronegativities are equalized, but the vibration continues, the interaction changes, and the electron stays at the new site; it cannot go back, unless conditions again become favorable during another vibration.

"Nonsymmetric" Reactions

For "nonsymmetric" reactions the potential energy curves will again appear as in Figure 7.8, except that now the difference between the minima will be larger. At the intersection the acceptor will be more "attractive" and/or the donor more "repulsive," and the efficiency for electron transfer will be enhanced. Accordingly, adiabatic paths are now more likely.

Another factor, also related to the relative depths of the minima, is the fractional population of the vibrational levels at the height of the intersection (Fig. 7.9).

(a)   (b)

Figure 7.9. Vibrational levels at the height of the intersection in (*a*) are less populated than the corresponding vibrational levels in (*b*), because the potential energy curve in (*a*) is deeper.

Orbital Combinations and Effective Charge

In one-electron transfer reactions between octahedral transition metal ion complexes, the following combinations can be distinguished:

1. Transfer from an e_g^* antibonding orbital of the donor to an e_g^* antibonding orbital of the acceptor. The driving force in this case is not due to the bonding, nonbonding, or antibonding character of the donor and acceptor orbitals, but rather to the difference in effective charge on the central ion.

Consider a self-exchange reaction between a d^4 and a d^3 metal ion. Schematically,

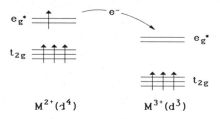

The increased positive charge for M^{3+} stabilizes the occupied t_{2g} orbitals and at the same time increases the splitting between the t_{2g} and e_g^* states. This is reflected, for example, in the values of the ionic radii. For first-row transition elements, the ionic radii of M^{3+} ions are about 20–30% smaller than those of the M^{2+} ions. Also, in going from M^{2+} to M^{3+}, the Δ_0 value (the splitting between t_{2g} and e_g^*) is increased by ~50%. If the combined effect of these two factors still leaves the e_g^* of M^{3+} below that of M^{2+}, there is a driving force for the transfer of the electron—a driving force which in last analysis is due to the difference in the effective charges on the two ions.

2. Transfer from a t_{2g} nonbonding orbital to a t_{2g} nonbonding orbital (t_{2g} is actually slightly antibonding). The situation is similar to case 1. Schematically,

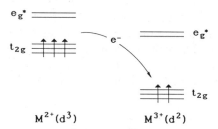

3. Transfer from an antibonding e_g^* orbital to a nonbonding t_{2g} orbital. In this case there is stabilization of the donor, which contributes to the driving force of the reaction. For a transfer from a d^4 ion to a d^2, schematically:

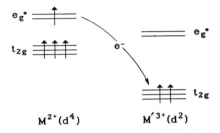

Here M and M' are different, and there is generally a net change in free energy.

4. Transfer from a nonbonding to an antibonding orbital, again between metal ions of different elements:

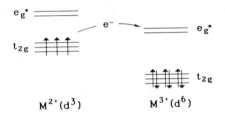

Reactions of this type may be unfavorable.

Inverted Donor–Acceptor Sites

For an effective electron transfer, the minimum of the acceptor potential energy curve must be lower than that of the donor (Fig. 7.8a). However, the equilibrium bond lengths of the acceptor are not necessarily smaller than those of the donor (as in Fig. 7.8). The potential energy for the "inverted" situation will then look as in Figure 7.10.

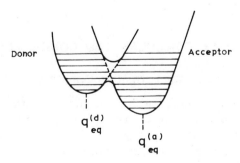

Figure 7.10. The equilibrium distance of the donor, $q_{eq}^{(d)}$, is smaller than the equilibrium distance of the acceptor, $q_{eq}^{(a)}$.

The situation is clearly less favorable than the "normal" one, since the donor is at the intersection on the "attractive" part of its own potential energy curve, and the acceptor is on the "repulsive" side. Transition coefficients are expected to be small.

Summary

How many factors can slow down an electron-transfer reaction? The following can be listed:

1. Small formation constant for the precursor complex, for example, because of strong coulombic repulsion
2. Slow rate of the electron transfer itself, due to large differences in bond lengths, and a high Franck–Condon barrier
3. Short time spent at the right configuration, due to unfavorable interactions
4. Insufficient driving force because the electron has to be transferred from a nonbonding to an antibonding orbital
5. Insufficient driving force because at the intersection the donor is "attractive" and the acceptor "repulsive"
6. Other factors, not examined here explicitly, such as changes in spin multiplicity.

Outer-Sphere, Inner-Sphere, and Intramolecular Electron Transfer

The donor–acceptor model described applies to all three categories, except that what is considered donor or acceptor varies. In outer-sphere reactions, the donor and acceptor are whole molecules; in inner-sphere and intramolecular electron transfer, the donor is the part of the molecule that includes the reductant, and the acceptor the part that includes the oxidant. In inner-sphere electron transfer there is an additional factor, which is absent in outer-sphere and intermolecular transfer reactions: the necessity for substitution, which in fact may even be the slow rate-determining step.

7.14. SOME USEFUL FORMULAS AND CORRELATIONS

Formulas for the Precursor Formation Constant

A number of useful formulas will be given here without derivation. For derivations the student should consult more specialized texts (see the reference list at the end of the chapter).

The rate of a bimolecular electron-transfer reaction can be expressed as the product of the equilibrium constant for the formation of the precursor complex and a first-order rate constant for electron transfer within this complex:

$$k = K_{pr} k_{eltr}$$

The equilibrium constant is often calculated from the formula

$$K_{pr} = \frac{4\pi N r^2 \delta r}{1000} \exp\left(-\frac{w(r)}{RT}\right)$$

where $w(r)$ is the work required to bring the reactants to the separation distance r, δr is a range of distances around r over which the probability of electron transfer is appreciable, and N is Avogadro's number.

Another expression for K_{pr}, derived by Marcus, is

$$K_{pr} = \frac{NhZ\rho}{RT} \exp\left(-\frac{w(d)}{RT}\right)$$

where $w(d)$ is the work required to bring the reactants to a distance, $d = r_A + r_D$, equal to the sum of the radii of the reactants; Z is the collision frequency; and ρ is proportional to the range of distances, δr, over which electron transfer can occur.

A third expression for K_{pr}, derived by Fuoss and Eigen, is the following:

$$K_{pr} = \frac{4\pi N d^3}{3000} \exp\left(-\frac{w(d)}{RT}\right)$$

Electron Transfer within the Precursor

There are also several expressions for k_{eltr}, based on different theories.

Ordinary transition state theory leads to the expression

$$k_{eltr} = \left(\frac{\sum v_i^2 \lambda_i}{\sum \lambda_i}\right)^{1/2} \exp\left(-\frac{\Delta G^{\ddagger}}{RT}\right)$$

which is applicable at ordinary temperatures. In this expression $\Delta G^{\ddagger}$ is the free energy required to achieve the activated complex configuration, and $v_i = (f_i/\mu_i)^{1/2}/2\pi$ is the harmonic frequency, where μ_i is the reduced mass associated with the activation.

A quantum-mechanical treatment yields the following expression at the high-temperature limit:

$$k_{eltr} = \frac{2H_{AB}^2}{h} \left(\frac{\pi^3}{E_\lambda RT}\right)^{1/2} \exp\left(-\frac{(E_\lambda + \Delta E_0)^2}{4E_\lambda RT}\right)$$

where H_{AB} is the coupling matrix element, a measure of the strength of interaction within the precursor complex, E_λ is the reorganization energy having inner- and outer-sphere components, and ΔE_0 is the exothermicity of the reaction.

Linear Free Energy Relations

In the theory worked out by Marcus, the free energy of activation for an outer-sphere electron-transfer reaction is given by

$$\Delta G^\ddagger = w_R + m^2\lambda$$

where

$$m = -\left(\frac{1}{2} + \frac{\Delta G^\circ + w_P - w_R}{2\lambda}\right)$$

w_R is the work required to bring the reactants together, w_p the work required for the departure of the products, ΔG° is the free energy of the reaction, and λ includes the free energy required for the necessary reorganization of the first coordination sphere of the reactants and also of the solvent molecules surrounding the reactant molecules.

If ξ, which is defined by the relation

$$\xi = \frac{\Delta G^\circ + w_P - w_R}{2\lambda}$$

is small ($\xi \ll 1$), we can neglect its square and obtain the approximate expressions:

$$m^2 \simeq \tfrac{1}{4} + \xi$$

and

$$\Delta G^\ddagger \cong w_R + \lambda\left(\frac{1}{4} + \frac{\Delta G^\circ + w_P - w_R}{2\lambda}\right)$$

or

$$\Delta G^\ddagger \cong \frac{\lambda}{4} + \frac{w_P + w_R}{2} + \frac{\Delta G^\circ}{2}$$

In other words, under these conditions, $\Delta G^\ddagger$ *depends linearly on* ΔG°. In a series of reactions between similar molecules, λ and $w_P + w_R$ may not vary appreciably, and then $\Delta G^\ddagger$ depends linearly only on ΔG°.

The Marcus Cross-Correlation

The rate constant k_{12} of the forward reaction

$$D + A \underset{k_{21}}{\overset{k_{12}}{\rightleftharpoons}} D^+ + A^- \qquad \text{(Equilibrium constant } K_{12})$$

is related to the rate constants of the self-exchange reactions:

$$D + D^+ \xrightarrow{k_{11}} D^+ + D$$

$$A + A^- \xrightarrow{k_{22}} A^- + A$$

by the Marcus cross-correlation relation

$$k_{12} = (k_{11}k_{22}K_{12}f)^{1/2}$$

Under specific conditions f can be calculated from the equation

$$\log f = \frac{(\log K_{12})^2}{4 \log(k_{11}k_{22}/Z^2)}$$

where Z is the collision number.

Examples of Calculations

Using the data for self-exchange given in Table 7.2 and the value $Z = 2.5 \times 10^{11} \text{ s}^{-1}$, calculate the rate constant for the reaction

$$Ce^{IV} + Mo(CN)_6^{4-} \xrightarrow{k_{12}} Ce^{III} + Mo(CN)_6^{3-}$$

From Table 7.2, $E_{12}^\circ = 1.44 - 0.80 = 0.64 \ V$. Thus,

$$K_{12} = \exp(nFE^\circ/RT) = \exp(16.92)(2.303)(0.64) = 6.8 \times 10^{10}$$

TABLE 7.2. Rate Constants for Self-Exchange

Couple	Reduction Potential	$k, M^{-1} s^{-1}$
Ce(III)/Ce(IV)	1.44	4.4
$W(CN)_8^{3-}/W(CN)_8^{4-}$	0.54	7×10^4
$Fe(CN)_6^{3-}/Fe(CN)_6^{4-}$	0.68	3×10^2
$Mo(CN)_8^{3-}/Mo(CN)_8^{4-}$	0.80	3×10^4
$IrCl_6^{2-}/IrCl_6^{3-}$	0.93	2×10^5

We can also calculate

$$\log f = \frac{(10.83)^2}{4 \log \left[\dfrac{(3 \times 10^4)(4.4)}{6.25 \times 10^{22}} \right]} = \frac{117.29}{4 \log (2.11 \times 10^{-18})} = -1.66$$

Thus,

$$f = 2.2 \times 10^{-2}$$

and

$$k_{12} = [(4.4)(3 \times 10^4)(6.8 \times 10^{10})(2.2 \times 10^{-2})]^{1/2}$$

$$k_{12} \sim 1.4 \times 10^7 \ M^{-1} \ s^{-1}$$

The experimentally observed value is $1.4 \times 10^7 \ M^{-1} \ s^{-1}$. Other values of calculated cross-reaction rate constants based on the data of Table 7.2 are given in Table 7.3 from Reynolds and Lumry. The student should verify them.

Correlation between Thermal and Optical Electron Transfer

Consider the potential diagram of a mixed valence complex (Fig. 7.11) with weak interaction between the two metal ion centers. First assume that M^{II}, M^{III} are different oxidation states of the same metal ion.

Diversion 1

We have used two kinds of potential energy curves, (1) those of the donor and acceptor separately (M^{II} and M^{III} separately in Fig. 7.10) and (2) those of the combined system ($[M^{II}, M^{III}]$, $[M^{III}, M^{II}]$ in Fig. 7.11).

TABLE 7.3. Cross-Reaction Rate Constants

Reaction	Observed k_{12}, $M^{-1} \ s^{-1}$	Calculated k_{12}, $M^{-1} \ s^{-1}$
$Ce(IV) + W(CN)_8^{4-}$	5×10^8	2.8×10^8
$Ce(IV) + Fe(CN)_6^{4-}$	1.9×10^6	8.7×10^6
$IrCl_6^{2-} + W(CN)_8^{4-}$	6.1×10^7	8.8×10^7
$IrCl_6^{2-} + Fe(CN)_6^{4-}$	5.8×10^5	7.2×10^5
$IrCl_6^{2-} + Mo(CN)_8^{4-}$	1.9×10^6	8.8×10^5
$Mo(CN)_8^{3-} + W(CN)_8^{4-}$	5.0×10^6	4.8×10^6
$Mo(CN)_8^{3-} + Fe(CN)_6^{4-}$	3.0×10^4	2.9×10^4
$Fe(CN)_6^{3-} + W(CN)_8^{4-}$	4.3×10^4	6.3×10^4

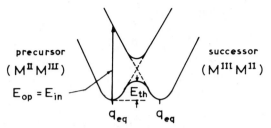

precursor
$(M^{II} M^{III})$

$E_{op} = E_{in}$

successor
$(M^{III} M^{II})$

E_{th}

q_{eq} q_{eq}

Figure 7.11. Schematic presentation of the crossing of the potential energy curves of $[M^{II}, M^{III}]$ (precursor) and $[M^{III}, M^{II}]$ (successor). For a self-exchange reaction, the equilibrium distances of the two are equal.

In the first case, one curve is for M^{II} the other for M^{III}, and the minimum for the former is lower than that for the latter, even for a self-exchange reaction. The equilibrium distances (bond lengths) are generally different.

In the second case, one curve is for the precursor complex, and the other is for the successor complex, that is, the mixed valence combinations $[M^{II}, M^{III}]$ and $[M^{III}, M^{II}]$. For a self-exchange reaction, only the oxidation states of the two partners have been interchanged, but this has no effect on equilibrium properties. At equilibrium, $[M^{II}, M^{III}]$ and $[M^{III}, M^{II}]$ are operationally identical. Interatomic distances in particular are the same. The fact that the curve for the successor is drawn to the right of the other (Fig. 7.11) does not mean that its equilibrium distance is measured along an x axis or that it is larger.

Diversion 2

In our figures, potential energy curves look like parabolas. This is a crude simplification, even for diatomic molecules. For polyatomic molecules or, even worse, for more complex combinations of such molecules (precursor, acceptor) with not just one bond length, not just one minimum, not just one barrier, this simplification is indeed considered arbitrary. Nevertheless, it is still used, but not as a realistic description. Rather, it is used as a visual two-dimensional aid for discussing certain concepts.

Let us return, however, to Figure 7.11. E_{th} in this figure is the barrier for the thermal transfer of the electron, and E_{op} is the energy for the light-induced electron transfer from M^{II} to M^{III} within the precursor.

Hush has shown that for a symmetric case such as that depicted in Figure 7.11, E_{th} and E_{op} are related by the simple equation

$$E_{th} = \cong \tfrac{1}{4} E_{op}$$

Light absorption by the intervalence compound leading to this kind of electron transfer corresponds to the *intervalence band,* a strong band with a maximum at a frequency $(\nu_{iv})_{max}$ given by

$$E_{op} = h(\nu_{iv})_{max}$$

The bandwidth at half-intensity, $(\Delta\nu_{iv})_{1/2}$, depends on $(\nu_{iv})_{max}$. At high temperatures ($h\nu \ll kT$), this dependence is simple:

$$(\Delta\nu_{iv})_{1/2} = 2310(\nu_{iv})_{max} \ cm^{-1}$$

What the relation between E_{th} and E_{op} means is that spectroscopic data can be correlated with kinetic data—that E_{op} can be determined by measuring the rate of the thermal reaction or E_{th} by taking the spectrum.

Figure 7.11 refers only to the intervalence compound itself and implies that only the barrier within this compound contributes to E_{op}. In reality, for solution reactions there is also a solvent barrier, and E_{op} should rather be expressed as a sum of two terms:

$$E_{op} = E_{in} + E_{out}$$

where E_{in} refers to the intervalence compound itself and E_{out} to the solvent.

A *nonsymmetric* mixed valence compound of the [II, III] class can have the following forms:

$$L_xM^{II}BM'^{III}L_y \qquad M \neq M'$$

1

or

$$L_xM^{II}BM^{III}L'_y, \qquad L \neq L'$$

2

with B being the bridging ligand. Upon electron transfer, **1** becomes **3**:

$$L_xM^{II}BM'^{III}L_y \underset{Thermal}{\overset{h\nu}{\rightleftharpoons}} L_xM^{III}BM'^{II}L_y$$

1   **3**

An analogous reaction can be written for **2**.

Suppose that **1** is considerably more stable than **3**. Then, at equilibrium there is little **3**, and there is no measurable spontaneous thermal reaction from left to right. The reaction can be driven to the right only by supplying energy externally, for example, as light. The reverse reaction is spontaneous.

A "profile" for the potential energy diagram for this reaction is given in Figure 7.12. The deeper well in Figure 7.12 corresponds to the most stable

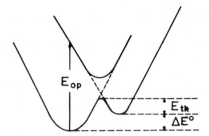

Figure 7.12. Potential energy "profile" for an electron transfer within a nonsymmetric mixed valence compound.

form **1**. It is obvious that the exothermicity $\Delta E°$ must be added to E_{th}, in order to obtain E_{op}:

$$E_{op} \simeq 4E_{th} + \Delta E°$$

The bandwidth in the nonsymmetric case is given by the relation

$$(\Delta \nu_{iv})_{1/2} = [(\nu_{iv})_{max} - 2310\Delta E°]^2 \text{ cm}^{-1}$$

GENERAL REFERENCES

G. C. Allen and N. S. Hush, Intervalence-Transfer Absorption. Part 1. Qualitative Evidence for Intervalence-Transfer Absorption in Inorganic Systems in Solution and in the Solid State, *Progr. Inorg. Chem.*, **8**, 357 (1967).

D. Benson, *Mechanisms of Inorganic Reactions in Solution: An Introduction,* McGraw-Hill, London, 1968.

R. D. Cannon, *Electron Transfer Reactions,* Butterworths, London, 1980.

J. P. Collman and L. S. Hegedus, *Principles and Application of Organotransition Metal Chemistry,* University Science Books, California, 1980.

J. O. Edwards, Ed., Inorganic Reaction Mechanisms, *Progr. Inorg. Chem.* **13**, 1970.

J. Halpern, Formation of Carbon–Hydrogen Bonds by Reductive Elimination, *Accts. Chem. Res.* **15**, 332 (1982).

J. Halpern, Binuclear Oxidative Addition–Reductive Elimination Reactions, *Inorg. Chim. Acta* **62**, 31 (1982).

R. Hoffmann, M. M.-L. Chen, and D. L. Thorne, Qualitative Discussion of Alternative Co-ordination Modes of Diatomic Ligands in Transition Metal Complexes, *Inorg. Chem.* **16**, 503 (1977).

N. S. Hush, Intervalence-Transfer Absorption. Part 2. Theoretical Considerations and Spectroscopic Data, *Progr. Inorg. Chem.* **8**, 391 (1967).

D. L. Kepert, Isopolytungstates, *Progr. Inorg. Chem.* **4**, 199 (1962).

J. K. Kochi, *Organometallic Mechanisms and Catalysis,* Academic, New York, 1978.

W. L. Reynolds and R. W. Lumry, *Mechanisms of Electron Transfer,* Ronald Press, New York, 1966.

M. B. Robin and P. Day, Mixed Valence Chemistry—A Survey and Classification, *Adv. Inorg. Chem. Radiochem.* **10**, 247 (1967).

K. Saito and Y. Sasaki, Substitution Reactions of Oxo-metal Complexes, *Adv. Inorg. Bioinorg. Mech.* **1**, 179 (1982).

A. G. Sykes, *Kinetics of Inorganic Reactions*, Pergamon, London, 1966.

H. Taube, Observations on Atom-Transfer Reactions, D. B. Rorabacher and J. F. Endicott, eds., in *Mechanistic Aspects of Inorganic Reactions*, ACS Symp. Ser. Vol. 198, ACS, Washington, DC, 1982.

H. Taube, *Electron Transfer Reactions of Complex Ions in Solution*, Academic, New York, 1970.

K. H. Tytko and O. Glemser, Isopolymolybdates and Isopolytungstates, *Adv. Inorg. Chem. Radiochem.* **19**, 239 (1976).

PROBLEMS

1. If in the reaction

$$NO_2^- + HOCl \xrightarrow{\text{H}_2\text{O}} ONO_2^- + HCl$$

the hypochlorous acid is enriched in ^{18}O, then $^{18}ONO_2^-$ is obtained. Propose a mechanism consistent with this observation.

2. The rate law for the reaction

$$I^- + OCl^- \longrightarrow IO^- + Cl^-$$

is given by the equation

$$\frac{d[IO^-]}{dt} = \frac{k[I^-][OCl^-]}{[OH^-]}$$

Propose a mechanism. Can the proposed mechanism be tested by isotopic labeling?

3. It has been shown by labeling with radioactive chlorine that chlorine exchange between ClO_2 and $HClO_3$ in aqueous solutions is slow. What conclusions can be drawn on the basis of this observation about the rate by which the following equilibrium is established?

$$2ClO_2 + H_2O \rightleftharpoons HClO_2 + H^+ + ClO_3^-$$

Is this conclusion contradicted by the fact that chlorine exchange between ClO_2 and $HClO_2$ is fast? Explain.

4. Phosphinic acid, HPH_2O_2, is slowly oxidized by diiodine:

$$HPH_2O_2 + I_2 + H_2O \longrightarrow H_2PHO_3 + 2H^+ + 2I^-$$

The rate law is

$$-\frac{d[HPH_2O_2]}{dt} = k[HPH_2O_2]$$

Give an appropriate mechanistic interpretation.

5. Electron exchange between $VO^{2+}(aq)$ and $VO_2^+(aq)$ has been studied with the NMR peak-broadening technique and found to obey the rate law

$$\text{Rate} = k[VO^{2+}][VO_2^+]^2$$

Propose a mechanism.

6. The oxidation of iodide ion by chromate ion in dilute acid solution,

$$2HCrO_4^- + 6I^- + 14H^+ \longrightarrow 2Cr^{3+} + 3I_2 + 8H_2O$$

and the oxidation by Fe^{3+}

$$2Fe^{3+} + 2I^- \longrightarrow 2Fe^{2+} + I_2$$

are slow reactions. Under similar conditions the reaction

$$HCrO_4^- + 3Fe^{2+} + 7H^+ \longrightarrow Cr^{3+} + 3Fe^{3+} + 4H_2O$$

is fast.

If $HCrO_4^-$, Fe^{2+}, and I^- are mixed, iodide ion is oxidized rapidly. Propose an explanation, and suggest a detailed mechanism.

7. The reaction

$$U^{4+} + 2PuO_2^{2+} + 2H_2O \longrightarrow UO_2^{2+} + 2PuO_2^+ + 4H^+$$

proceeds with a rate given by the expression

$$-\frac{d[U^{4+}]}{dt} = \frac{k[U^{4+}][PuO_2^{2+}]}{k' + [H^+]}$$

Propose a mechanism.

8. The rate constant for electron exchange between $V^{2+}(aq)$ and $V^{3+}(aq)$ is observed to depend on hydrogen ion concentration:

$$k_{obs} = a + b[H^+]$$

Propose a mechanism. Express a and b in terms of the rate constants of the proposed mechanism. *Hint*: $V^{3+}(aq)$ hydrolyzes more than $V^{2+}(aq)$.

9. Anhydrous $CrCl_3$ dissolves slowly in water. However, in the absence of air, this process can be accelerated by adding a small amount of Cr^{II}. Give an interpretation. What is the original geometry around Cr^{III}? What is the geometry after the chloride has been dissolved?

10. Electron exchange between $Cr^{2+}(aq)$ and $Cr^{3+}(aq)$ is accelerated by various ligands, in the following order:

$$EDTA > P_2O_7^{4-} > citrate \simeq PO_4^{3-} > F^- > tartrate > SCN^- > SO_4^{2-}$$

In every case the ligand is found to be attached to Cr^{III}. Propose a mechanism, and try to (qualitatively) explain the observed order.

11. The oxidation of Cr^{III} by Ce^{IV} follows the rate law

$$\frac{d[Ce^{IV}]}{dt} = \frac{k[Ce^{IV}]^2[Cr^{III}]}{[Ce^{III}]}$$

Propose a mechanism, and show that this mechanism is consistent with the empirical rate law.

12. The rate law for the reaction

$$Hg_2^{2+} + Tl^{3+} \longrightarrow 2Hg^{2+} + Tl^+$$

is

$$-\frac{d[Hg_2^{2+}]}{dt} = \frac{k[Hg_2^{2+}][Tl^{3+}]}{[Hg^{2+}]}$$

Propose a mechanism consistent with this rate law.

13. The following data have been reported in the literature for the electron (self) exchange reactions between Fe^{2+} and Fe^{3+} and between Cr^{2+} and Cr^{3+}.

Couple	$E°$, V	$\Delta H^{\ddagger}$, kJ mol^{-1}	$\Delta S^{\ddagger}$, J mol^{-1} deg^{-1}
$Fe^{3+} - Fe^{2+}$	$+0.771$	39	-104
$Cr^{3+} - Cr^{2+}$	-0.41	88	-33

Utilize the Marcus theory to determine the forward and reverse rate of the cross reaction at 25°C:

$$Fe^{3+} + Cr^{2+} \rightleftharpoons Fe^{2+} + Cr^{3+}$$

Calculate $\Delta G^{\ddagger}$ for this reaction.

14. Postulate a reasonable mechanism involving water and proton or hydroxide ion transfer steps, and explaining the quantitative transfer of an oxygen atom from ClO_3^- to SO_2.

15. The overall redox reaction

$$M^{II} + L \longrightarrow M^{III} + L^-$$

is studied under pseudo-first-order conditions using a large excess of L. Under these conditions M^{II} is first transformed quickly and quantitatively into a complex $M^{II}L_3$ (charges are omitted), which then reacts with L according to the equation:

$$M^{II}L_3 + L \longrightarrow M^{III}L_3 + L^-$$

with a slower, measurable rate. Predict the rate law, expressing it in terms of the initially added total concentrations of M^{II} and L.

16. In the reaction of Cr^{2+} with complexes A, B, and C, Co^{III} is reduced to Co^{II}. The reaction of complexes A and B involves formation of an inner-sphere activated complex, that of C an outer-sphere complex. What conclusion can be drawn regarding the site of attack by Cr^{2+}? Does Cr^{2+} attack the carboxyl adjacent to Co^{III} or the remote one?

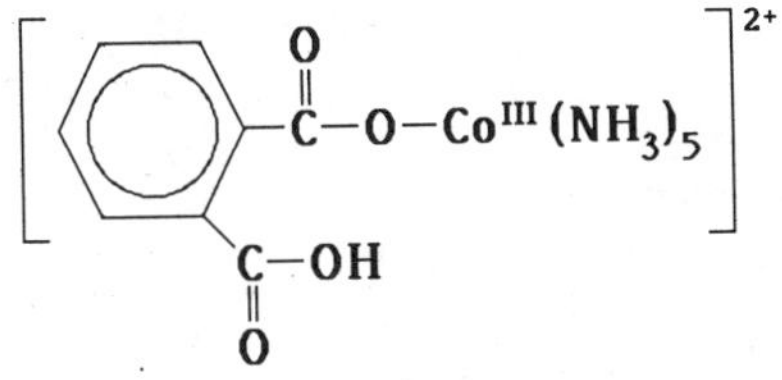

(A)

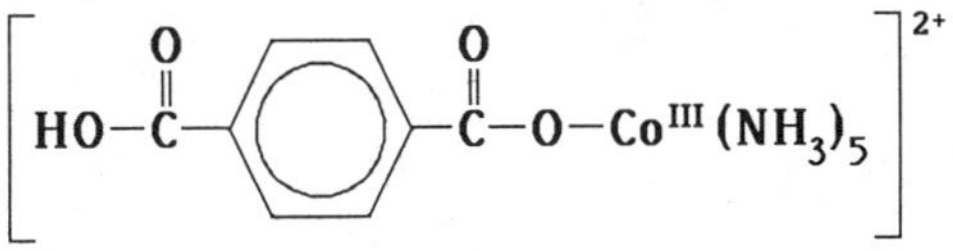

(B)

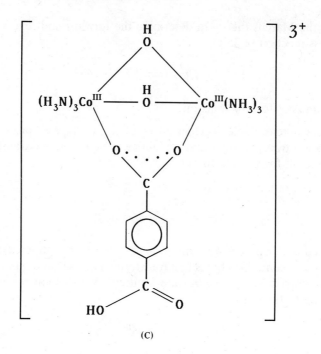

(C)

8

CATALYSIS

Several catalytic processes have already been mentioned. Here catalysis is examined more systematically. Since the objective is to restrict this book to solution chemistry, only homogeneous catalysis is discussed. "Heterogenized" homogeneous catalysts will also be discussed, but heterogeneous catalysis on metal and other surfaces will be mentioned only occasionally.

8.1. GENERAL CONSIDERATIONS

The Catalyst

The minimum requirements for a substance to be characterized as a catalyst for a given reaction are

1. It must be involved in the reaction and must increase the rate by providing a new path with a lower activation energy. It is therefore necessary to have an "encounter" of the catalyst with at least one of the substrate molecules.
2. The catalyst itself returns to its original form. With respect to the catalyst, the process is *cyclic*. The substrate eventually reacts and changes.

Each of these two characteristics represents *at least* one elementary reaction. Hence, the mechanism of a catalytic reaction consists of at least two such reactions. In fact, the mechanism often contains more than two reactions, because either one of these two processes (encounter and regeneration) may be complex.
Schematically:

$$C + S \longrightarrow CS \qquad\qquad 8.1$$

$$CS \longrightarrow C + P \qquad\qquad 8.2$$

where C is the catalyst, S the substrate, and P the products. The energetics are summarized in Figure 8.1.

The combination of reactions 8.1 and 8.2 is generally faster than the uncatalyzed reaction 8.3.

$$S \longrightarrow P \qquad\qquad 8.3$$

It is clear that the catalyst increases the rate of the reaction, but

It does not change the position of equilibrium between substrate and products.

Either it does not appear in the overall (stoichiometric) chemical equation, or it appears on both sides of the equation.

Homogeneous and Heterogeneous Catalysis

If the catalytic reaction takes place in only one phase, the catalysis is said to be *homogeneous*; if it takes place on an interface, it is *heterogeneous*. For the catalytic reaction to be considered homogeneous, it is not necessary that all components be in one phase; only the catalytic reaction itself must take place homogeneously. For example, in the decomposition of aqueous hydrogen peroxide, which is catalyzed by some metal ions, the reactant and the catalyst are in the aqueous phase, but one of the products is gaseous dioxygen. Descriptively, we may say that homogeneous catalysis is a three-dimensional process and that heterogeneous catalysis is two-dimensional.

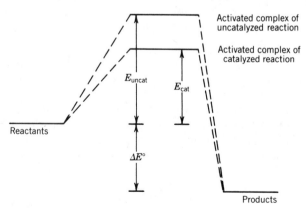

Figure 8.1. The activated complex of a critical step in the catalytic reaction corresponds to smaller activation energy (E_{cat}) compared to the uncatalyzed reaction (E_{uncat}).

Promoters, Inhibitors, Poisons, and Precursors

Promoters are substances that increase the effectiveness of the catalyst. *Inhibitors* inactivate the catalyst, reactants, or intermediates, and in effect they slow down the reaction. *Poisons* combine with the catalyst and inactivate it also; in a sense they are a special class of inhibitors.

The *active* form of the catalyst, that is, the form (or forms) actually participating in the catalytic cycle, is often different from the form that was added initially; the initial form is called the *precursor*. In hydroformylation, for example, one of the active forms is a carbonyl hydride of cobalt. The precursor, however, can be some cobalt(II) salt, such as $CoCl_2$, which reacts with the substrates (carbon monoxide and dihydrogen) to give the active form.

Catalysis and Excitation

The increased rate of a catalytic cycle (e.g., reactions 8.1, 8.2) compared to the noncatalytic reaction 8.3 is related to a decrease in the activation energy. Alternatively, it can be regarded as the result of some kind of excitation (activation) of the substrate, caused by the catalyst. The catalyst "prepares" the reactants to react faster and in so doing is modified itself, but it is then regenerated.

If a catalyst C removes electron density from a bonding region of a substrate molecule or increases the electron density at an antibonding region, or both, it is essentially energizing and weakening the bond. Schematically,

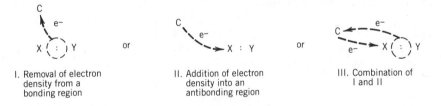

I. Removal of electron density from a bonding region

or

II. Addition of electron density into an antibonding region

or

III. Combination of I and II

Since the bonding region corresponds to lower energy and the antibonding region to higher energy, any of these processes is conceptually equivalent to excitation, and the excited substrate is, of course, expected to react faster. If, in addition, at the end of this reaction C is not consumed, the reaction is catalytic.

Ways of Achieving Activation

There are essentially two general ways to achieve the activation just described:

1. With acids and bases
2. With redox couples

(a) *Acid-base catalysis.* A proton attached to a molecule can remove electron density from a bonding region; a base can add electron density into an antibonding region.

In aqueous solutions, catalysis by $H^+(aq)$ and $OH^-(aq)$ is called *specific acid–base catalysis.* Catalysis by other Lewis acids and bases is called *general acid–base catalysis.* In nonaqueous solutions, specific catalysis is caused by the conjugate acid or base of the solvent.

In catalysis by an acid, a base must also be present (and vice versa) to remove the acid attached to the substrate and to complete the catalytic cycle.

The rate constant of a reaction catalyzed by $H^+(aq)$ and $OH^-(aq)$ can be written in the general form

$$k = k_0 + k_1[H^+] + k_2[OH^-] = k_0 + k_1[H^+] + \frac{k_2 K_w}{[H^+]}$$

Metal ions (complexes) also act as (Lewis) acids or bases (back-bonding). In fact, transition metal ions are generally more versatile than the proton or the hydroxide ion, because of their metal d orbitals and electrons and the variety of ligands to which they can be bound.

A case of activation by metal ions was given in Chapter 6; a transition metal ion attracts electrons into one of its d orbitals and simultaneously gives back electrons from another d orbital into the antibonding orbital of the substrate. This synergic action, in the last analysis, amounts to a transfer of electron density from an occupied bonding orbital to an unoccupied antibonding orbital—it is essentially an excitation.

(b) *Catalysis by redox couples.* If the electrons are not simply displaced, but are neatly transferred, in a way involving change in a formal oxidation number of the catalyst, we may talk about catalysis by a redox couple.

EXAMPLE

Catalysis of the decomposition of hydrogen peroxide by the Br_2/Br^- couple:

$$Br_2 + H_2O_2 \longrightarrow 2Br^- + 2H^+ + O_2$$

$$\underline{2Br^- + H_2O_2 + 2H^+ \longrightarrow Br_2 + 2H_2O}$$

$$2H_2O_2 \longrightarrow 2H_2O + O_2$$

Some transition metal ion couples can also have a similar effect. ☐

Two general categories of redox reactions are particularly important in catalysis: *oxidative addition* and *reductive elimination* (see Sections 7.6 and 7.7).

The 16- or 18-Electron Rule

A very useful rule for identifying intermediates in catalytic reactions can be stated as follows: *Diamagnetic organometallic or analogous complexes with 16 or 18 electrons around the central atom have increased stability and are the likely intermediates or products.*

The 18 electrons correspond to the configuration of the next noble gas. The loss of the two *s* electrons results in the 16-electron configuration. The rule applies to the diamagnetic complexes of the transition metal families IV to VIII, with ligands such as CO, N_2, CN^-, RCN, PR_3, $P(OR)_3$, olefins, acetylene derivatives, acetyls, SiR_3^-, R^-, H^-, allyl, cyclopentadienyl, and aryls.

EXAMPLES

$$IrCl(CO)(PPh_3)_2 \xrightarrow{+BF_3} IrCl(CO)(PPh_3)_2BF_3$$

(16)   (16)

$$Co(CN)_5(OH_2)^{2-} \xrightarrow{-H_2O} Co(CN)_5^{2-} \xrightarrow{+X^-} Co(CN)_5X^{3-}$$

(18)   (16)   (18)

$$HIr(CO)(PPh_3)_3 \xrightarrow{-PPh_3} HIr(CO)(PPh_3)_2 \xrightarrow{+HSi(OEt)_3}$$

(18)   (16)

$$H_2Ir[Si(OEt)_3](CO)(PPh_3)_2$$

(18)

Reactions of the Energized Substrate

The reactions the energized substrate will undergo may in principle be no different than those of the nonenergized species. What has changed is the facility by which they proceed. Breaking the X—Y bond in processes I, II, and III is easier than breaking this bond in free X—Y. Also, intra- or inter-molecular nucleophilic attack on Y or electrophilic attack on X in processes I and II may be easier than in free X—Y.

Focusing on the Catalyst

So far the discussion has been centered mainly on the substrate: how it is energized and how it reacts further. Let us now focus on the catalyst.

It is self-evident, but often overlooked, that what happens to the catalyst and what happens to the substrate are interrelated but drastically different. From the minimum requirements of a catalytic cycle it is obvious that the *active catalyst must have at least two forms:*

The "free" form (without substrate)
The "bound" form (bound to substrate)

During the reaction the catalyst changes back and forth between these forms.

For a proton the two forms are the proton in the bulk of the solution and the proton bound to a basic site of a substrate. For a metal ion complex these two forms may be two different oxidation states or two different coordination numbers. Often, one of these states is considerably less stable than the other, in which case we may talk about *unstable oxidation states* or *coordinatively unsaturated intermediates* or *coordinatively unstable adducts* or *outer-sphere adducts*. In any case the catalytic function of a metal ion complex is intimately related to its redox and substitution behavior.

Thus, the active catalyst is not *one* molecule, *it is at least a pair of molecules.* In practice it is often even more than that—it is *n* different molecular species with $n \geq 2$.

Simple Cycles, Catalytic Cycles

The basic catalytic mechanism (equations 8.1 and 8.2) can also be presented in the form of a cycle (Fig. 8.2). The basic rules for constructing such a cycle are

Keep the catalyst and its various metamorphoses (C, CS) on the cycle
Feed in the substrate (S, reactants)
Get out products (P)

Most cycles are much more complex than the one illustrated, but for the purpose of our discussion, this *basic* or *elementary cycle* suffices. To describe it again with words: The substrate is fed in, and an intermediate is formed, which then gives back products and the original form of the catalyst.

In a similar way, the historic Solvay process for the manufacturing of soda (Na_2CO_3) is shown in Figure 8.3. Ammonia and its metamorphosis ammonium

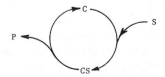

Figure 8.2. The elementary catalytic cycle.

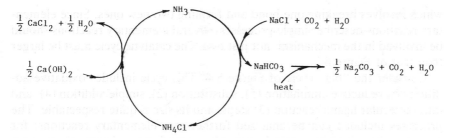

Figure 8.3. The Solvay cycle.

chloride are again parts of the cycle, and various substances are fed in and gotten out. What is the difference, then, between the cycles in Figures 8.2 and 8.3?

The main feature that qualitatively differentiates these two cycles is heat, which must be supplied at some stage in the Solvay process but not in the catalytic process. The energy used in the catalytic process comes from the reactants themselves, where it is stored in the form of chemical energy. The catalytic process is thermodynamically spontaneous. In the Solvay cycle the energy must be supplied externally.

The difference does not seem to be particularly serious, but traditionally processes like the Solvay process are not considered catalytic, and they will not be examined here.

Mystical Connotations

The "philosophers' stone" of the alchemists was a mystical substance, which in small quantities was said to act as a ferment, a yeast, to transform large amounts of base metals to gold or silver. In solution, this same substance was said to become the "elixir" of life, restoring youth and providing everlasting life.

The concept of the catalyst has many similarities. Chemical transformation of base metals to gold is not possible, but many other chemical transmutations are, including transmutations in the body, catalyzed by enzymes. Catalysis seems to have inherited some mystery and some aura from alchemy.

8.2. BIG CYCLES, SMALL CYCLES

The elementary catalytic cycle of Figure 8.2 consists of elementary reactions. If the number of elementary reactions is larger, the cycle is bigger and there are more metamorphoses of the catalyst.

Consider the following overall catalytic reaction:

$$X\!-\!Y + S \xrightarrow{\text{Catalyst}} X\!-\!S\!-\!Y \qquad\qquad 8.4$$

which involves breaking one bond and forming two new ones. Since elementary reactions describe simple processes, several elementary reactions should be involved in the mechanism, not just two. The catalytic cycle must be larger than that of Figure 8.2.

Consider the "big" cycle of Figure 8.4. This cycle includes oxidative addition (1), reductive elimination (5), substitution (2), simple addition (4), and intramolecular ligand reaction (3) steps, and its size is quite respectable. The processes included can be analyzed further into elementary reactions; for example, the substitution step is certainly a complex reaction itself—it is not elementary.

If the whole mechanism were included, the cycle would have become overly big. "Condensation" is a common practice in presenting complex catalytic processes. Side reactions are usually omitted.

What we are generally trying to keep in the presentation are the transmutations of the catalyst (five in Fig. 8.4).

The Kinetics of a Small Cycle

Consider a mechanism consisting of reactions 8.1 and 8.2 and the corresponding back-reaction for equation 8.1:

$$C + S \underset{k_{-1}}{\overset{k_1}{\rightleftharpoons}} CS$$

$$CS \overset{k_2}{\longrightarrow} P + C$$

This is a common catalytic mechanism, especially for enzymatic reactions.

To simplify the treatment, all stoichiometric coefficients are taken equal

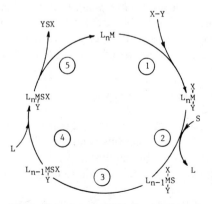

Figure 8.4. A typical "big" catalytic cycle.

to 1. The rate of product formation is given by the equation

$$\frac{d[P]}{dt} = k_2[CS]$$ 8.5

The concentration of the intermediate, [CS], can be obtained by applying the steady-state approximation:

$$k_1[C][S] - k_{-1}[CS] - k_2[CS] = 0$$ 8.6

It is assumed, of course, that this approximation is justified ($k_2 \gg k_1$).

To express the rate in terms of the initial concentrations, the stoichiometric relations are used:

$$[S]_0 = [S] + [CS] + [P]$$ 8.7

$$[C]_0 = [C] + [CS]$$ 8.8

Initially $[P] \simeq O$, and equation 8.7 becomes

$$[S]_0 \simeq [S] + [CS]$$ 8.9

The concentration of the catalyst is usually small compared to the concentration of the substrate. Therefore, even if most of the catalyst has been transformed into CS, the following inequality holds:

$$[CS] \ll [S]$$ 8.10

and reaction 8.9 is simplified further to

$$[S]_0 \simeq [S]$$ 8.11

Substituting reaction 8.8 and 8.11 into 8.6, we obtain

$$k_1[S]_0 \left([C]_0 - [CS] \right) - k_{-1}[CS] - k_2[CS] = 0$$

and solving for [CS]:

$$[CS] = \frac{k_1[S]_0[C]_0}{k_{-1} + k_2 + k_1[S]_0}$$

Equation 8.5 becomes

$$\left(\frac{d[P]}{dt} \right)_{initial} = \frac{k_1k_2[S]_0[C]_0}{k_{-1} + k_2 + k_1[S]_0}$$

or

$$\left(\frac{d[P]}{dt}\right)_{\text{initial}} = \frac{k_2[S]_0[C]_0}{K_m + [S]_0} \qquad 8.12$$

where $K_m = (k_{-1} + k_2)/k_1$.

For enzymatic reactions, K_m is called the *Michaelis–Menten constant*.

Comments on Equation 8.12

First, we summarize the assumptions and simplifications:

1. The time after mixing is relatively short, and equation 8.12 holds for the initial rate.
2. It is assumed that the intermediate CS is in a steady state.
3. The concentration of the catalyst is small compared to the concentration of the substrate.
4. The mechanism is as given (reactions 8.1, 8.2, and the corresponding back-reaction).

It is also noted that even though the reverse of the first reaction has been included, the intermediate CS is generally not in equilibrium with C and S. However, if $k_{-1} \gg k_2$, the concentration [CS] approaches the equilibrium concentration, in which case

$$K_m = \frac{k_{-1} + k_2}{k_1} \simeq \frac{k_{-1}}{k_1} = \frac{1}{K}$$

where K is the equilibrium constant. Moreover, since it has been assumed also that $k_2 \gg k_1$ (the condition for the steady-state approximation), it follows that

$$k_{-1} \gg k_2 \gg k_1$$

and

$$K = \frac{k_1}{k_{-1}} \ll 1$$

In this case equation 8.12 becomes

$$\left(\frac{d[P]}{dt}\right)_{\text{initial}} \simeq \frac{k_2 K[S]_0[C]_0}{1 + K[S]_0} \qquad 8.13$$

If, in addition, $K[S]_0 \ll 1$,

$$\left(\frac{d[P]}{dt}\right)_{\text{initial}} \simeq k_2 K[S]_0[C]_0$$

that is, the reaction is initially second-order. In contrast, if $K \ll 1$ but $K[S]_0 \gg 1$, equation 8.13 becomes

$$\left(\frac{d[P]}{dt}\right)_{\text{initial}} \simeq k_2[C]_0$$

and the initial rate depends only on the initial concentration of the catalyst.

In many cases the intermediate complex exists in several forms in equilibrium with each other. It also often happens that these forms are not all equally effective in accelerating the reaction.

Consider, for example, the following scheme (the charges on the various species are omitted):

$$
\begin{array}{ccccc}
\text{CH}_2 & \underset{}{\overset{-\text{H}^+}{\rightleftharpoons}} & \text{CH} & \underset{}{\overset{-\text{H}^+}{\rightleftharpoons}} & \text{C} \\
\updownarrow\, \text{S} & & \updownarrow\, \text{S} & & \updownarrow\, \text{S} \\
\text{SCH}_2 & \rightleftharpoons & \text{SCH} & \rightleftharpoons & \text{SC} \\
& & \downarrow & & \\
& & \text{CH+P} & &
\end{array}
$$

With such a scheme the rate will depend on the hydrogen ion concentration (Fig. 8.5), and there will be an optimum pH value for which this rate will be maximized.

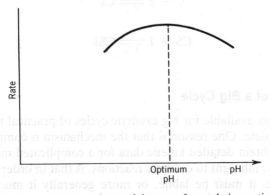

Figure 8.5. Plot of the dependence of the rate of a catalytic reaction on pH.

In a catalytic reaction like the one considered in this section, an inhibitor can act antagonistically or nonantagonistically or both. The following are simple models of such effects:

1. Antagonistic action of an inhibitor I:

$$C + S \underset{k_{-1}}{\overset{k_1}{\rightleftharpoons}} CS$$

$$CS \overset{k_2}{\longrightarrow} P + C$$

$$C + I \underset{k_{-3}}{\overset{k_3}{\rightleftharpoons}} CI$$

The inhibitor binds to the catalyst, in competition with the substrate.

2. Nonantagonistic action of an inhibitor I:

$$C + S \underset{k_{-1}}{\overset{k_1}{\rightleftharpoons}} CS$$

$$CS \overset{k_2}{\longrightarrow} P + C$$

$$CS + I \underset{k_{-3}}{\overset{k_3}{\rightleftharpoons}} CSI$$

In this case, there is no competition with the substrate, but the inhibitor binds (and deactivates) the intermediate.

3. Combined action:

$$C + S \underset{k_{-1}}{\overset{k_1}{\rightleftharpoons}} CS$$

$$CS \overset{k_2}{\longrightarrow} P + C$$

$$C + I \underset{k_{-3}}{\overset{k_3}{\rightleftharpoons}} CI$$

$$CS + I \underset{k_{-4}}{\overset{k_4}{\rightleftharpoons}} CSI$$

The Kinetics of a Big Cycle

The kinetic data available for big catalytic cycles of practical importance are usually incomplete. One reason is that the mechanism is complicated, and it is difficult to obtain detailed kinetic data for a complicated mechanism. Another difficulty, inherent to catalytic reactions, is that in order for the system to be catalytic, it must be labile, or more generally it must have readily interchangeable forms. But these are exactly the features that make experimental study difficult. It is easier to study stable structures and nondynamic (static) systems.

Yet in spite of the difficulties, attempts are constantly being made to collect and interpret data for complex systems.

It should be obvious that for complex systems only numerical methods can be of general use, that is, methods based on the fitting of the data obtained for the entire reacting system to the proposed mechanism, using assumed or independently determined values for the rate and equilibrium constants. However, as the number of elementary reactions in the mechanism increases, the fitting becomes questionable, because the same data with the associated experimental error can be fitted to more than one mechanism and to several values of the parameters (rate and equilibrium constants) with equal reliability (or lack of it).

Alternative Methods for the Study of Big Cycles

If the system cannot be studied as a whole, the obvious alternative is to break it down and study the pieces, hoping that these pieces behave separately in the same way as they behave within the system—an assumption by no means always justifiable.

There are two aspects in such a stepwise search.

(*a*) *Characterization of the intermediates.* This can be done in the following ways:

1. In solution by techniques such as NMR and EPR.
2. After *isolation* and/or *synthesis.*
3. By *simulation.* This is a very popular method. Another compound is prepared that resembles in some respects (usually structurally) the real one but is easier to study. The expectation is that the "impersonator" sufficiently resembles the "original."

(*b*) *The study, including a kinetic study, if possible, of the individual reactions of the mechanism.* Generally, these reactions are not cyclic, and their separate study is possible only if the reactants (or their models) are available independently. Examples of big cycles will be given in Sections 8.6–8.12.

8.3. SELECTIVITY

In mixtures of alkyne, alkene, and dihydrogen, in the presence of polynuclear rhodium complexes as catalysts, the alkyne is hydrogenated selectively. This is just one example of selectivity. There are many, many more, especially in enzymatic reactions.

Definitions

The *integral selectivity*, S_P, for a product P is the fraction of an initial substance transformed into this product up to time t. S_P is defined by the equation

$$S_P(t) = \frac{fP_t}{A_t} = \frac{f\int_0^t W_P\, dt}{\int_0^t |W_A|\, dt} \qquad 8.14$$

where P_t is the amount of the product P formed, W_P the rate of its formation, A_t the amount of reactant A at time t, $|W_A|$ the absolute rate of its consumption, and f a normalizing factor related to the stoichiometric coefficients of the chemical equation.

Integral selectivity is related to the *yield*, ϕ_P, through the relation

$$S_P(t) = \frac{\phi_P}{C_A} \qquad 8.15$$

where $C_A = A_t/A_0$, the conversion fraction of A. By combining 8.14 and 8.15 we obtain

$$\phi_P = \frac{fP_t}{A_0} \qquad 8.16$$

The *differential selectivity*, σ_P, is defined by the relation

$$\sigma_P = -\frac{f\, dP}{dA} = -\frac{f\, dP/dt}{dA/dt} = -\frac{fW_P}{W_A} \qquad 8.17$$

which is more useful in flow systems.

It is obvious from the definition that selectivity depends on the rate and therefore on the mechanism and the values of the rate constants.

EXAMPLE I. PARALLEL FIRST-ORDER REACTIONS

Consider the case:

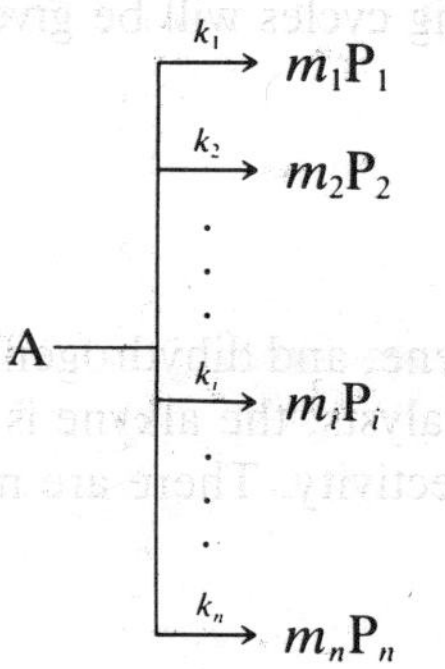

$$
A \;
\begin{cases}
\xrightarrow{k_1} & m_1 P_1 \\
\xrightarrow{k_2} & m_2 P_2 \\
\vdots & \\
\xrightarrow{k_i} & m_i P_i \\
\vdots & \\
\xrightarrow{k_n} & m_n P_n
\end{cases}
$$

The rate of consumption of A is

$$-\frac{d[A]}{dt} = \sum_{i=1}^{n} k_i[A]$$

Hence, the integral selectivity for product P_j is

$$S_{P_j}(t) = \frac{k_j}{\sum\limits_{i=1}^{n} k_1} \qquad\qquad 8.18$$

In this case, there is no dependence on concentration or time. The selectivity depends only on the rate constants of the mechanism considered. In the general case, the selectivity also depends on both time and concentration.

In the limiting case, in which only one of the k's is different than zero, for example, $k_j \neq 0$, equation 8.18 gives

$$S_{P_j} = 1 \qquad S_{P_i} = 0 \qquad (i \neq j)$$

This case is approached in highly selective catalytic reactions, with a catalyst accelerating one reaction only.

EXAMPLE II. RELATIVE SELECTIVITY OF TWO SMALL CYCLES

Consider the action of a catalyst C on a mixture of substrates S_1 and S_2. Furthermore, consider that the catalytic cycle for each of these substrates is simple, consisting of a fast catalyst–substrate adduct formation equilibrium and a subsequent rate-determining step:

Cycle I	Cycle II
$C + S_1 \underset{}{\overset{K_1}{\rightleftharpoons}} CS_1$ (fast)	$C + S_2 \underset{}{\overset{K_2}{\rightleftharpoons}} CS_2$ (fast)
$CS_1 \xrightarrow{k_1} P_1 + C$ (slow)	$CS_2 \xrightarrow{k_2} P_2 + C$ (slow)

It is also assumed that all stoichiometric coefficients are equal to 1. The corresponding rates are

$$W_{P_1} = k_1 K_1[C][S_1] \qquad \text{and} \qquad W_{P_2} = k_2 K_2[C][S_2]$$

The *relative integral selectivity* can be defined either relative to S_1 or relative to S_2. In the first case,

$$S_{P_1}, S_1 = \frac{\int_0^t W_{P_1}\,dt}{\int_0^t W_{S_1}\,dt} = \frac{\int_0^t W_{P_1}\,dt}{\int_0^t W_{P_1}\,dt} = 1$$

$$S_{P_2}, S_1 = \frac{\int_0^t W_{P_2}\,dt}{\int_0^t W_{S_1}\,dt} = \frac{k_2K_2}{k_1K_1}\frac{\int_0^t [\mathrm{C}][\mathrm{S}_2]\,dt}{\int_0^t [\mathrm{C}][S_1]\,dt}$$

Initially ($t \rightarrow 0$),

$$S_{P_1}, S_1 = 1$$

$$S_{P_2}, S_1 \simeq \frac{k_2K_2}{k_1K_1}\frac{[\mathrm{S}_2]_0}{[\mathrm{S}_1]_0}$$

The initial relative selectivity depends on the ratios of the rate and equilibrium constants, but it also depends on the ratio of the initial concentrations of the substrates.

Selectivity of Big Cycles

Big cycles may contain smaller cycles, parallel reactions, and so forth. Only a sufficient knowledge of the mechanism and of the rate law can lead to an identification of the factors that determine selectivity.

Stereoselectivity

Asymmetric catalysis can be regarded as a special kind of selectivity; in this catalysis one of the enantiomers is obtained preferentially. A necessary condition is that the catalyst itself be chiral.

The drug for Parkinson's disease, L-DOPA, is among the substances commercially synthesized by asymmetric catalysis:

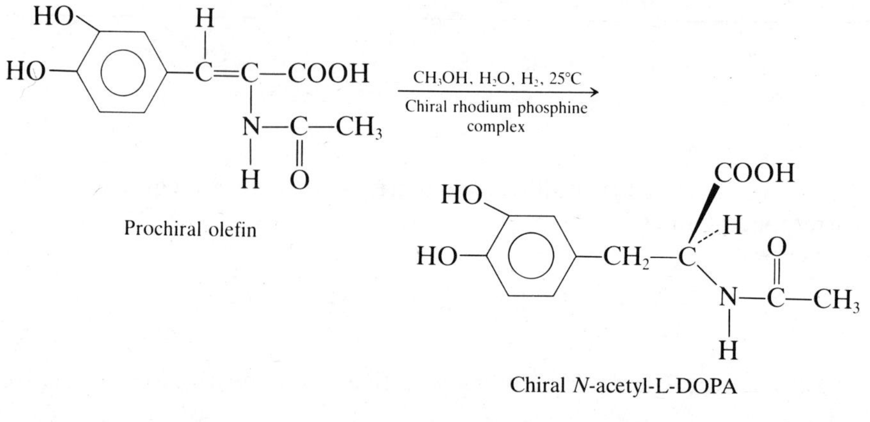
HO
HO
H
C=C—COOH
N—C—CH3
H O
CH3OH, H2O, H2, 25°C
Chiral rhodium phosphine complex
Prochiral olefin
HO
HO
CH2—C
COOH
H
O
N—C—CH3
H
Chiral N-acetyl-L-DOPA

The efficiency of an asymmetric synthesis is usually expressed in terms of the percentage excess of one enantiomer over the other. In the synthesis of L-DOPA this excess approaches 100%.

The synthesis of L-DOPA is an example of asymmetric hydrogenation catalyzed by a chiral metal complex, but the scope of such catalysis is not restricted to hydrogenation. Asymmetric catalysis is also known in hydroformylation, hydrosilation, olefin coupling, and so on. Moreover, there are many purely organic systems with organic chiral catalysts, and of course there are a wide assortment of asymmetric enzymes that can convert chiral substrates into chiral products or one chiral molecule into another.

The factors affecting stereoselectivity are generally not expected to be much different from those affecting selectivity in general. In any case, a careful study of the mechanism is necessary before any attempt is made to identify these factors.

8.4. MONO- AND POLYNUCLEAR CATALYSTS

The common-sense advantage of having bi-, tri-, . . . , polynuclear complexes instead of mononuclear species as catalysts seems to be that the many nuclei can do what each of them can do separately, plus some more. The less obvious disadvantage is that two, three, or more nuclei may get in each other's way and interfere with each other destructively.

A related question is: What really is the function of the metal ion center in catalysis? This question has already been discussed, but it is worth recalling. The main functions of the metal ion center are:

1. To activate the reactant molecules.
2. To hold the activated reactants together long enough for reaction to occur.

Referring to these functions, we will try to compare mononuclear species to polynuclear ones and eventually to metal surfaces. To do that we must first review briefly some relevant aspects of the molecular basis of selectivity.

Some Sources of Nonselectivity

Two kinds of nonselectivity have been implied in Section 8.3:

1. Nonselectivity due to more than one mode of activation of the same substrate.
2. Nonselectivity due to competitive activation of different substrates by the same catalyst species.

Where do these nonselectivities originate?

Consider first a simple mononuclear catalyst ML. In solution this catalyst dissociates quickly. Lability is one of the prerequisites of catalytic action. It then follows that species will be present in solution that have had one or more ligands replaced by solvent, reactant, or product molecules, or even dimeric and higher order species. In principle, each one of these can give a different distribution of products. Moreover, each one may be nondiscriminating toward different reactants. The end result is the formation of mixtures.

With a polynuclear complex, these variations still exist. In addition, there are collective effects involving two or more metal centers, that is, new modes of binding and activation, which can lead to further loss of selectivity.

The variations can result in very subtle effects indeed. However, with a mononuclear complex, nonselectivity can be narrowed down to a minimum, and the catalytic action can be matched closely to the desired effect, by the proper choice of metal ions, ligands, solvent, and conditions.

With a polynuclear catalyst, the achievement of such selectivity is more difficult because of the inherent collective effects, which add tremendously to the scope of homogeneous catalysis by providing new opportunities, but at the same time they also add new sources of nonselectivity.

Collective Modes of Binding and Activation

One of the collective (cooperative) modes of activation is the formation of bridged species. There are a very large number of bridged structures known or suspected to be intermediates. Typical examples are included in Table 8.1. The number of metal atoms held together varies (in these examples) from two to four and includes both one- or two-atom bridges.

The reduction of dinitrogen to form ammonia requires six electrons, and since usually only one electron is transferred at a time, it requires six elementary redox steps. In addition, to form ammonia, six hydrogen atoms must be obtained from three hydrogen molecules. If the substitution steps are also counted, the number of elementary steps increases tremendously. It is not surprising that one mononuclear complex cannot do this complicated job. At least, no such complex has yet been found.

Similar remarks can be made for reactions such as the hydrogenation of carbon monoxide, the rearrangement of n-C_6H_{14} to give a mixture of alkanes and cycloalkanes, and many others. However, such complex reactions can be catalyzed by clusters, for example, by $Os_3(CO)_{12}$ or $Ir_4(CO)_{12}$. One reason is that clusters can *bind different reactant molecules on neighboring sites* and give them the opportunity to stay close together long enough to react. This is a possibility beyond any offered by single metal centers. It is perhaps too much to expect from a mononuclear catalyst to keep holding together, for as long as it is needed, all the species involved—for example, in the hydrogenation of carbon monoxide. With clusters, the chances are better.

From Polynuclear Species to Metal Surfaces

Polynuclear catalysts, especially cluster compounds, have been regarded as simple models of metal surfaces, as little pieces of metal (very little indeed!),

TABLE 8.1. Bridge-Structured Intermediates

Metal Centers	Bridging Atoms	Examples
2	1.	
2	1, 2	
3	1	
3	1, 2	
4	1	
4	2	

with ligands attached to them. The gap between clusters and solid catalysts might seem smaller if it is remembered that in heterogeneous catalysis the common practice is to use microcrystals to maximize the surface area.

A cluster like $Rh_6(CO)_{16}$ has six rhodium atoms bound to each other, but it also has 16 carbon monoxide ligands (Fig. 8.6). It therefore resembles not a clean metal surface, but a metal surface carrying chemisorbed species. On the metal surface each metal atom is in contact with other metal atoms of the same kind. On the surface of an alloy, a metal atom may be in contact

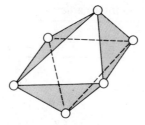

Figure 8.6. Schematic presentation of a cluster, $Rh_6(CO)_{16}$, used in catalysis. [CO + $(1/2)O_2 \longrightarrow CO_2$ at 100°C.] The vertices of the octahedron are occupied by Rh atoms bound to two CO molecules and to each other. In addition, four CO molecules cap the four trigonal faces (the three shaded and the one in the back indicated by the dashed lines).

with an atom of a different metal. Thus, while the environment of a metal atom in a cluster can be changed practically at will by changing the ligands, the changes of the environment of a metal atom on a metal surface are rather limited. As a result, close adaptation of a metal catalyst to the requirements of a specific reaction is also limited, and selectivity is low. In industrial applications this selectivity becomes even lower, because the heterogeneous catalyst is usually a mixture of active components on some carrier, and there are different active sites (including defects in the crystal structure), not all of which lead to the desired products.

Another significant difference is that the metal surface has one "face" only. Metal atoms on a surface are accessible from one side only, while metal clusters are "three-dimensional." In addition, even the smallest piece of metal, provided it contains more than about 100 atoms, has collective metallic properties such as conductivity.

The combination of these factors makes metal surfaces not only less selective but also less effective in lowering the activation energy. It is often forgotten, but it is a fact, that heterogeneous catalysis usually requires quite vigorous conditions, whereas homogeneous catalysis is often carried out under very mild conditions.

8.5. HETEROGENIZED HOMOGENEOUS CATALYSTS

The advantages of the mono- or polynuclear homogeneous catalysts have been discussed at some length. However, heterogeneous catalysts also have advantages, the main one being the ease with which they can be separated from the reaction mixture without additional costly procedures. Another advantage that is very important in large-scale applications is that a solid catalyst generally has higher thermal stability and is easier to regenerate.

Thus, a compromise seems to be in order: heterogenize the homogeneous catalyst. Use the same complexes as in homogeneous catalysis, but hold them firmly in place on a solid support, such as a polymer. Heterogenized homogeneous catalysts (also called "anchored," "immobilized," "supported," or "hybrid phase" catalysts) retain the advantages of the homogeneous catalysts, and at the same time separation is easy.

EXAMPLE

Consider the following case:

Homogeneous

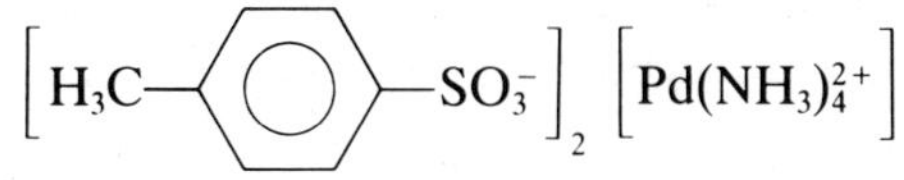

$$\left[H_3C\!-\!\!\left\langle\bigcirc\right\rangle\!\!-\!SO_3^- \right]_2 \left[Pd(NH_3)_4^{2+} \right]$$

Heterogenized

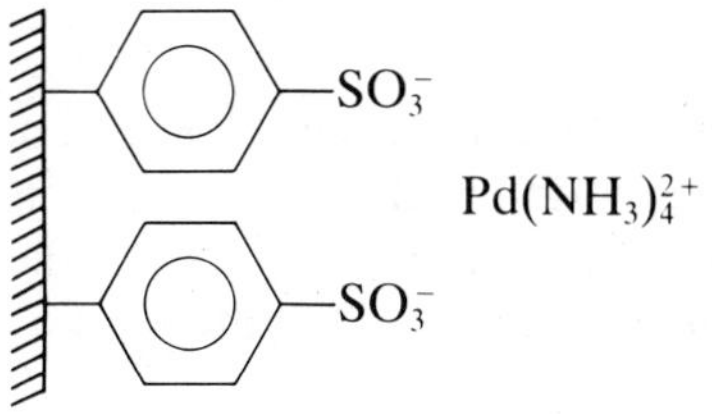

$$Pd(NH_3)_4^{2+}$$

   In this case the replacement of the methyl group by the polymeric support does not appreciably change the catalytic activity. It is generally desirable to have the homogeneous catalyst supported without markedly affecting its catalytic properties.
   Among the supports that have been used are the following:

Inorganic: Silica, zeolites, glass, γ-alumina, clay.

Organic: Polystyrene, polyamines, silicones, polyvinyls, polybutadiene, poly(amino acids), acrylic polymers, cellulose.

The catalyst can be bound to an inorganic support by first attaching the ligand to the support and then attaching the metal complex fragment, or vice versa. □

EXAMPLES

First, attach the ligand and then the metal complex:

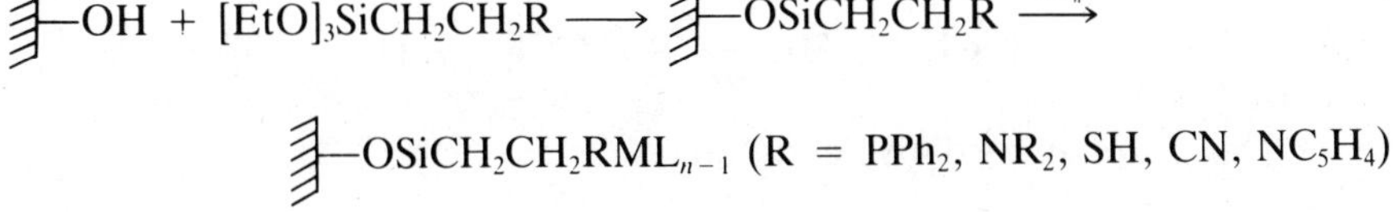

$$\text{≥}\!-\!OH + [EtO]_3SiCH_2CH_2R \longrightarrow \text{≥}\!-\!OSiCH_2CH_2R \xrightarrow{\ ML_n\ }$$

$$\text{≥}\!-\!OSiCH_2CH_2RML_{n-1} \ (R = PPh_2,\ NR_2,\ SH,\ CN,\ NC_5H_4)$$

Alternatively, first make the metal complex with the appropriate ligand and then attach it to the support.

$$(EtO)_3SiCH_2CH_2PPh_2 \xrightarrow{ML_n} (EtO)_3SiCH_2CH_2PPh_2ML_{n-1} \xrightarrow{\text{}\text{—OH}}$$

$$\text{—OSiCH}_2CH_2PPh_2ML_{n-1}$$

With organic polymeric supports, the metal complex or the appropriate ligand can also be introduced before polymerization. □

8.6. COMMON FEATURES IN THE HYDROGENATION OF OLEFINS AND RELATED COMPOUNDS

The catalytic cycle for the hydrogenation of an olefinic double bond is not unique. It is generally different for different olefins and different catalysts, but there are some common features, one of them being that at some stage, in some form, both the olefin and hydrogen are coordinated and activated.

Activation of Dihydrogen

Dihydrogen is activated in one of the following ways:

1. Two-electron oxidative addition, for example,

$$Ir^ICl(CO)(PPh_3)_2 + H_2 \longrightarrow Ir^{III}H_2Cl(CO)(PPh_3)_2$$

Four- and five-coordinate d^8 complexes, like Vaska's compound $IrCl(CO)(PPh_3)_2$, undergo cis addition.

2. *Homolytic cleavage,* for example,

$$2Co(CN)_5^{3-} + H_2 \longrightarrow 2CoH(CN)_5^{3-}$$

If the product is regarded as a hydride, the reaction should be considered a one-electron oxidative addition. If the hydrogen in the product is regarded as an atom, the reaction is a homolytic cleavage that does not involve change in the oxidation state of cobalt.

3. *Heterolytic cleavage,* for example,

$$Ru^{III}Cl_6^{3-} + H_2 \longrightarrow Ru^{III}HCl_5^{3-} + H^+ + Cl^-$$

This *is not* an oxidative addition.

Cases 1 and 2 involve homolytic cleavage of H—H bonds, requiring 435 kJ mol^{-1}. This highly endothermic process is partly balanced by the exothermic formation of M—H bonds. The strength of the latter is typically ~240 kJ mols^{-1} each.

Activation of the Olefin

After the addition of the hydride ion, some ligands are labilized and eventually replaced by an olefin. Thus, hydride ion and olefin become coordinated to the same metal ion (only mononuclear complexes are considered here). The olefin is usually coordinated in an η^2 fashion (the Dewar–Chatt–Duncanson model):

This model involves a combination of synergistic bonding and back-bonding shown schematically in Figure 8.7, where the dashed regions represent occupied orbitals. The lengthening of the C—C bond in the coordinated olefin is exaggerated to draw attention to the activation involved. This activation is associated with a lengthening and weakening of the carbon–carbon bond, but the main interest here is in the eventual transformation of the olefin into an alkyl radical. What is the relation between these aspects?

Consider a complex containing both a hydride ion and an olefin in cis

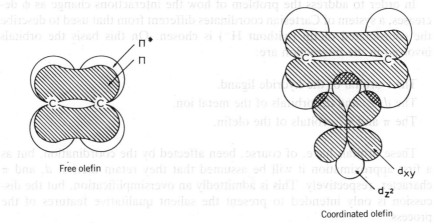

Figure 8.7.

position to each other. There are two "extreme" configurations of this hydride–metal–olefin system, the "coplanar" and the "perpendicular".

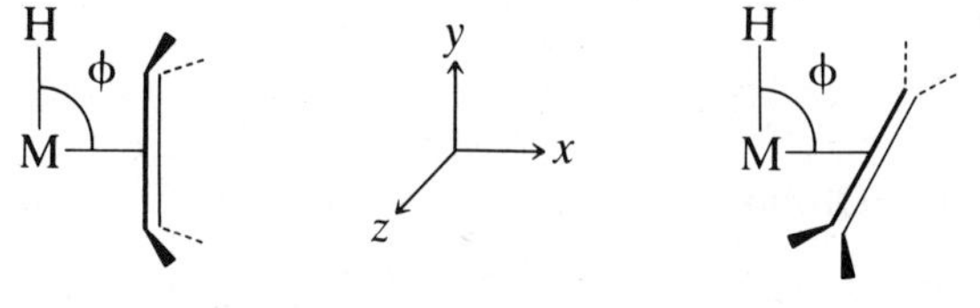

1	System of	**2**
"Coplanar"	cartesian	"Perpendicular"
	coordinates	

Consider as an example a d^8 complex of Pt^{II}. In the "coplanar" geometry, H and the two carbons are in one plane, which also contains the metal ion. In the "perpendicular" configuration, the olefin is perpendicular to the plane defined by its own center, H, and M.

For reaction to occur, H$^-$ and olefin must interact, and this can be achieved by a decrease in the angle ϕ, for example,

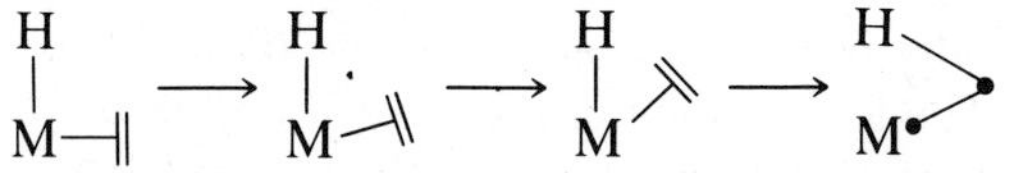

The angle ϕ can be considered the reaction coordinate.

The questions are then: What are the interactions involved in this pathway, and what is the significance of the olefin activation that preceded this process?

The significance of the activation of dihydrogen, namely, of the breaking of the H—H bond, is obvious and needs no comment.

In order to address the problem of how the interactions change as ϕ decreases, a system of Cartesian coordinates different from that used to describe the coordinated olefin (without H$^-$) is chosen. On this basis the orbitals involved in the interaction are:

The s orbital of the hydride ligand.
The $d_{x^2-y^2}$ and d_{xy} orbitals of the metal ion.
The π and π^* orbitals of the olefin.

These orbitals have, of course, been affected by the coordination, but as a first approximation it will be assumed that they retain their s, d, and π character, respectively. This is admittedly an oversimplification, but the discussion is only intended to present the salient qualitative features of the process.

The initial and an intermediate phase in the interaction between s, $d_{x^2-y^2}$, π, and π^* can be schematically presented as in Figure 8.8. In going

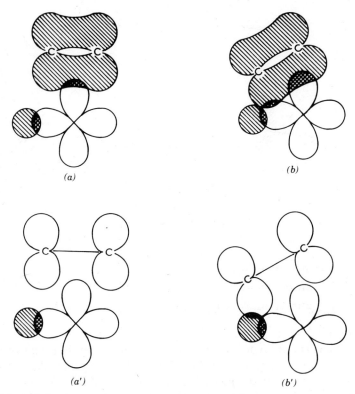

Figure 8.8. $(a)(a')$ Initial state of interaction. $(b)(b')$ Intermediate state of interaction. (The filled areas symbolize occupied orbitals.)

from (a) to (b) the repulsion between the occupied s and π orbitals increases, but this increase is partly offset by the increased attractive overlap between s and the empty π^* [from (a') to (b')]. The latter is important in forming the new H—C bond, and since it involves population of the antibonding π^* orbital, it causes further lengthening of the already longer carbon–carbon double bond and its gradual transformation into a single bond. The displacement of electrons into the π^* orbital also causes weakening of the M—H bond and eventually causes it to break. At the same time, the η^2 mode of coordination of the metal ion changes smoothly into an η^1 mode. In this process some back-bonding between the occupied d_{xy} and the empty π^* is lost [(a'') and (b'') in Fig. 8.9]. However, this loss is probably offset by an increased attractive σ overlap between $d_{x^2-y^2}$ and π [(a) and (b) of Fig. 8.8].

In summary, the main forces that change during the insertion of the olefin into the M—H bond are

1. $s-\pi^*$ attraction, which increases
2. $s-\pi$ repulsion, which increases

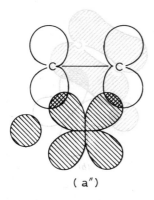

(a″)

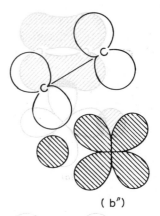

(b″)

Figure 8.9.

3. $d_{x^2-y^2}-\pi$ attraction, which increases
4. $d_{x^2-y^2}-s$ attraction, which decreases
5. $d_{xy}-\pi^*$ attraction, which decreases

Changes 1, 3, and 4 favor insertion, while 2 and 5 oppose it and contribute to the activation energy required. In fact, with the "perpendicular" geometry, the unfavorable factors dominate and insertion does not take place. This is the reason only the coplanar approach is considered.

What Happens after Insertion?

The metal–alkyl species formation in the insertion reaction is only half the story. Hydrogenation of the double bond requires the addition of two hydrogen atoms. How does the second enter?

The answer to this question depends on whether we start with a dihydride or with a monohydride. In the former case the two hydrogen atoms can be successively added to the olefin monomolecularly. In the latter case this is not possible.

The *intramolecular* reaction of a coordinated hydride with a coordinated alkyl group and the detachment of a hydrocarbon is a reductive elimination reaction:

$$L\!-\!Pt^{II}\!\!-\!CH_3 \longrightarrow Pt^0L_2 + CH_4 \qquad (L = PPh_3)$$
with H and L above/below Pt

This is an elementary reaction. The product PtL_2 reacts further. Intramolecular eliminations of this kind are generally fast, especially if the eliminated ligands are cis to each other. Elimination of trans ligands is probably preceded by *trans*-to-*cis* rearrangement.

Intermolecular reductive elimination, by a mechanism not involving organic free radicals, is exemplified by the reaction

$$RMn(CO)_5 + HMn(CO)_5 \xrightarrow{\text{In benzene}} RH + Mn_2(CO)_{10} \qquad 8.19$$

where R = p-$CH_3OC_6H_4CH_2$. For this reaction, the following mechanism has been proposed:

$$RMn(CO)_5 \underset{k_{-1}}{\overset{k_1}{\rightleftharpoons}} RMn(CO)_4 + CO$$

$$RMn(CO)_4 + HMn(CO)_5 \longrightarrow RH + Mn_2(CO)_9$$

$$Mn_2(CO)_9 + CO \longrightarrow Mn_2(CO)_{10} \quad \text{(fast)}$$

This mechanism involves a coordinatively unsaturated intermediate, $RMn(CO)_4$, and reductive elimination (the second reaction of the mechanism) occurs via a binuclear intermediate (two manganese atoms).

A *free radical mechanism* has been postulated for the reaction

$$RMn(CO)_4P + HMn(CO)_4P \xrightarrow{\text{In benzene}} RH + Mn_2(CO)_8P_2 \qquad 8.20$$

where R = p-$CH_3OC_6H_4CH_2$ and P = $(p$-$CH_3OC_6H_4)_3P$. The reactants in 8.20 differ from those in 8.19 only in the replacement of one CO ligand by a phosphine ligand, but the mechanism differs radically. The mechanism for 8.20 is

$$RMn(CO)_4P \longrightarrow R^· + Mn(CO)_4P$$

$$R^· + HMn(CO)_4P \longrightarrow RH + Mn(CO)_4P$$

$$2Mn(CO)_4P \longrightarrow Mn_2(CO)_8P_2$$

For mononuclear catalysts, the simultaneous coordination of a hydride ion and an alkyl group is a necessary condition for an intramolecular mechanism, but it is not sufficient. Even if the complex contains the group M—R, the
$$\overset{H}{\underset{|}{}}$$
mechanism may still be intermolecular.

Extensions

Other unsaturated groups that are known to be hydrogenated in the presence of homogeneous catalysts are

$$-C\equiv C-, \quad {>}C{=}O, \quad -N{=}N-, \quad -CH{=}O, \quad -NO_2$$

Aromatic and heterocyclic rings are also hydrogenated, but less effectively.

The rates of reduction of such groups vary widely, depending on the catalyst and on the conditions. In fact, the catalysis may be highly selective. However, mechanisms are generally similar. Some of these groups—for example, acetylenes, CO_2, and allenes—possess a second empty π^* orbital, but this does not seem to change the mechanism to a marked extent.

Hydrosilation and hydrocyanation reactions are frequently mentioned under the heading "Hydrogenation Reactions." In hydrosilation, one hydrogen atom and one SiR_3 radical are added to the double bond (instead of two hydrogen atoms):

In hydrocyanation, one hydrogen atom and one CN group are added:

The commercially important hydrocyanation of butadiene belongs in this last category:

$$CH_2{=}CHCH{=}CH_2 \ + \ 2HCN \ \xrightarrow{\ Ni^0 \ catalyst\ } \ NCCH_2CH_2CH_2CH_2CN$$

1,3-Butadiene   Adiponitrile

The activation is analogous to that in hydrogenation. The reaction

$$M^{II}L_n \ + \ HSiR_3 \ \longrightarrow \ HM^{IV}(SiR_3)L_n$$

is an oxidative addition reaction, analogous to the oxidative addition of dihydrogen.

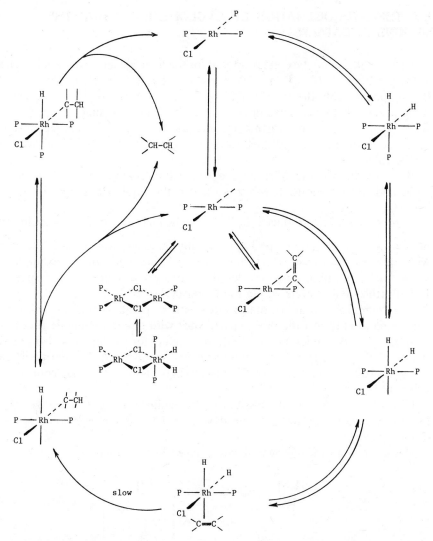

Figure 8.10. The mechanism for the hydrogenation of cyclohexene using Wilkinson's catalyst. Cyclohexene is symbolized as $\diagup C=C \diagdown$, the hydrogenated product as $\diagup CH-CH \diagdown$, and the phosphine ligands (PPh₃) as P.

8.7. THE HYDROGENATION OF CYCLOHEXENE USING THE WILKINSON CATALYST

A version of the catalytic cycle proposed for the hydrogenation of cyclohexene in the presence of $RhCl(PPh_3)_3$ (Wilkinson's catalyst) is given in Figure 8.10. For simplicity, only the intermediates are indicated; the species with which these intermediates react are implied. It has also been assumed that all substitutions involve only one ligand and take place by a dissociative mechanism. Associative and multiligand substitution steps are not included, but they cannot be ruled out.

It is seen that what is called a cycle is actually a complex network of cyclic paths and side reactions. It is also seen that the catalyst undergoes several metamorphoses. The product ($CHCH$) can be reached from any point in the network by a multitude of paths. Generally, among parallel paths the one with the least resistance—the fastest one—will dominate. In this example it seems to be the path that includes the coordinatively unsaturated species $RhClP_2$ (the smaller cycle within the larger cycle).

The "communication" of the product with all points in the network provides a method to test the mechanism: Start with any intermediate that can be prepared independently, and look for product formation. It is also noted that because of the irreversible steps we cannot get back to the intermediates from the products, but there are other points in the network that are connected reversibly.

Common features of all possible paths leading to product are the key intermediate at the bottom and the slow, rate-determining insertion reaction it undergoes.

The expression for the overall rate is given by the equation

$$[k_{\text{catalysis}}]^{-1} = \frac{[\text{Rh}]_{\text{total}}}{\text{Rate}} = \frac{1}{k_{\text{slow}}} + \frac{[\text{P}]}{k_1 \left[\overset{\diagup}{\underset{\diagdown}{\text{C}}}=\overset{\diagdown}{\underset{\diagup}{\text{C}}} \right]} + \frac{[\text{P}]}{k_2[\text{H}_2]}$$

where k_{slow} is the constant for the rate-determining step, and k_1 and k_2 are combinations of rate and equilibrium constants.

8.8. REACTIONS INVOLVING CARBON MONOXIDE

Important catalytic reactions involving carbon monoxide are:

The *Fischer–Tropsch* conversion of $CO + H_2$ mixtures into hydrocarbons and other products.

The hydroformylation of olefins (the *Oxo process or Roelen process*):

$$RCH{=}CH_2 + CO + H_2 \xrightarrow[\text{or other catalysts}]{\text{Cobalt, rhodium,}} RCH_2CH_2\overset{\displaystyle O}{\overset{\|}{C}}H$$

The carbonylation of methanol to acetic acid (the *Monsanto process*):

$$CH_3OH + CO \xrightarrow{\text{Rhodium catalyst}} CH_3COOH$$

The first of these processes is heterogeneous, but it will be briefly examined here because of the special interest it presents.

The common feature of all these processes is that they involve activation of carbon monoxide. Therefore, before entering into their description, it is instructive to examine the ways in which carbon monoxide can be activated.

The activation modes of dihydrogen have already been described. They are the same in the Fischer–Tropsch process, in hydroformylation, or in olefin hydrogenation.

Activation of Carbon Monoxide

Usually carbon monoxide coordinates through carbon. This is called η^1 coordination, linear coordination, or end-on coordination. Pictorially, the bonding is

$$M{-}C{\equiv}O$$

There are other modes too, such as coordination through the triple bond or bridged coordination, but they will not be considered here.

The bonding in the end-on coordination consists of σ donation of carbon monoxide electrons to the metal and π back-donation (see Fig. 8.11). Back-donation involves charge transfer from the metal into an empty π^* orbital, but it is also affected by the occupied π orbital. As a result, there is a tendency for an accumulation of negative charge on carbon monoxide and particularly on oxygen. In a way, the d electrons push the π electrons away.

Thus, σ donation tends to create a positive effective charge on carbon and makes this atom susceptible to *nucleophilic attack,* for example,

$$M{-}\overset{\delta^+}{C}O + H^- \longrightarrow M{-}C\overset{\displaystyle O}{\underset{\displaystyle H}{\diagdown}}$$

On the other hand, π back-donation tends to restore the electron population on carbon and decrease its electrophilicity, while simultaneously increasing

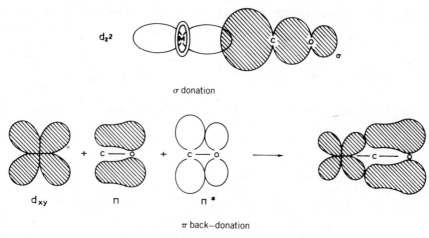

σ donation

π back–donation

Figure 8.11.

the nucleophilicity of oxygen. Oxygen then becomes susceptible to *electrophilic attack*:

$$M-\overset{\delta^-}{C}O + H^+ \longrightarrow [M-COH]^+$$

Attachment of an electrophile to oxygen makes the carbon more positive again, and the coordinated carbon monoxide easier to reduce. In short, η^1 coordination may make carbon positive enough to accept electrons, and this is assisted by the attachment of an electrophile on oxygen.

Insertion of carbon monoxide into a metal–hydrogen or metal–alkyl bond can be regarded as an intramolecular nucleophilic attack.

$$
\begin{array}{c}
M-C\equiv O \longrightarrow M-C\overset{\displaystyle O}{\underset{\displaystyle R}{\diagup}}\\
\;\;\;| \\
\;\;\;R
\end{array}
$$

The salient features of the mechanism are similar to those of olefin insertion.

Activation of carbon monoxide sometimes leads to complete *CO bond cleavage*. This is the case in the stoichiometric reduction of CO to CH_4 and H_2O by some cluster compounds in strong protic media.

The Fischer–Tropsch Method

A mixture of CO and H_2 is converted catalytically into a mixture of hydrocarbons and oxygenated products. The special importance of the method lies in the fact that the $CO + H_2$ mixture (synthesis gas, abbreviated *syn-gas*) can be obtained from coal. The goal is to obtain liquid fuels.

$$Coal \xrightarrow[\text{heat}]{+\ H_2O} CO + H_2 \longrightarrow H_2C{=}O \xrightarrow[-\ \text{Ketones}]{-\ \text{Aldehydes}} CH_3OH \xrightarrow[-\ \text{Alcohols}]{-\ \text{Olefins}}$$

$$CH_4 + H_2O \longrightarrow \text{liquid fuel}$$

In the course of the reaction there is carbon–carbon and carbon–hydrogen bond formation. Coal is a mixture containing organic macromolecules. It already has such bonds. It may then look strange that we break these macromolecules completely to form carbon monoxide, and then try to glue the carbons together again. This "gasification" makes more sense if it is regarded as an unusual purification method for getting rid of undesired components (e.g., minerals). In this context it is noted that there are other methods for coal utilization, such as partially breaking the organic macromolecules in the presence of hydrogen, where there is no need for gasification. At any rate, gasification is used, and therefore it is useful to quote some details.

Some Energetics

The overall reactions

$$2H_2 + 2CO \longrightarrow CH_4 + CO_2$$

$$3H_2 + CO \longrightarrow CH_4 + H_2O$$

are thermodynamically favored (negative $\Delta G°$) only at temperatures below 600°C. At these temperatures the reactions need activation.

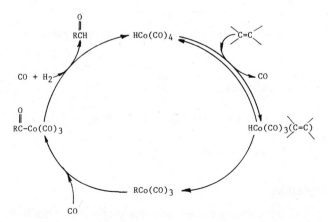

Figure 8.12. A simplified cycle for the hydroformylation reaction.

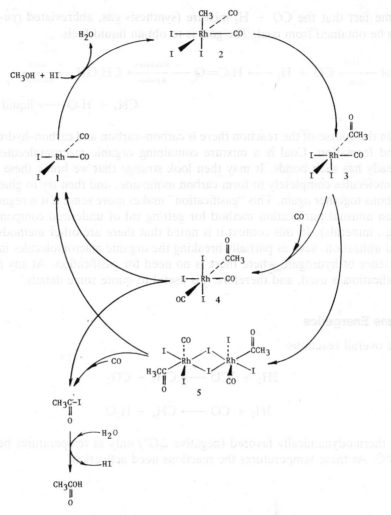

Figure 8.13. Mechanism of the carbonylation of methanol.

For the general reaction

$$2nH_2 + nCO \longrightarrow C_nH_{2n+1}OH + (n-1)H_2O$$

$\Delta G°$ is negative for all $n \neq 1$ below 325°C.

Hydroformylation

Hydroformylation is catalyzed homogeneously by cobalt, rhodium, and other complexes. A simplified catalytic cycle of a hydroformylation reaction is given in Figure 8.12. The steps in this simplified cycle are complex; they are not elementary. In addition, there are parallel nonproductive steps.

In a practical situation the temperature ranges from 140 to 180°C and the pressure from 200 to 300 atm. Cobalt can be introduced in various forms (precursors) such as $CoCl_2$. In the reactor the precursor is converted into the active forms, which participate in the cycle. The replacement of one of the carbon monoxide ligands by an olefin probably proceeds through a coordinatively unsaturated species, $HCo(CO)_3$.

Other catalysts used in hydroformylation reactions are hydridocobalt or rhodium carbonyls, also containing phosphine ligands.

Carbonylation of Methanol

Carbonylation of methanol is achieved using rhodium catalysts. The pressure of CO in practical situations is 30–40, atm and the temperature is ~180°. Iodide is also added as a promoter. The mechanism proposed is summarized in Figure 8.13. The catalyst **1** reacts with methanol and HI to give the product of oxidative addition **2**, which in turn undergoes intramolecular rearrangement (insertion of CO into the M—CH_3 bond) and is transformed into the 5-coordinate species **3**. This coordinatively unsaturated complex has the tendency to dimerize (**5**) but also the tendency to pick up another CO ligand (**4**) to become 6-coordinate again. Both **4** and **5** give back the original catalyst and acetyl iodide, which hydrolyzes to acetic acid.

Methyl iodide is formed from the reaction of methanol and HI as a by-product. The first reaction (**1 → 2**) is rate-determining. Iodide ion can be replaced by other ligands, such as bromide ion, which is, however, less effective.

8.9. THE WACKER PROCESS

The overall reaction for the Wacker process can be represented as

$$CH_2{=}CH_2 + \tfrac{1}{2}O_2 \xrightarrow{Pd^{II}/Cu^{II}} CH_3\overset{\displaystyle O}{\overset{\displaystyle \|}{C}}H$$

The reaction can also be carried out with olefins of the type RCH=CHR' or RCH=CH_2.

In the absence of copper(II), the olefin reacts with palladium(II) stoichiometrically to give palladium metal:

$$CH_2{=}CH_2 + PdCl_2 + H_2O \longrightarrow CH_3CHO + Pd^0 + 2HCl \qquad 8.21$$

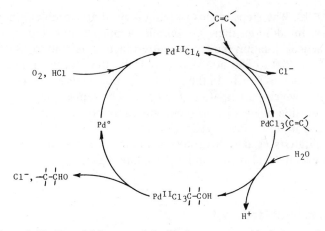

Figure 8.14. Essential features of the Wacker process, with charges omitted.

In the presence of copper(II), however, palladium is reoxidized:

$$Pd^0 + 2Cu^{II} \longrightarrow Pd^{II} + 2Cu^I$$

and the reaction becomes cyclic, because Cu^I is in turn oxidized back to Cu^{II} by O_2:

$$2Cu^I + \tfrac{1}{2} O_2 + 2H^+ \longrightarrow 2Cu^{II} + H_2O$$

The formation of the aldehyde per se [equation 8.21] does not require activation of O_2. It only requires activation of the alkene, which is then aquated. The precursor of the catalyst is usually $PdCl_4^{2-}$. The alkene is co-ordinated to the metal in the usual η^2 manner, replacing one chloride ion, and then it is hydrated, presumably by nucleophilic attack by a noncoordinated water molecule (or OH^-).

The essential features of the Wacker process can be represented by the cycle illustrated in (Fig. 8.14). The last step is not direct; it is carried out with the mediation of the Cu^{II}/Cu^I couple.

The oxidation of Pd^0 by Cu^{II} probably proceeds by way of a chloride-bridged activated complex, and oxidation of the resulting Cu^I by O_2 proceeds by a free radical mechanism with HO_2 and HO as intermediates.

8.10. ISOMERIZATION

Only migration of double bonds within an alkene molecule will be examined. This migration is equivalent to a transfer of a hydrogen atom from a carbon bound to one of the double-bond carbons to the other carbon of the double bond:

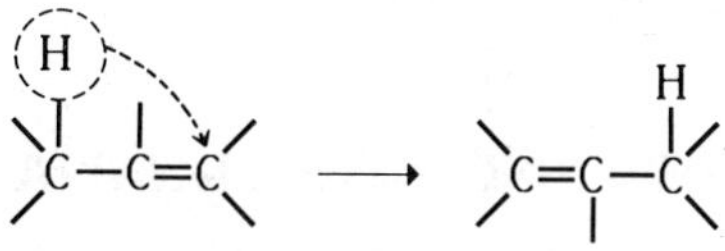

This shift is catalyzed by metal complexes mainly by two mechanisms:

Through formation of a metal–alkyl intermediate.
Through formation of an allyl-type intermediate.

Both of these mechanisms involve a metal hydride intermediate.

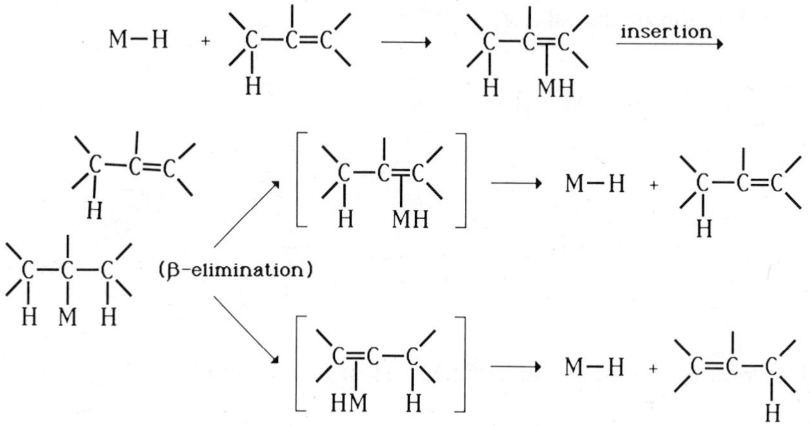

Scheme I

In the β-elimination, either of the two hydrogens can be transferred back to the metal. Consequently two alkenes are formed, the original one and its isomer. The hydride (the catalyst) is either added as such or it is formed in situ. Isomerization is catalyzed by many complexes, including complexes of group VIII elements.

The second mechanistic scheme has been established with $Fe_3(CO)_{12}$ as a catalyst:

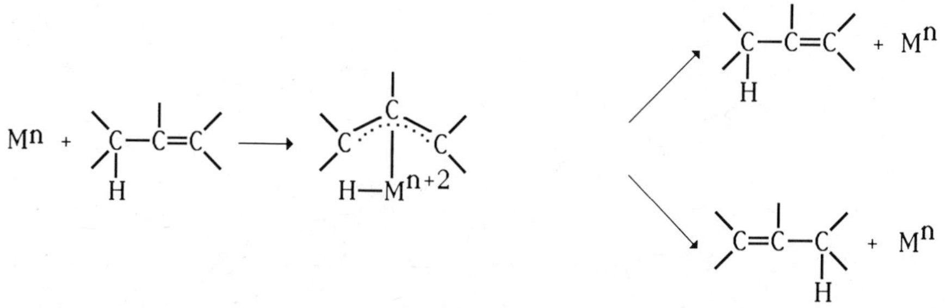

Scheme 2

The formation of the allyl complex can be characterized as an oxidative addition and/or as a hydride abstraction.

8.11. OLIGOMERIZATION AND ZIEGLER–NATTA POLYMERIZATION OF ALKENES

The Mechanism

The basic steps of the widely accepted mechanism for oligomerization and Ziegler–Natta polymerization reactions are the following:

Formation of a metal hydride M—H (in oligomerization) or of a metal alkyl M—R (in Ziegler–Natta polymerization)
Coordination of the alkene:

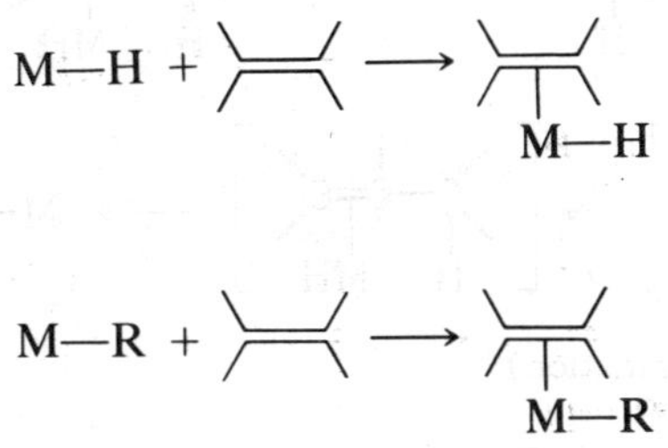

$$M—H + \quad \longrightarrow \quad M—H$$

$$M—R + \quad \longrightarrow \quad M—R$$

Insertion of the alkene into the M—H or M—R bond:

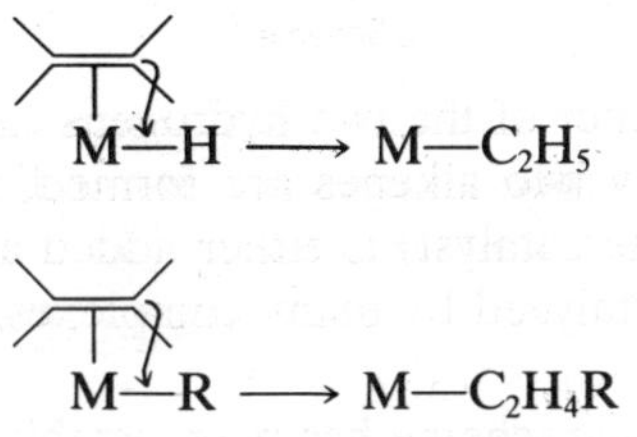

$$M—H \longrightarrow M—C_2H_5$$

$$M—R \longrightarrow M—C_2H_4R$$

Coordination of another alkene:

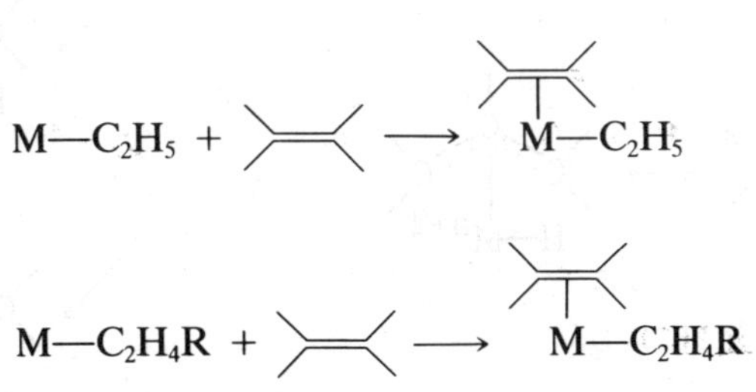

$$M—C_2H_5 + \quad \longrightarrow \quad M—C_2H_5$$

$$M—C_2H_4R + \quad \longrightarrow \quad M—C_2H_4R$$

Another insertion into the metal–carbon bond:

$$M-C_2H_5 \longrightarrow M-CH_2CH_2CH_2CH_3$$

$$M-C_2H_2R \longrightarrow M-CH_2CH_2CH_2CH_2R$$

These double steps of olefin coordination and insertion may continue, but now in competition with β elimination:

$$M-CH_2CH_2CH_2CH_3 \longrightarrow M-H + CH_2{=}CHCH_2CH_3$$

The distribution of products depends on this competition, which in turn depends on the catalyst system used. With nickel compounds (also Pd^0 and Rh^{III} compounds), the products obtained are oligomeric. The distribution among linear, branched, and cyclic oligomers is further tailored by the proper choice of the other ligands and of the oxidation state of the metal.

With Ziegler–Natta catalysts the products are polymeric. Ziegler–Natta catalysts contain titanium, zirconium, vanadium, or chromium complexes, and an alkylaluminum as an alkyl source for the formation of the initial $M-CH_3$ complex.

Carbene Intermediates

An alternative to the mechanism just described involves carbene intermediates that have metal–carbon double bonds, M=C:

$$M^I-CH_2CH_3 \rightleftharpoons M^{III}{=}C\begin{smallmatrix}H\\CH_3\end{smallmatrix}$$

These carbene hydrides can react with an olefin to give an adduct, and then the coordinated alkene, carbene, and hydride react to give the higher alkyl:

$$M^{III}{=}C\begin{smallmatrix}H\\CH_3\end{smallmatrix} + \longrightarrow M{=}C\begin{smallmatrix}H\\CH_3\end{smallmatrix} \longrightarrow M^I-CH_2CH_2CH_2CH_3$$

8.12. NITROGEN FIXATION

A cycle for nitrogen fixation is given in Figure 8.15. A dinitrogen complex is formed, which undergoes a sequence of one-proton, one-electron transfer reactions. The first dinitrogen complex was synthesized in 1965. Since then many such complexes have been studied, and there is a wealth of evidence indicating that in the vast majority of cases dinitrogen binds to the metal end-on and that in this position addition of a proton is facilitated. The binding to the metal polarizes the ligand, increases its basicity, and facilitates the addition of the first proton and the formation of the diazenido ligand —NNH. The activation of dinitrogen is similar to the activation of the isoelectronic carbon monoxide.

In principle, the diazenido ligand in mononuclear complexes can adopt one of the following geometries:

1   2   3   4

The known diazenido complexes assume either the singly bent (2) or the doubly bent (3) configuration. Protonation of the singly bent complex occurs on the remote nitrogen, while protonation of the doubly bent species occurs on the nitrogen that is bound to the metal. In the first case, a *hydrazido* $(2-)$ complex is formed ($MNNH_2$), and in the second case a *diazene* complex ($MNHNH$). None of the known diazene complexes has been shown to yield ammonia. Nevertheless, some investigators believe the diazene is a key intermediate in nitrogen fixation. The participation of the hydrazide complex in the cycle(s) for nitrogen fixation is better documented. Configurations 5 and 6 are possible.

5
Two resonance forms of
the linear configuration.

and

6
The bent
configuration

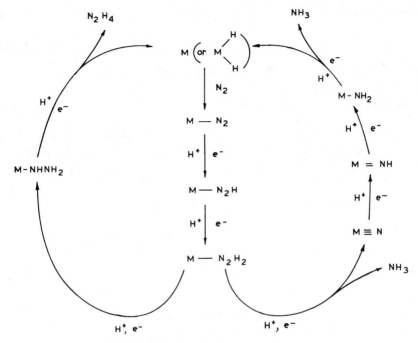

Figure 8.15. A cycle for nitrogen fixation.

Protonation of the hydrazido ligand can lead to ammonia or hydrazine, for example, by reactions such as

$$2Mo(NNH_2)Br(PPh_3)(triphos)^+ \xrightarrow{\text{2HBr + 2Br}^-}$$

$$2MoBr_3(triphos) + 2NH_3 + N_2 + 2PPh_3$$

where triphos is the tridentate ligand $C_6H_5P[CH_2CH_2P(C_6H_5)_2]_2$, bis(2-diphenylphosphinoethyl)phenylphosphine. In this reaction the coordinated hydrazido ligand disproportionates, and there is no need for an external reductant.

8.13. AN OVERVIEW

Comparisons

Two categories of activation have been discussed:

Activation of multiple bonds (e.g., double bonds, carbon monoxide).
Activation of single bonds (e.g., H—H, M—H, M—C)

The first category is associated with synergistic σ and π^* (back) bonding, the second with oxidative addition $\left(\text{e.g.,} \quad M^I + H_2 \rightarrow M^{III} \overset{H}{\underset{H}{\diagup}} \right)$, and either homolytic cleavage (e.g., $2M + H_2 \rightarrow 2M—H$) or heterolytic cleavage (e.g., $M^+ + H_2 \rightarrow M—H + H^+$).

In the category of single-bond activation, one should include the activation of C—H and C—C bonds. Homogeneously catalyzed oxidation and isomerization reactions of alkanes have not yet been developed on an industrial scale, but many efforts are made in this direction. What is particularly interesting is the activation of the simple saturated hydrocabons contained in abundance in natural gas.

In comparing the reactions

$$M + \overset{\diagup}{\underset{\diagdown}{C}}—C\overset{\diagup}{\diagdown} \longrightarrow M^{II} \overset{\diagup}{\underset{\diagdown}{C\overset{\diagup}{\diagdown}}} \qquad 8.22$$

$$M + \overset{\diagup}{\underset{\diagdown}{C}}—H \longrightarrow M^{II} \overset{\diagup}{\underset{H}{C\overset{\diagup}{\diagdown}}} \qquad 8.23$$

to the reaction

$$M + H—H \longrightarrow M^{II} \overset{H}{\underset{H}{\diagup}} \qquad 8.24$$

we note that the strength of the M—C bonds is somewhere in the 80–120 kJ mol^{-1} range, whereas that of the M—H bond is $\sim$240 kJ mol^{-1}. The strength of the C—C, C—H, and H—H bonds is 347, 414, and 431 kJ mol^{-1}, respectively. These values show that reactions 8.22 and 8.23 (especially reaction 8.22) are thermodynamically less favorable than reaction 8.24.

Classification Based on Interactions

Instead of classifying the activation processes as multiple- and single-bond activations, the classification can be based on the direct interactions involved.

Three categories of such interactions can be discerned:

1. The catalyst C interacts simultaneously with the reacting molecules A and B, but A and B do not interact with each other. Schematically,

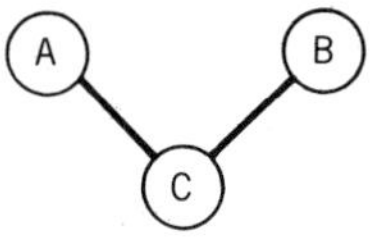

I 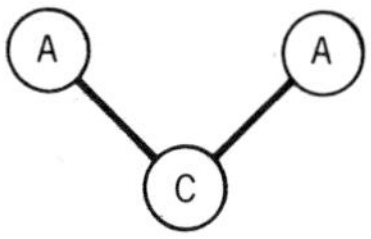   Ia

2. All three interact with each other:

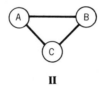

II

3. A and B interact with each other, but C interacts only with one of them, for example,

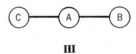

III

It is obvious that the first case does not lead directly to reaction between A and B. However, it is often the necessary first step. This is the case in hydrogenation, for example, where both the olefin and the hydride ion must first coordinate to the metal center (**I**) and then interact (**II**). Accordingly, the HOMOs and LUMOs for this interaction must be considered first, but then in following the orbital metamorphoses as the system goes from the reactant into the product state, the A–B interactions must also be considered.

Case 3 involves activation of a preexisting bond (single or multiple). It is the first step in many oxidative additions, in which case it develops into **I** or **Ia,** for example,

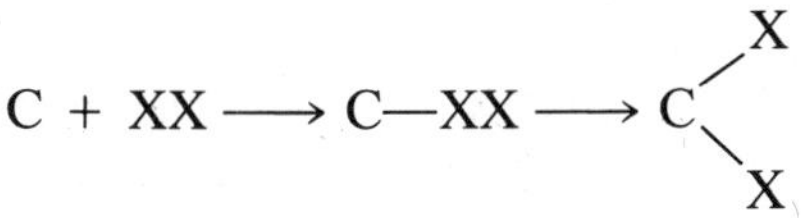

$$C + XX \longrightarrow C{-}XX \longrightarrow C\begin{smallmatrix} \nearrow X \\ \searrow X \end{smallmatrix}$$

It is also the first step in the activation for attack by another reagent, (nucleophilic or electrophilic) such as with η^1 coordinated carbon monoxide or dinitrogen.

So the recurring sequences in catalytic reactions are the following:

1. $C + A + B \longrightarrow C\begin{smallmatrix} A \\ B \end{smallmatrix} \longrightarrow C\begin{smallmatrix} A \\ | \\ B \end{smallmatrix} \longrightarrow$ Products

                        I   **II**

2. $C + A—B \longrightarrow C—A—B \longrightarrow C\begin{smallmatrix} A \\ B \end{smallmatrix}$ (or **Ia**) $\longrightarrow C\begin{smallmatrix} A \\ | \\ B \end{smallmatrix} \longrightarrow$ Products

              III   **I**   **II**

3. $C + A—B \longrightarrow C—A—B \xrightarrow[\text{another reagent D}]{\text{Direct attack by}}$ Products

              III

A large cycle may contain combinations of such sequences under various "disguises."

GENERAL REFERENCES

H. Alper, ed., *Transition Metal Organometallics in Organic Synthesis,* Vols. 1 and 2, Academic, New York, 1976, 1978.

D. C. Bailey and S. H. Langer, Immobilized Transition-Metal Carbonyls and Related Catalysts, *Chem. Rev.* **81**, 109 (1981).

Y. Chauvin, D. Commereuc, and F. Dawans, Polymer Supported Catalysts, *Progr. Polym. Sci.* **5**, 95 (1977).

N. M. Emanuel, Problems of the Selectivity of Chemical Reactions, *Russ. Chem. Rev.* **47**, 705 (1978).

J. Falbe and H. Bahrman (transl. by H. G. Gilde), Homogeneous Catalysis—Industrial Applications, *J. Chem. Educ.* **61**, 961 (1984).

P. C. Ford, ed., *Catalytic Activation of Carbon Monoxide (ACS Symp. Ser.* Vol. 152), Am. Chem. Soc., Washington, DC, 1981.

K. Fukui and S. Inagaki, An Orbital Interaction Rationale for the Role of Catalysts, *J. Am. Chem. Soc.* **97**, 4445 (1975).

J. Halpern, Mechanistic Aspects of Homogeneous Catalytic Hydrogenation and Related Processes, *Inorg. Chim. Acta* **50**, 11 (1981).

J. Halpern, Mechanism and Stereoselectivity of Asymmetric Hydrogenation, *Science* **217**, 401 (1982).

J. Halpern, Formation of Carbon–Hydrogen Bonds by Reductive Elimination, *Acct. Chem. Res.* **15**, 332 (1982).

R. Henderson, G. J. Leigh, and C. J. Pickett, The Chemistry of Nitrogen Fixation and Models for the Reactions of Nitrogenase, *Adv. Inorg. Chem. Radiochem.* **27**, 197 (1984).

J. K. Kochi, *Organometallic Mechanisms and Catalysis,* Academic, New York, 1978.

V. I. Labunskaya, A. D. Shebaldova, and M. L. Khidekel, Catalysis of Symmetry Disallowed Reactions, *Russ. Chem. Rev.* **43**, 1 (1974).

C. Masters, *Homogeneous Transition-Metal Catalysis: A Gentle Art,* Chapman and Hall, London, 1981.

E. L. Muetterties, Molecular Metal Clusters, *Science* **196**, 839 (1977).

G. W. Parshall, Industrial Applications of Homogeneous Catalysis. A Review, *J. Mol Catal.* **4**, 243 (1978).

C. U. Pittman, Jr., and R. C. Ryan, Metal Cluster Catalysts, *Chemtech.* **8**, 170 (1978).

G. N. Schrauzer, ed., *Transition Metals in Homogeneous Catalysis,* Dekker, New York, 1971.

D. F. Shriver, Activation and Reduction of Carbon Monoxide, *Chem. Br,* 19, 482 (1983).

D. L. Thorn and R. Hoffmann, The Olefin Insertion Reaction, *J. Am. Chem. Soc.* **100**, 2079 (1978).

E. Tsuchida and H. Nishide, Polymer-Metal Complexes and their Catalytic Activity, *Adv. Polym. Sci.,* **24**, 1 (1977).

J. Tsuji, *Organic Synthesis by Means of Transition Metal Complexes,* Springer-Verlag, New York, 1975.

R. Ugo, ed., *Aspects of Homogeneous Catalysis,* Vols. 1–3, D. Reidel, Publ., Dordrecht, Neth., 1970, 1974, 1977.

R. Ugo, ed., *Catalysis,* Specialist Periodical Reports (Chemical Society), Vols. 1 and 2, London, 1977, 1978.

PROBLEMS

1. Describe a perpetual motion device that could be constructed if a catalyst accelerated the reaction

$$N_2 + 3H_2 \longrightarrow 2NH_3$$

but not its reverse, while another catalyst accelerated only the reverse.

2. The reaction

$$C_2O_4^{2-} + S_2O_8^{2-} \longrightarrow 2CO_2 + 2SO_4^{2-}$$

is catalyzed by the couple $A + 2e^- = B$, which participates in the cycle

$$A + C_2O_4^{2-} \xrightarrow{k_1} B + 2CO_2$$

$$B + S_2O_8^{2-} \xrightarrow{k_2} A + 2SO_4^{2-}$$

Assume that these reactions are elementary and that B is in a steady state. Derive the rate law in terms of the total concentration of the catalyst and the concentrations $[C_2O_4^{2-}]$ and $[S_2O_8^{2-}]$. What form does the rate law take if $[S_2O_8^{2-}] \gg [C_2O_4^{2-}]$, and what form if $[S_2O_8^{2-}] \ll [C_2O_4^{2-}]$? Using the mechanism, explain what these limiting forms mean.

3. The reaction of a substance S is catalyzed by the hydrogen ions that originate in the dissociation of a weak acid HA. The empirical rate law is

$$\text{Rate} = k[S][HA]^{1/2}$$

Propose a mechanism.

4. The empirical rate law for the reaction

$$H_2O_2 + 3I^- + 2H^+ \longrightarrow I_3^- + 2H_2O$$

is

$$\frac{d[I_3^-]}{dt} = k[H_2O_2][I^-] + k'[H_2O_2][I^-][H^+]$$

That is, the reaction proceeds by two parallel paths, one catalyzed and the other not catalyzed by hydrogen ion.

(a) Propose a mechanism for both pathways.

(b) The reaction is also catalyzed by VO_2^+, which can be regarded as a Lewis acid. What will be the form of the rate law in the presence of VO_2^+? What is the mechanism for the VO_2^+-catalyzed path?

5. It has been found that the decomposition of paraldehyde to aldehyde is slow in the absence of catalysts but can be accelerated by several substances. Among them, in order of increased efficiency, are $ZnCl_2 < HCl \ll HBr < BCl_3 < SnCl_4 < FeCl_3 < AlCl_3 < FeBr_3 < TiCl_4$. Correlate and interpret these data.

6. The decomposition of nitramine,

$$H_2NNO_2 \longrightarrow H_2O + N_2O$$

is influenced by bases B. The following results were obtained:

B	K_b	k (M min^{-1})
HPO_4^{2-}	2.0×10^{-7}	86
OAc^-	5.5×10^{-10}	0.50
$C_2O_4^{2-}$	2.2×10^{-10}	0.104
H_2O	$\sim 2 \times 10^{-16}$	$\sim 8 \times 10^{-6}$

Use K_b as a measure of the strength of the base, and comment on these results.

7. $[(CH_3)_4N]^{15}NO_3$ equilibrates isotopically with liquid N_2O_4 in 36 hours. The isotope exchange reaction is catalyzed by various bases. Propose a detailed mechanism.

8. The following observations were made:

   (a) The reactions of $S_2O_8^{2-}$ with Ce^{III} and Cr^{III} are catalyzed by Ag^+.

   (b) In both cases the rate is independent of the concentration of the trivalent ion:

   $$Rate = k[Ag^+][S_2O_8^{2-}]$$

   where $[Ag^+]$ represents the initial Ag^+ concentration.

   What conclusions can be drawn?

9. The reactions of H_2O_2 with various reducing agents is catalyzed by Mo(VI) but not by S(VI), in spite of the fact that these elements are in the same group in the periodic table (but in different families). Give a reasonable explanation.

10. Propose a mechanism consistent with all the following observations:

    (a) The reaction between Br_2 and $H_2C_2O_4$ *is not* catalyzed by $Mn^{2+}(aq)$ unless a small part of $Mn^{2+}(aq)$ is first oxidized to $Mn^{3+}(aq)$.

    (b) $Mn^{3+}(aq)$ is reduced rapidly by $H_2C_2O_4$. Yet in the presence of both $H_2C_2O_4$ and Br_2, $Mn^{3+}(aq)$ does not react until one of the two (Br_2 or $H_2C_2O_4$) is consumed. It is also known that Br_2 (with or without $H_2C_2O_4$) does not oxidize $Mn^{2+}(aq)$ to $Mn^{3+}(aq)$.

    (c) The rate of the reaction between Br_2 and $H_2C_2O_4$ in the presence of $Mn^{2+}(aq)$ and $Mn^{3+}(aq)$ is independent of the $Mn^{2+}(aq)$, Br_2, and $H_2C_2O_4$ concentrations, but it is first-order in the concentration of $Mn^{3+}(aq)$.

    (d) The addition of HF causes a decrease in the rate.

    (e) For the oxidation of Mn^{II} to Mn^{III}, oxidizing agents like Ce^{IV}, Co^{III}, and Mn^{VII} are usually used. Mn^{III} can also be prepared photochemically using blue light.

    Suggest further experiments to check the proposed mechanism.

11. The value of the rate constant for the hydrolysis of urea at 100°C is 4.2×10^{-5} s^{-1}. The enzyme-catalyzed reaction (urease, 5.1×10^{-10} M) at pH 8.0 and 20.8°C has a rate constant 3×10^4 s^{-1}. The corresponding activation enthalpies are 134 and 44 kJ mol^{-1}. Calculate the decrease in

the free energy of activation caused by urease. Comment on the effect of the enzyme on the enthalpy and entropy of activation.

12. In the catalytic hydrogenation of a mixture of compounds A_1 and A_2, the rates are proportional to the corresponding concentrations:

$$-\frac{d[A_1]}{dt} = k_1[A_1] \qquad -\frac{d[A_2]}{dt} = k_2[A_2]$$

What is the relation between the concentrations of the remaining A_1 and A_2 at time t?

13. The reaction between β-D-glucose and oxygen to form hydrogen peroxide and δ-gluconalactone is catalyzed by glucose oxidase. The proposed mechanism is

$$E + G \xrightarrow{k_1} RL$$

$$RL \xrightarrow{k_2} R + L$$

$$R + O_2 \xrightarrow{k_3} E \cdot H_2O_2$$

$$E \cdot H_2O_2 \xrightarrow{k_4} E + H_2O_2$$

where E is the oxidized enzyme, R is the reduced enzyme, G is β-D-glucose, and RL and E · H_2O_2 are complexes.

(a) Sketch a catalytic cycle for the process.

(b) Assuming steady state, derive an expression for E_t/rate as a function of [G] and [O_2], where E_t is the total concentration of the enzyme.

(c) Modify this expression for E_t/rate by including the step

$$RL + G \xrightarrow{k_5} R + L + G$$

9

INORGANIC PHOTOCHEMISTRY

Each molecule has only one ground state but innumerable excited states. In principle, the chemistry of the excited states is limitless, and it will probably receive more and more attention as the various technical and theoretical tools improve.

These studies are important in themselves, but they can also help gain more insight about the reactivity of the ground state. It must be recalled that the activation needed for the ground-state molecules to be transformed into products can be regarded as a mixing of the ground-state wavefunction with wavefunctions of excited states.

A nice clean way to produce excited states is by the absorption of light. The focus here will be on electronic excitation.

9.1. GENERAL CONSIDERATIONS

The main differences between the ground and the excited states lie in their

Energy
Molecular geometry
HOMO–LUMO symmetry

These three aspects are, of course, interrelated.

Differences in ground- and excited-state reactivities have already been discussed within the context of the Woodward–Hoffmann rules and in relation to the symmetry of the HOMOs and LUMOs. Another example of a simpler species in the gas phase is provided by the reactivity of atomic oxygen produced photochemically from O_2 or O_3.

$$O_2 \xrightarrow{h\nu} 2O$$

$$O_3 \xrightarrow{h\nu} O_2 + O$$

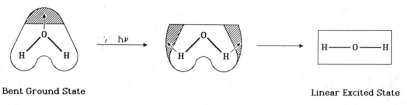

Bent Ground State   Linear Excited State

Figure 9.1. An illustration of the correlation between molecule geometry and excitation.

The ground state of atomic oxygen is ^3P ($L = 1$, $S = 1$) and corresponds to the configuration

The first excited state is ^1D ($L = 2$, $S = 0$), and its configuration is

For atomic oxygen, the ^1D state, in addition to being energetically at a higher level, has a stronger "acidic" (electrophilic) character because of its empty p orbital (its LUMO). Thus, with water, O(^1D) reacts readily:

$$O(^1D) + HOH \longrightarrow OH + OH$$

whereas O(^3P) does not.

The correlation between molecular geometry and excitation can be illustrated with the water molecule as in Figure 9.1. Excitation is equivalent to the transfer of electron density from the top of the molecule to the sides. This results in the appearance of new forces that act on the hydrogen nuclei and tend to linearize the molecule.

In the rest of this chapter we will concentrate on transition metal ion complexes in solution, and we will have the chance to see again the interplay between energy, molecular geometry, and orbital symmetry.

9.2. EXCITATION MODES IN TRANSITION ELEMENT COMPLEXES

Three fundamentally different electronic transitions can be observed in transition metal ion complexes

From a metal-centered to a metal-centered orbital. These are called *d–d transitions.*

From a metal-centered orbital to a ligand-centered one or vice versa. These are the *charge-transfer (CT) transitions,* which can be either ligand-to-metal (CTLM) or metal-to-ligand (CTML).

From a ligand-centered to a ligand-centered orbital (*intraligand transition*).

In this classification it is assumed that we can specify whether the orbitals are mostly metal- or ligand-centered, and this is indeed the case for most complexes. There are also cases in which such an assumption is not justified, but they will not be dealt with here.

The d–d Transitions

There are three kinds of *d–d* transitions: (1) promotional, (2) intraconfigurational, and (3) combinations of the two. These transitions are briefly described here for octahedral complexes. In the *promotional transitions,* one or more electrons are promoted from a t_{2g} orbital to an e_g^* orbital without change in spin multiplicity, for example,

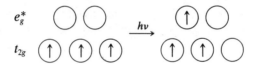

Absorptivities for transitions of this kind usually range between 10 and 100 M^{-1} cm^{-1}, and the spectra are broad. It is obvious that the position of a band corresponding to a $t_{2g} \rightarrow e_g^*$ promotion should depend on the energy difference between these levels, Δ_0, and consequently be sensitive to the nature of the ligands.

Intraconfigurational transitions involve spin flipping and pairing within a set of degenerate *d* orbitals, such as

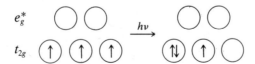

The bands are sharper than in the promotional transitions and are insensitive to the nature of the ligands. The absorptivities are very small; the bands are sometimes too weak to be observed, especially with 3d elements. For 4d and 5d elements, absorptivities are higher because of spin–orbit interaction.

The third kind of *d–d* transition is a *combination* of the other two: the electron is promoted, and simultaneously its spin changes. The energy of such transitions should be higher than the other two, because it is the sum of the promotional and pairing energies.

Bond Length and Bond Energy Changes Associated with *d–d* Transitions

If we recall that t_{2g} orbitals are less antibonding than the e_g^* (in accordance with common practice, t_{2g} orbitals are symbolized without an asterisk), we readily reach the conclusion that promotion of an electron from t_{2g} to e_g^* must be associated with bond lengthening and weakening. Relative to the ground state, the promotional excited state is generally distorted. This is illustrated in Figure 9.2, in which potential energy profiles of the ground and the lower excited state of an octahedral Cr^{III} complex are pictured.

It is seen in this figure that the minimum of the potential energy of the state T_{2g}, which involves promotion of one electron from t_{2g} to e_g^*, is displaced to the right. It is instructive to emphasize that these very electrons—the *d* electrons—that are of enormous interest to spectroscopists and photochemists actually thermodynamically *destabilize* the complexes—the e_g^* electrons more than the t_{2g} electrons, but the effect of both on the ligands is repulsive. The t_{2g} orbitals are also antibonding, and generally they should not be called nonbonding unless their antibonding character is compensated by back-bonding into empty π^* orbitals of the ligand.

Intraconfigurational excitations do not involve promotion into a more antibonding orbital. Hence, bond lengths and bond strengths in this case are expected to remain essentially the same. This is shown graphically in Figure 9.2 for the state ^2E_g of Cr^{III}.

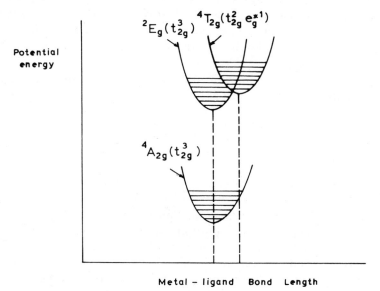

Figure 9.2. "Profiles" of potential energy curves for the ground state and the lower excited states of Cr^{3+} in an octahedral environment.

Charge-Transfer Transitions

Many ligands have π and π^* orbitals with energies comparable to those of the t_{2g} and e_g^* metal-centered orbitals. This is, for example, the case of the complex of Ru^{II} (t_{2g}^6) with bipyridine,[1] $[Ru(bipy)_3]^{2+}$. The qualitative energy diagram for this complex is given in Figure 9.3. The π and π^* orbitals are ligand-centered; the t_{2g} and e_g^*, metal-centered. The transition to the lower empty orbital is of the charge-transfer type, $(t_{2g})^6 \rightarrow (t_{2g})^5(\pi^*)^1$ (or simply $d \rightarrow \pi^*$). This transition is in a sense a cooperative one; it has no counterpart in Ru^{II} alone or in free bipyridine.

Metal-to-ligand charge-transfer transitions like the one just described weaken the bonds of the ligand but tend to strengthen metal–ligand bonds. Ligand-to-metal charge transfer transitions tend to have the opposite effect. This is the case, for example, for the absorption at 250 nm of the complex $Co(NH_3)_5Br^{2+}$, which has an absorptivity $\epsilon_{max} = 20,000 \ M^{-1} \ cm^{-1}$.

Charge-transfer transitions generally have high absorptivities, typically in the range of 10^3–$10^4 \ M^{-1} \ cm^{-1}$ or higher. They are also strongly affected by the solvent.

Intraligand Transitions

The energy-level diagram of $Ru(bipy)_3^{3+}$ resembles that of $Ru(bipy)_3^{2+}$ (Fig. 9.3). However, in $Ru(bipy)_3^{3+}$ the filled π_1 lies higher than the filled t_{2g}, and the lowest energy transition is $\pi_1 \rightarrow \pi_1^*$. Both of these orbitals are ligand-centered, and the transition essentially weakens only the bonding within the ligand. Accordingly, the physical and chemical behavior of the excited state is mainly related to the ligand, and the metal ion merely plays an indirect role, as it modifies the ligand orbitals.

Selection Rules for Electronic Transitions

For dipole transitions, the probability is related to the intensity, which depends on the magnitude of the integral:

$$R = \int \phi_j^* \overline{\mu} \phi_k d\tau \qquad 9.1$$

where ϕ_j, ϕ_k are the wavefunctions of the two stationary states involved in

---

[1] 2,2'-bipyridine, abbreviated bipy:

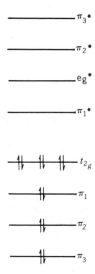

Figure 9.3. Qualitative energy-level diagram of $[Ru(bipy)_3]^{2+}$.

the transition and

$$\bar{\mu} = \sum q_i \bar{r}_i$$

where $\bar{r}_i$ is the vector from the origin to the point charge q_i. The operator $\bar{\mu}$ is antisymmetric, that is, it changes sign when $\bar{r}_i$ is replaced by $-\bar{r}_i$. If both wavefunctions ϕ_k, ϕ_j are symmetric or both are antisymmetric, their product does not change sign when $\bar{r}_i \rightarrow -\bar{r}_i$. Then the integrand in 9.1 is an odd function of $\bar{r}_i$ and the integral vanishes. The corresponding transitions are *forbidden*. This is the *Laporte rule*. Transitions from symmetric to antisymmetric states and vice versa are *allowed*. In transition metal complexes, d–d transitions are Laporte-forbidden, whereas charge-transfer and intraligand transitions of high intensity are allowed.

In an atom the state is even if Σl_i is even, odd if this sum is odd, where l_i is the angular momentum quantum number of the occupied orbitals.

The spin eigenfunctions of the two states are not functions of the space coordinates, and they are not affected by the dipole moment operator. Consequently, their product can be removed from the integral 9.1. However, if these spin functions correspond to different spin multiplicities, they are orthogonal and their product vanishes. Transitions involving change in spin (e.g., triplet $\rightarrow$ singlet) are forbidden.

Also forbidden are the one-electron transitions for which $\Delta l \neq 1$ (for weak Russell–Saunders coupling). Generally, the photon emitted or absorbed has an angular momentum quantum of $\hbar$ and since total angular momentum must be conserved, a compensating change in the angular momentum of the absorbing or emitting system is also necessary.

9.3. A SIMPLIFIED ELECTROSTATIC DESCRIPTION OF PROMOTIONAL EXCITATION

The process will be illustrated by describing a complex with octahedral geometry. The conventions adapted are given in Figure 9.4. The ligands are assumed to be identical.

The orbital shown in Figure 9.4 is the d_{xy}; the d_{yz} and d_{xz} orbitals are similar but lie between the y, z and x, z axes, respectively. As is well known, these three orbitals constitute the t_{2g} set. The $d_{x^2-y^2}$ orbital has higher energy and is directed toward the ligands 2, 3, 4, and 5, and the d_{z^2} is directed toward the ligands 1 and 6. These two orbitals constitute the e_g^* set. It must also be recalled that d_{z^2} is a linear combination of the $d_{x^2-z^2}$ and $d_{y^2-z^2}$ orbitals, which are analogous to $d_{x^2-y^2}$, except that they are directed toward other ligands. In the description that follows, $d_{x^2-z^2}$ and $d_{y^2-z^2}$ will be used, in order to find the degeneracy of the various states. We will also use d_{x^2}, d_{y^2}, which are analogous to d_{z^2}.

Configuration d^1

In the ground state, the electron can be placed in orbital d_{xy} or in orbital d_{yz} or in orbital d_{xz}: these three choices are equivalent. Hence, the state is triply degenerate, with spin multiplicity $2(1/2) + 1 = 2$. That is, the ground state is ^2T (T denotes triplet). In fact, it is a $^2T_{2g}$ state, because it is antisymmetric to a C_2 axis (x or y) which is perpendicular to the principal axis, and gerade (g), that is, symmetric with respect to the center of inversion.

Upon excitation, the electron can be transferred to d_{z^2} or to $d_{x^2-y^2}$. In the absence of interelectronic interactions (for the one-electron case only), these two positions are equivalent, and the excited state is doubly degenerate (^2E_g).

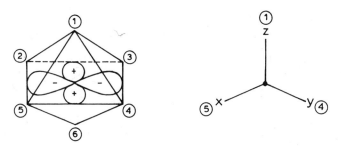

Figure 9.4. Labeling of ligands and axes in octahedral geometry.

For the same reason, that is, the lack of interelectronic repulsion, different positions within the same configuration are not differentiated energetically, and such transitions (interconfigurational) are not observed. Only one peak is expected, as in Ti^{III} complexes.

Configuration d^2

There are three equivalent ways to place the two electrons in the three t_{2g} orbitals:

$$d_{xy}d_{yz} \qquad d_{yz}d_{xz} \qquad d_{xz}d_{xy}$$

The ground state is triply degenerate, with spin multiplicity 3, $^3T_{1g}$. The subscript 1 means symmetry with respect to a C_2 axis perpendicular to the principal axis.

The *first* spin-allowed promotional excited state can be pictured to result from the ground state by a 45° rotation of one of the occupied orbitals as shown in Figure 9.5. Three such rotations are possible, giving a triplet first excited state ($^3T_{2g}$):

$$d_{xy}d_{y^2-z^2} \qquad d_{yz}d_{x^2-z^2} \qquad d_{xz}d_{x^2-y^2}$$

It is obvious that the rotation shown in Figure 9.5 should be associated with a deformation of the octahedron, because the 2345 plane expands (d electron–ligand repulsion), while the 2146, 1365 planes become rhombic (Fig. 9.6). In fact, there is also a difference between these last two planes, because one of them contains an electron while the other does not.

During this transition ($^3T_{1g} \rightarrow {}^3T_{2g}$) the repulsion *between* the d electrons does not change appreciably, because these electrons remain in different planes. Accordingly, this spin-allowed absorption depends only on the energy difference, Δ_0, between the t_{2g} and e_g^* orbitals. It does not depend on the Racah parameters, which are related to interelectronic repulsion.

The *second* spin-allowed promotional excited state can be pictured to result

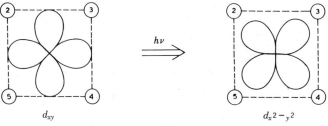

Figure 9.5. The $d_{x^2-y^2}$ orbital can be pictured to result from d_{xy} by a 45° rotation within the xy plane.

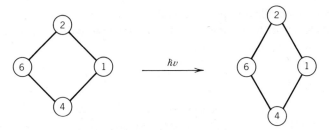

Figure 9.6. Deformation of the 2146 plane after absorption of a photon.

from a displacement of one electron from its plane in the ground state to the axis perpendicular to this plane. This means that the electron is put on the same plane with the other electron, but directed toward the axial ligands. Three combinations are possible; the resulting state is triply degenerate ($^3T_{1g}$) but different from the ground state (which is also $^3T_{1g}$):

$$d_{xy}d_{x^2} \qquad d_{yz}d_{y^2} \qquad d_{xz}d_{z^2}$$

These transitions ($^3T_{1g} \rightarrow {}^3T_{1g}$ excited) cause an increase in bond length along the axis where the electron goes and a shrinkage of the square where the electron came from. Generally, there is again a deformation of the octahedron. However, in this case there is also a change in interelectronic repulsion; the electrons are forced to reside in the same plane, and mutual repulsion increases. This adds to the energy required for the transition and also makes this transition dependent on the Racah parameters.

The *third* spin-allowed promotional excited state corresponds to a transition of both electrons from the t_{2g} to the e_g^* level. The two electrons go into the two e_g^* orbitals with parallel spins, and only one configuration is possible, $d_{z^2}d_{x^2-y^2}$. The state is $^3A_2$. In conclusion, three spin-allowed promotional excitations have been described for a d^2 ion, as for example in V^{III} complexes.

Configuration d^3

For the ground state there is only one choice, $d_{xy}d_{xz}d_{yz}$. This is a nondegenerate $^4A_{2g}$ state.

The *first* spin-allowed promotional excited state is a triplet, $^4T_{2g}$. It is equivalent to a rotation of one of the three orbitals by 45° (as shown in Fig. 9.5). The three resulting configurations are

$$d_{xy}d_{yz}d_{x^2-z^2} \qquad d_{xy}d_{y^2-z^2}d_{xz} \qquad d_{x^2-y^2}d_{yz}d_{xz}$$

Again, as in the first excitation of a d^2 ion, the electrons remain in separate planes and the interelectronic repulsion is not affected. The transition will depend on Δ_0, but not on the Racah parameters.

The *second* spin-allowed promotional excited state is also a triplet, $^4T_{1g}$. It corresponds to a transfer of one of the three electrons to the axis perpendicular to the plane it occupied originally.

$$d_{xy}d_{yz}d_{y^2} \qquad d_{xy}d_{x^2}d_{xz} \qquad d_{z^2}d_{yz}d_{xz}$$

This electron will then lie in the plane of either of the other two, which means increased interelectronic repulsion, dependence on the Racah parameters, and higher energy.

The *third* spin-allowed promotional excited state again involves, as in the d^2 case, transfer of two electrons, and the resulting state is again a triplet, $^4T_{1g}$:

$$d_{xy}d_{x^2-y^2}d_{z^2} \qquad d_{yz}d_{y^2-z^2}d_{x^2} \qquad d_{xz}d_{x^2-z^2}d_{y^2}$$

Two electrons now occupy higher levels, and the interelectronic repulsions are stronger. The state is energetically higher than the other two excited states.

Three spin-allowed peaks are expected for a d^3 ion, as in Cr^{III}.

A Useful Generalization

On the basis of the electrostatic description just given, one can anticipate that in the second and third excited states of a d^3 ion (sometimes designated as $^4L_2$ and $^4L_3$, where L means "ligand field" state), the interelectronic repulsions will tend to push the electrons away from each other. The electrons are separated from each other in the first excited state ($^4L_1$), where the interelectronic repulsion is minimized, and in the ground state. Thus, a tendency toward the first excited state, and of course eventually toward the ground state, is predicted.

It is, then, not surprising that most of the observed photochemical and photophysical events originate in the lower excited state.

Configuration d^4 (High Spin)

The configuration of the ground state can be either

$$d_{xy}d_{yz}d_{xz}d_{x^2-y^2} \qquad \text{or} \qquad d_{xy}d_{yz}d_{xz}d_{z^2}$$

It is, therefore, a doublet, ^5E_g.

In the *first* spin-allowed promotional excited state one electron goes from d_{xy} or d_{yz} or d_{xz} to the empty d_{z^2} (or $d_{x^2-y^2}$). The state is therefore a triplet, $^5T_{2g}$, and it obviously involves increased d electron–ligand and d electron–d electron repulsions.

TABLE 9.1. Complementary Pairs

	d^1	d^9	d^2	d^8
Third excited state			$^3A_{2g}$	$^3T_{1g}$
Second excited state			$^3T_{1g}$	$^3T_{1g}$
First excited state	^2E_g	$^2T_{2g}$	$^3T_{2g}$	$^3T_{2g}$
Ground state	$^2T_{2g}$	^2E_g	$^3T_{1g}$	$^3A_{2g}$

Simultaneous excitation of two electrons, without change in spin multiplicity, is not possible. Only one spin-allowed absorption peak is expected. An example would be Cr^{II}.

Configuration d^5 (High Spin)

The ground state is $^6A_1$:

$$d_{xy}d_{yz}d_{xz}d_{z^2}d_{x^2-y^2}$$

There is no way to transfer one or more electrons to another orbital without changing the spin. No spin-allowed transitions are possible. An example would be Mn^{II}, which has only weak spin-forbidden transitions.

Configurations d^6, d^7, d^8, d^9 (High Spin)

These configurations are treated like the d^4, d^3, d^2, and d^1 cases, respectively, but by considering the holes rather than the electrons; the order of the ground and the last excited state is reversed as in Table 9.1.

Low-Spin Configurations

The ground-state configurations are

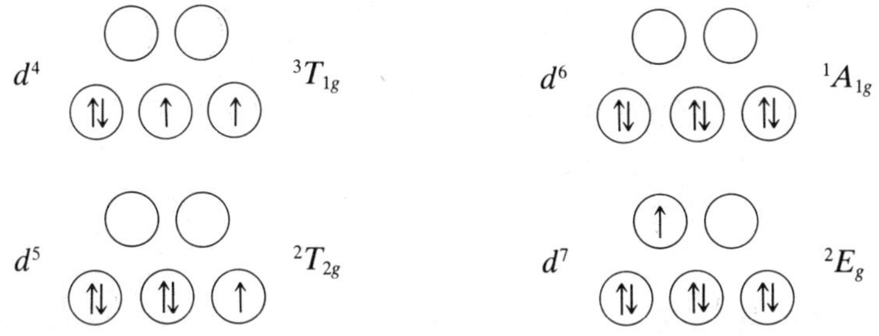

d^4 $^3T_{1g}$

d^5 $^2T_{2g}$

d^6 $^1A_{1g}$

d^7 ^2E_g

Finding the excited states is more complicated than in the high-spin cases and will not be discussed here.

9.4. WHAT HAPPENS AFTER THE ELECTRONIC EXCITATION

A molecule cannot just stay excited. It has to return to "peace": back to its own ground state or to the ground state of a chemically transformed species. Only the modes of deactivation vary.

Physical Modes of Deactivation

The molecule gives the extra energy it acquired upon excitation to its environment or to one particular part of the environment. This is usually done:

In "big lumps"
In "small pieces"

In the first case, we talk about photon emission, fluorescence, phosphorescence, and intermolecular energy transfer. In the second case, we have vibrational relaxation, energy degradation, internal conversion, and intersystem crossing. The meaning of each of these terms is explained next.

Photon Emission, Fluorescence, Phosphorescence

The excited state emits a photon and returns to the ground state. If the excited state and the ground state have the same spin multiplicity (same number of unpaired electrons), the emission is called fluorescence; if there is a change in spin multiplicity it is called phosphorescence.

The energy of the emitted photon, $E = h\nu$, is invariably smaller than that of the absorbed photon, because emission is always preceded by some kind of energy degradation. The energy that has been degraded is not available for photon emission.

Photon emission is unimolecular, a first-order process, that is, first-order in the concentration of the excited state. The rate constants for fluorescence are typically of the order of 10^8 s^{-1}. For phosphorescence, the rate constants are as a rule much smaller. Common rate constants for phosphorescence in transition metal complexes are in the range 10^2–10^5 s^{-1}.

Intermolecular Energy Transfer

A large part of the energy of the electronically excited (*donor*) molecule is transferred to another (*acceptor*) molecule.[2] The donor returns to the ground state or to a lower excited state, and the acceptor is excited. However, energy matching is never perfect. The energy received by the acceptor is, as a rule, less than that given by the donor. The rest of the energy is again degraded and is spread over the environment. The reverse, that is, the process in which

---

[2]"Donor" or "acceptor" of energy, not of electrons as in other parts of this book.

the acceptor receives energy both from the donor and from the environment, is possible in principle, but inefficient.

The energy can be transferred as a photon: the donor emits it, and the acceptor absorbs it. This is the "trivial" mechanism. Of more interest is the "contact" mechanism, which requires an encounter between donor and acceptor, first-order in the concentration of the excited donor and first-order in the concentration of the acceptor. However, the overall energy transfer is usually diffusion-controlled. The slow step is not the second-order transfer itself, but the diffusion. Accordingly, rate constants are of the order of 10^8–10^9 M^{-1} s^{-1}.

Vibrational Relaxation, Energy Degradation

If the excited state is not distorted relative to the ground state, absorption of light is not expected to produce a vibrationally excited state. This case is schematically presented in Figure 9.7a. If the excited state is distorted (Fig. 9.7b), the electronically excited state is also vibrationally excited. This is a consequence of the Franck–Condon principle, which in effect dictates that the transition should be "vertical."

In solution, not only the distorted excited state itself, but also the surrounding solvent molecules suddenly find themselves in an arrangement that does not correspond to the new situation. The adjustment to the new situation is rapid, but by no means infinitely fast; it takes several picoseconds.

Additional insight into the process can be obtained by recalling that the d-electron transitions that lead to distortion are the promotional ones from

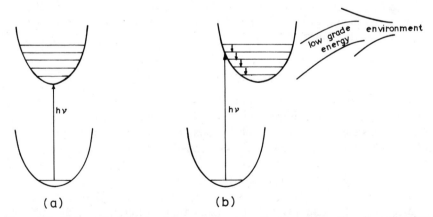

Figure 9.7. If the excited state has the same geometry as the ground state, there is no vibrational excitation. If it is distorted, the resulting electronically excited state is also excited vibrationally. In (a) the absorption maximum corresponds to the difference between the minima, in (b) it is larger than this difference. The relaxation to a vibrationally equilibrated excited state (the "thexi" state), which is associated with dissipation of heat to the environment, is fast.

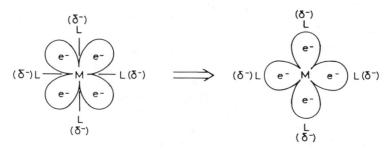

Figure 9.8. Electrostatic model for $t_{2g} \longrightarrow e_g^*$ transitions.

t_{2g} to e_g^*. These transitions correspond to an intramolecular displacement of the d electron from the space between the ligands onto a metal–ligand axis.

The time required for this transition is several orders of magnitude less than the time required for the ligands (L) to move. Thus, the ligands are found at first in the position they were in before the transition, but now new repulsive forces have been created between the d electron and the effective negative charge of the ligands (Fig. 9.8). Under the influence of these forces, L starts moving away from M toward the new equilibrium position. The molecule will now vibrate around this position, at first with a large amplitude, which is gradually damped, until it reaches thermal equilibrium (a Boltzmann distribution). This "thermally equilibrated excited" state has been called by Adamson the "thexi" state.

Vibrational relaxation results in a decrease in the energy of the system, but, of course, energy is not lost. It is simply dissipated to the environment (Fig. 9.7b) in the form of "low-grade" energy, energy of low availability. This is why the emitted photon has less energy than the absorbed photon, and also why in energy transfer the acceptor usually receives less energy than that given by the donor.

Internal Conversion, Intersystem Crossing

Energy dissipation to the environment does not take place only during the relaxation of the nonequilibrium vibrational distribution. It also occurs, often with equal efficiency, even after the establishment of vibrational equilibrium, and it is associated with a radiationless transition from one electronic state to another.

If the transition is from one electronically excited state to another state of the same spin multiplicity, the term *internal conversion* is used. If it is from one electronically excited state to another state of different spin multiplicity, it is called *intersystem crossing* (Fig. 9.9).

The rate constants for internal conversion from one excited state to another excited state are, like those of vibrational relaxation, of the order of 10^{12} s^{-1}. If other processes are to be observed (e.g., fluorescence), the internal con-

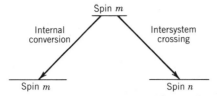

Figure 9.9. In internal conversion there is no change of the spin; in intersystem crossing the spin changes.

version from the lowest excited state to the ground state must be considerably slower. The rate constants of intersystem crossing in inorganic systems range between 10^9 and 10^{12} s^{-1}.

Internal conversion is characteristic of polyatomic molecules. Isolated smaller molecules do not show internal conversion, but if they are not isolated, as for example in solution where they are solvated, they can also be regarded as polyatomic aggregates. In such polyatomic aggregates, there is so much complexity in the vibrations and in the distribution of the electrons that it becomes possible for the distribution of one energy state to "look like" the distribution of another state. Distributions are not static; they are not even steady. They constantly change into a multitude of patterns, and it may very well happen that the patterns of different states will at some phases be the same. Then the system just does not "know" at what state it is, and even the slightest random fluctuation may push it one way or another.

This is a rough, qualitative, mechanistic description of internal conversion and intersystem crossing. However, even from this qualitative description it becomes apparent that, among the factors influencing internal conversion, the following should be considered:

1. The difference between the energy levels. It is reasonable to expect that with smaller differences the matching will be easier. This seems to be the reason why cascading to the lower excited state is fast, while internal conversion from this state to the ground state may be slow. The energy "distance" between the ground and the first excited states is always larger than the differences between excited states.

   In intersystem crossing, energy differences are usually small. Thus, in spite of the fact that these crossings involve violation of the spin conservation rules, they are often fast.
2. Higher molecular complexity favors faster internal conversion. There are more chances for coincidence.

Within this context it should be mentioned again that in multielectron, multinuclear systems there are a large number of minima resulting from configurational interactions, vibronic coupling, and so on. For instructional purposes, a "state" may be pictured as a single "potential profile" and a single energy value may be assigned to this "state." But it should always be remembered that there are a large number of minima, which are clustered around the deeper ones or even around the "global" one (the ground state), and that this should certainly facilitate internal conversion.

Understanding these and other factors is of the utmost importance, because internal conversion is by far the most frequent mode of deactivation, and intersystem crossing the best way of populating the state that will react chemically or emit light. A huge number of photons come in every day from the sun—a number that can satisfy more than the present energy needs of all humanity—but most of these photons are lost because of effective internal conversion to the ground state and ineffective intersystem crossing and because they are degraded to heat.

Photochemical Processes

Another "mechanism" of energy loss is chemical change. The excited state undergoes a chemical transformation, which effectively leads to at least partial dissipation of the absorbed energy.

The classification of the chemical reactions of the excited states into unimolecular and bimolecular is very important in practice, because unimolecular reactions *can be* (although they are not necessarily) as fast or faster than and competitive with vibrational relaxation, internal conversion, intersystem crossing, or light emission. On the other hand, bimolecular reactions cannot be faster than the rate of the (diffusion-controlled) encounter of the two reactants.

Reactions of the first category include substitution processes of dissociative character and proton transfer over a small distance.

Among the bimolecular processes are substitution reactions of an associative character and electron-transfer reactions.

As a rule, unimolecular processes are exergodic (dissipative). They cannot be used for storing energy. Their study is nevertheless important because it gives information on how to minimize losses, but also for developing new synthetic methods. The synthetic importance comes from the fact that light-driven dissociative processes often lead to different products than do the corresponding thermal reactions.

The distinction between dissociative and associative processes in ground-state chemistry usually requires extensive investigations and comparisons that are not always conclusive. In excited state chemistry the situation is even more difficult, because the short lifetimes of the excited states constitute a serious obstacle in obtaining meaningful data. General arguments can be formulated favoring either dissociation or association:

1. The increased energy of the excited state favors a dissociative mechanism, and this is indeed the case in fast substitutions, which are competitive with physical processes. An additional factor of a general nature

supporting the view that dissociative mechanisms dominate in substitution is the exergonic nature of most photochemical substitutions and the associated dissipation of heat. It seems easier to accept that this "energy scattering" is associated with "matter scattering" (dissociation) rather than with "matter organization" (association).

2. Excitation creates low-energy vacancies in the d orbitals of the transition metal complex and enhances its acceptor properties, which is of course a general factor favoring association.

Thus, the question of dissociation versus association is still open.

9.5. THREE CASES: PHOTOSUBSTITUTION, PHOTOSTEREOCHEMISTRY, AND PHOTOCHEMICAL ELECTRON TRANSFER

Photosubstitution in Rh(NH$_3$)$_5$X^{2+} (X = I, Br, Cl)

The ground state of the d^6 low-spin $Rh(NH_3)_5X^{2+}$ complexes is $^1A_{1g}$. Promotional photoexcitation into ligand field states (d–d excitation) is followed by rapid intersystem crossing to an orbitally doubly degenerate state with two unpaired electrons (spin multiplicity 3). This state, which is designated as ^3E, then undergoes substitution:

$$(^3E)\ \ Rh(NH_3)_5X^{2+} + S \begin{cases} \longrightarrow Rh(NH_3)_5S^{3+} + X^- & \quad 9.2 \\ \\ \longrightarrow \textit{trans-}Rh(NH_3)_4(S)X^{2+} + NH_3 & \quad 9.3 \end{cases}$$

where S is a solvent molecule.

With ground-state $Rh(NH_3)_5X^{2+}$, reaction 9.2 is the only one observed. With the excited state (^3E), both paths are observed. Another difference is that substitution from this excited state is more than 14 orders of magnitude faster than that from the ground state. This tremendous increase in lability can be attributed at least partly to the ^3E configuration. One electron has been promoted from the less antibonding t_{2g} to the more antibonding e_g^*:

$$t_{2g}^6\, e_g^{*0} \longrightarrow t_{2g}^5\, e_g^{*1}$$

and hence the σ bonding becomes weaker. A dissociative mechanism is widely accepted for this substitution.

Stereochemistry of Photosubstitution: Cr^{III} Complexes

Consider as an example the complex $Cr(NH_3)_5Cl^{2+}$. In the ground state, Cr^{III} (d^3) has one electron in each of the t_{2g} orbitals. As we have seen, the first promotional excitation is equivalent to a 45° rotation of one of these partly occupied orbitals. This is the excited state that undergoes substitution.

Suppose that the rotation occurs in the xz plane (Fig. 9.10a). In the notation of Figure 9.10, the z axis passes through the chloride ion ligand, and only three of the ammonia ligands are shown; the other two are out of the plane of the paper.

Thus, the excitation causes the negatively charged electron "cloud" to turn toward the negatively charged ligand atoms. The repulsive forces created lead

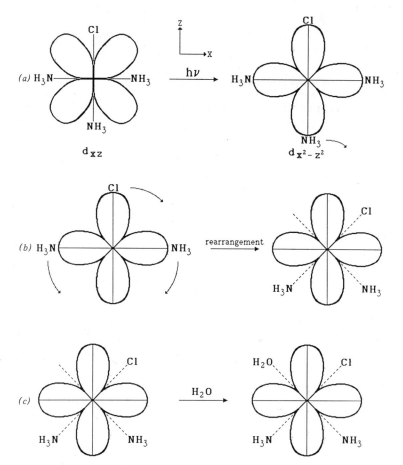

Figure 9.10. Schematic representation of photosubstitution. The absorption of the photon leads to the removal of the ammonia trans to Cl, the remaining ligands move away from the electron "cloud," and a water molecule enters cis to chlorine.

to removal of the ammonia molecule trans to the chloride ion, which is the more negative (repulsive) of all ligand atoms. In the resulting coordinatively unsaturated species, repulsions have not all been relieved. The chloride ion and the remaining ammonias are still repelled, but a further decrease in repulsion is achieved by the movement of these three ligands away from the d electron, as indicated in Figure 9.10b. Movement outside this plane is inhibited by the presence of the other two d electrons in the xy and yz planes.

The incoming group (e.g., H_2O) should necessarily enter into the vacancy, which is cis in relation to the chloride ion ligand (Fig. 9.10c). This simple electrostatic model predicts cis (to Cl) substitution products, as is indeed observed experimentally.

Redox Reactions of Excited Ru(bipy)$_3^{2+}$

At this point, tris(2,2'-bipyridine)ruthenium(II), Ru(bipy)$_3^{2+}$, is perhaps the complex that has been studied photochemically more than any other.

In contrast to the ground state, in which all spins are paired, the reactive excited state of Ru(bipy)$_3^{2+}$ has unpaired electrons. However, because of strong spin–orbit coupling it is inaccurate to attach spin labels to this excited state, that is, to call it a triplet. At any rate, return to the ground state by photon emission is spin-forbidden and slow, which allows time for participation in bimolecular reactions.

The energy of this reactive state corresponds to 18,000 cm^{-1} (550 nm) and the half-life (at 77 K) is 13.9 μs. It is also interesting that this level is populated very rapidly with 100% efficiency, even if the initial excitation is to a higher level. It is also important to note that this excited state has charge-transfer character (metal-to-ligand), which means that compared to the ground state it is better both as an oxidant and as a reductant. In fact, this charge-transfer character seems to be quite a common requirement for excited states participating in electron-transfer reactions, although redox reactions not originating in a change-transfer state are also known.

The promotion of the electron to a higher orbital makes its removal easier; for example,

$$\text{Excited Ru(bipy)}_3^{2+} + Eu^{3+} \longrightarrow \text{Ru(bipy)}_3^{3+} + Eu^{2+}$$

but also, this same promotion creates a "hole," since it makes one orbital half-empty and makes the molecule a better acceptor:

$$\text{Excited Ru(bipy)}_3^{2+} + Eu^{2+} \longrightarrow \text{Ru(bipy)}_3^{+} + Eu^{3+}$$

Within this context it is also worth mentioning again that excitation does not simply involve an increase in the energy content; it is also accompanied by a change in the HOMOs and LUMOs and in the geometry of the molecule.

9.6. COMPETITION

The processes described in Section 9.5 are generally competitive. It is therefore instructive to consider some simple competition kinetics. Examine the following processes:

Process	Elementary Reaction	Rate
Absorption of light	$A + h\nu \longrightarrow A^*$	I_a
Emission of light (luminescence)	$A^* \longrightarrow A + (h\nu)_{\text{lum}}$	$k_{\text{lum}}[A^*]$
Unimolecular chemical reaction	$A^* \longrightarrow B$	$k_{\text{chem}}[A^*]$
Quenching	$A^* + Q \longrightarrow P + Q^*$	$k_q[A^*][Q]$

I_a represents the number of photons absorbed per unit time. The *quencher* Q in the last (bimolecular) reaction can either modify the excited absorber A^* chemically or simply take away its energy of excitation. In the latter case, we talk about energy transfer.

Continuous Illumination

With continuous illumination the excited molecule A^* will quickly reach a steady-state concentration given by the relation

$$0 = \frac{d[A^*]}{dt} = I_a - k_{\text{lum}}[A^*]_{\text{ss}} - k_{\text{chem}}[A^*]_{\text{ss}} - k_q[A^*]_{\text{ss}}[Q]$$

$$[A^*]_{\text{ss}} = \frac{I_a}{k_{\text{lum}} + k_{\text{chem}} + k_q[Q]}$$

The *quantum yield* of a process is defined as the ratio of the rate of this process divided by the rate of light absorption:

$$\phi = \frac{\text{process rate}}{\text{rate of light absorption}}$$

Thus, for a unimolecular chemical reaction,

$$\phi_{\text{chem}} = \frac{k_{\text{chem}}[A^*]}{I_a}$$

At steady state,

$$\phi_{\text{chem}} = \frac{k_{\text{chem}}}{k_{\text{lum}} + k_{\text{chem}} + k_q[Q]} \qquad 9.4$$

Analogously, the quantum yield for luminescence is

$$\phi_{lum} = \frac{k_{lum}}{k_{lum} + k_{chem} + k_q[Q]}$$

9.5

At zero quencher concentration,

$$\phi°_{chem} = \frac{k_{chem}}{k_{lum} + k_{chem}}$$

9.6

Dividing equation 9.6 by equation 9.4, we obtain

$$\frac{\phi°_{chem}}{\phi_{chem}} = \frac{k_{lum} + k_{chem} + k_q[Q]}{k_{lum} + k_{chem}} = 1 + K_{sv}[Q]$$

9.7

where

$$K_{sv} = \frac{k_q}{k_{lum} + k_{chem}}$$

Thus, according to equation 9.7, a plot of $[\phi°_{chem}/\phi_{chem}]$ versus the concentration of the quencher should be linear. This is the so-called Stern–Volmer plot, which essentially illustrates the concentration dependence of the competition between a first- and a second-order process.

Pulsed Illumination

If the sample is not illuminated continuously but absorbs a pulse of light of short duration, the concentration of the excited state is first built up quickly and then decays according to the equation

$$\frac{d[A^*]}{dt} = -k_{chem}[A^*] - k_{chem}[A^*] - k_q[A^*][Q]$$

which upon integration gives

$$[A^*]_t = [A^*]_0 \, e^{-k_d t}$$

where $k_d = \tau^{-1} = k_{lum} + k_{chem} + k_q[Q]$, and τ is the *lifetime* of the excited state. During the decay, which is fast, it is assumed that [Q] remains constant.
  In terms of τ, the yields (equations 9.4 and 9.5) become

$$\phi_{chem} = \tau k_{chem}$$

$$\phi_{lum} = \tau k_{lum}$$

Finally, it must be emphasized that the shortness of the lifetimes of the excited states has an important practical implication: these states can participate only in fast reactions. If such reactions are not possible, there is simply no photochemical change.

What to Measure

From what has been said so far it becomes obvious that in photochemical experiments we measure

The wavelength and intensity of the absorbed light
The amount and identity of the products
Rates of decay of the emission or of the excited state absorption

From these measurements, quantum yields, rate constants, and lifetimes are estimated.

More about these measurements can be found in specialized texts (see end-of-chapter reference listing).

Counting Losses

Take as an example the photochemical process

$$\text{Excited Ru(bipy)}_3^{2+} + \text{Cr(bipy)}_3^{3+} \longrightarrow \text{Ru(bipy)}_3^{3+} + \text{Cr(bipy)}_3^{2+}$$

Energetically, the ground-state products Ru(bipy)_3^{3+} and Cr(bipy)_3^{2+} are 146 kJ mol^{-1} higher than the ground-state reactants but 59 kJ mol^{-1} lower than excited Ru(bipy)_3^{2+} and ground-state Cr(bipy)_3^{3+} (Fig. 9.11). This means that

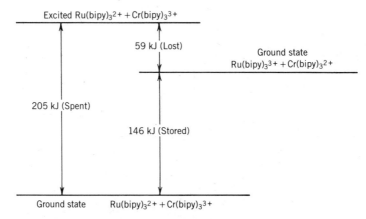

Figure 9.11. Energy diagram for the $\text{Ru(bipy)}_3^{2+} + \text{Cr(bipy)}_3^{3+}$ system (energies in kJ mol^{-1}).

at least 205 kJ mol^{-1} is needed[3] to excite Ru(bipy)$_3^{2+}$, but only 146 kJ mol^{-1} are stored in the products. The rest is lost as heat. But this is not all. If the excitation is to higher electronic or vibrational levels, the corresponding excited Ru(bipy)$_3^{2+}$ undergoes rapid internal conversion and/or thermal equilibration. Also, the back-reaction—that is, the reaction between Ru(bipy)$_3^{3+}$ and Cr(bipy)$_3^{2+}$—is exergodic and fast, and it also tends to decrease the yield and to contribute to the losses.

In summarizing, there are losses because

The minimum of excitation is higher than the level of storage.

The excitation may be at a level higher than the minimum required.

The back-reaction contributes to the dissipation of energy.

In addition, we may also have losses from:

Radiative transitions

Incomplete absorption of light

Scattering of light

Classical thermodynamics does not forbid the complete (100%) transformation of the energy of a photon into chemical energy. Yet this never happens. Light and matter never match the way we wish them to, and this creates entropy (disorder, waste) that we may call entropy of interaction or entropy of mismatching.

9.7. SOLAR ENERGY CONVERSION AND STORAGE

Alternatives

Solar energy can be converted into heat. This is easy, and it can be done with 100% efficiency. Heat then can be used directly, for example, for heating space, or it can be transformed back into other forms of energy, such as into chemical energy or into electricity, or it can be used to perform mechanical work.

However, if light is first transformed into heat and then into electricity, the efficiency is no longer 100%. This is dictated by the second law of thermodynamics.

Light can also be transformed into electricity directly, for example, using *photovoltaic cells*. These "solar cells" are quite efficient, although still ex-

---

[3]One mole of photons (6.023×10^{23} photons) is called an einstein. For 350-nm photons, this corresponds to 343 kJ mol^{-1}; for 400-nm photons, 300 kJ mol^{-1}; for 700-nm photons, 170 kJ mol^{-1}; and for 800-nm photons, 150 kJ mol^{-1}. It is also worth recalling that 1 eV = 8066 cm^{-1} = 23.06 kcal mol^{-1} = 96.48 kJ mol^{-1}.

pensive, and they are already in use in electronic devices, in space, and in remote areas on earth. As they become less expensive they will certainly become more competitive in relation to oil and coal.

Solar cells have two basic disadvantages:

1. A huge surface area is needed to collect appreciable amounts of energy. In a good sunny countryside during a sunny day at noon, we cannot collect more than ~500 W per square meter.
2. Electricity is not stored as such. It can be stored only after being transformed into some other form of energy, such as chemical energy, as in rechargeable batteries.

For the first of these disadvantages there is nothing that can be done. It is an inherent feature of solar energy. Nevertheless, things are not all that bad. The total amount of energy falling on earth in 15 days is equivalent to the energy contained in all of the fossil fuels available on earth. This is a huge amount, and although the whole surface of the earth cannot be used to collect it all, a good part of it can still be collected by using regions that are not useful for other purposes.

For the second disadvantage there are two alternatives. One is to produce electrical energy and then store it, for example, in the form of a dihydrogen–dioxygen mixture obtained by the electrolysis of water. The yield for this transformation can be as high ast 70%. The other alternative is to transform solar energy directly into chemical energy. Here, only this last alternative will be examined. In fact, we will discuss a special case—how to directly convert solar energy into chemical energy in homogeneous solutions.

As it stands now, most of the solar energy falling on earth is directly transformed into heat:

$$h\nu \longrightarrow \text{heat} \qquad\qquad 9.8$$

What we want is to interfere and to transform $h\nu$ into a more useful form of energy. Eventually, when this form is used, it will also end up as heat. We do not stop process 9.8, we merely delay it and make it happen in a more useful way.

Basically, there are two ways to transport energy: (1) as electricity and (2) as fuel. In the first case there is no need for large amounts of consumable raw materials, other than the materials needed for the installations. Electricity, however, cannot satisfy all our needs. Fuel is also needed to move cars and trains, run engines, and generate electricity, and this fuel is not recycled; it is scattered as exhaust gases and heat. Thus, the starting material for preparing this fuel must be cheap, readily available, and plentiful.

The most attractive candidate is water. Energy can be stored by splitting water into dihydrogen and dioxygen. This is an endergonic process, and dihydrogen is an excellent and clean fuel.

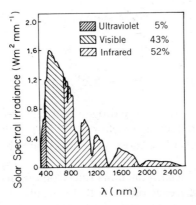

Figure 9.12. Solar spectrum at the surface of the earth.

Solar Light

The temperature inside the sun is of the order of a million kelvin. The surface is "cooler," only 6000 K! A typical spectrum of solar light at the surface of the earth is given in Figure 9.12. It is mostly in the infrared (52%) and the visible (43%). The UV portion is small (5%).

9.8. PHOTOCHEMICAL SPLITTING OF WATER

Basic Chemistry and Thermodynamics

For each dioxygen molecule formed, two water molecules are needed:

$$\overset{(+\mathrm{I})\ (-\mathrm{II})\ (+\mathrm{I})}{2\ \mathrm{H-O-H}} \longrightarrow \overset{(0)\quad(0)\qquad(0)\quad(0)}{2\ \mathrm{H-H} + \mathrm{O=O}}$$

This is the least that can be done: the splitting of two water molecules, which is a four-electron change; four hydrogens go from the $+\mathrm{I}$ oxidation state to the zero oxidation state, and two oxygens from $-\mathrm{II}$ to 0. There is a simultaneous reduction ($\mathrm{H^I} \to \mathrm{H^0}$) and oxidation ($\mathrm{O^{-II}} \to \mathrm{O^0}$).

The standard free energy for the water-splitting reaction is 474 kJ per mole of dioxygen produced, or half of this figure per mole of water split. The corresponding enthalpy is 572 kJ per mole of dioxygen. These values are for liquid water and gaseous products.

In the splitting process, water disproportionates. It is self-reduced and self-oxidized. However, reduction can also be brought about by another substance, not by water itself; for example,

$$\mathrm{H_2O} + \mathrm{V^{II}\text{-}cysteine(aq)} \longrightarrow \tfrac{1}{2}\,\mathrm{H_2} + \mathrm{V^{III}\text{-}cysteine(aq)} + \mathrm{OH^-}$$

In this reaction, only the oxidation state of hydrogen changes; the oxidation state of oxygen remains the same.

Similarly, there may be only oxidation, as in the reaction

$$2Ce^{4+}(aq) + H_2O \xrightarrow{\text{heat}} 2Ce^{3+}(aq) + 2H^+ + \tfrac{1}{2}O_2$$

Here it is the oxidation state of hydrogen that remains invariant.

The oxidation or reduction reactions of water with "external reagents" (also called "sacrificial") like those just quoted are thermal, but others can be assisted by light.

There are also sacrificial reductants (electron donors) and oxidants (electron acceptors), which react only photochemically. In any case, sacrificial donors and acceptors are not very useful for solar light conversion and storage, because they involve consumption of reagents, which require energy to prepare. Nevertheless, their reactions are simpler than the cyclic self-reduction and self-oxidation of water, and in this respect they are precious for giving information and insight about the mechanism. They are also useful in preparative chemistry.

It can also happen that the reduction product is not dihydrogen, that the electron is transferred to another compound, to form another energy-rich molecule. This, for example, happens in photosynthesis by green plants and green, red, or blue-green algae, where water is oxidized to dioxygen but is not at the same time reduced to dihydrogen. What is reduced instead is carbon dioxide:

$$CO_2 + H_2O \xrightarrow{h\nu, \text{ catalyst}} [CH_2O] + O_2$$

Number of Photons

The correspondence between the number of photons absorbed and the number of electrons is not necessarily 1 to 1. The number of photons can be fewer than the number of electrons. It is even conceivable that only one photon is used in the transfer of the four electrons. But this single photon must supply 474 kJ per mole of dioxygen, that is, it must have a wavelength in the UV range. Even if the overall four-electron process is driven by two photons, the wavelengths needed (counting inevitable losses) are smaller than 500 nm, and most of the solar light is still not usable. Three quanta for four electrons is a strange combination. Three-photon–four-electron processes have not been reported, but there is really no a priori reason why they should not happen.

Presently, the only processes that seem to have a good practical outlook are the four-photon–four-electron processes, and this is perhaps not too bad after all! Nature, in photosynthesis, does it more lavishly, with eight photons for the four electrons, that is, with two photons for each electron. But, of course, the plants do not carry out the photosynthesis in order to satisfy our needs for good yields. They photosynthesize to satisfy their own needs and to sustain their existence and growth; for this, high yields are not of such high priority.

Direct Routes

Two electrons are needed for each dihydrogen molecule, four for each dioxygen molecule. Therefore, solar protons must be absorbed and electrons (or "holes") must be transferred stepwise and somehow stored transiently until their number is sufficient to produce H_2 and/or O_2. Unfortunately, water itself is unsuitable for all these roles.

In the first place, water does not absorb in the solar wavelength range; it absorbs only below 200 nm. The energetics for a direct route, without mediators, are also unfavorable. Consider, for example, Figure 9.13, which is in effect the reverse of the reduction of dioxygen to water by one-electron steps (discussed in Chapter 5).

The average potential per electron is only -1.23 V, and if the four steps were equally spaced, the threshold wavelength required would be ~1000 nm. However, the steps are not equally spaced, and for the first step more than twice this potential is needed. It is estimated that even without counting losses, the photons needed to supply the energy required for the first step are in the vicinity of 440 nm. With the inevitable losses, the estimated wavelengths are much smaller, well into the UV region, and they are not available in the solar spectrum reaching the earth.

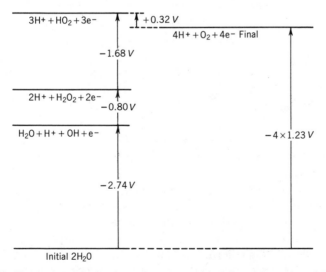

Figure 9.13. Energetics of the four one-electron steps in a "direct" formation of dioxygen from water for a 1 M H^+ ideal solution with ideal gases. With neutral solutions (pH 7), the potential (25°C) is $E^\circ_{neutral} = E^\circ_{1MH^+} + 0.059 \text{ pH} = E^\circ_{1MH^+} + 0.41$. The half-reactions are written as oxidation processes.

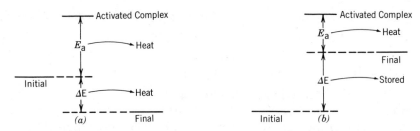

Figure 9.14. Energy losses in (a) exergonic and (b) endergonic processes.

It should also be recalled that whatever activation energy is required in any of the steps (whether endergonic or exergonic) is lost as heat (Fig. 9.14). For an exergonic step, there is also the loss of the energy of the reaction (Fig. 9.14a).

Another serious disadvantage of the direct route is that the photons must be absorbed successively by the intermediates (OH, H_2O_2, HO_2), which, however, are reactive and are present in low concentrations. In addition, in all these half-reactions electrons are given up, but to which species?

Are they given to single protons, successively?

$$e^- + H^+ \longrightarrow H$$

The energy penalty for this is high ($E° = -2.3$ V).

Are two electrons given to two protons, simultaneously?

$$2e^- + 2H^+ \longrightarrow H_2$$

There is no energy penalty here ($E° = 0.00$), but the probability for a simultaneous two-electron transfer process is small.

Perhaps they are given to water itself?

$$e^- + H_2O \longrightarrow H_2O^- \qquad E° = -2.7\ V$$

But water is a poor acceptor.

Still another obstacle is that the high-energy reactive intermediates considered can readily react in a number of ways other than the desired ones.

In short, water itself, without anything else added to it, and the intermediates derived from it (OH, H_2O_2, HO_2) are poor electron-storing (or hole-storing) species, and high thermodynamic barriers or kinetically unlikely steps are needed for dioxygen and dihydrogen formation. The use of direct schemes is not practical. The only practical approach seems to be to bypass completely the reactive water intermediates and invent schemes involving other, less energy demanding and kinetically more suitable intermediates.

Photosensitizers, Reversible Donors and Acceptors, Relays, Catalysts

Since water itself does not absorb solar wavelengths, it is necessary to use *photosensitizers*. The role of the photosensitizer is to absorb the photons and then to somehow transfer the energy to water or to intermediates in the route toward dihydrogen and dioxygen.

A photosensitizer must have the following properties. It must:

Be a good light absorber, in the wavelength range of the solar spectrum.
Have a sufficiently long-lived excited state to react bimolecularly (~1 ns for a diffusion-controlled reaction).
Undergo electron-transfer reactions in the excited state.
Not be destroyed by the irradiation.

Energy transfer from the excited state of the photosensitizer is associated with electron transfer, usually not directly to water but through *reversible donors and acceptors*. The necessity for reversible charge-storing donors or acceptors (*relays* as they are also called) comes from the multistep nature of the process and the fact that water itself and the intermediates derived from it are not efficient in charge storing. A relay may store just one charge, in which case many relay molecules are needed to split water; but it may store more than one charge—up to $+4$ for O_2 and up to -2 for H_2.

The transfer of the charges from the relays or from the oxidized or reduced photosensitizer to water is by no means a simple process, and it generally requires *catalysts,* ordinary catalysts for thermal reactions, which accelerate the slow steps and diminish losses due to high activation energy.

Some substances may play more than one role, but it is important to remember that the following three roles always exist:

Absorption of light
Charge transfer and storing
Catalysis

Reference to a specific hypothetical scheme will help clarify these points further. In this scheme, water is split photochemically by using a photosensitizer (S), catalysts (cat), a reversible donor (D), and a reversible acceptor (A).

The photosensitizer absorbs a photon:

$$S \xrightarrow{h\nu} S^*.$$

The excited photosensitizer reacts with the reversible acceptor:

$$S^* + A \longrightarrow S^+ + A^-$$

In the presence of a catalyst, dioxygen is formed:

$$4S^+ + 2H_2O \xrightarrow{\text{cat}_1} O_2 + 4S + 4H^+$$

The excited photosensitizer also reacts with the reversible donor:

$$S^* + D \longrightarrow S^- + D^+$$

and in the presence of a catalyst gives dihydrogen:

$$2S^- + 2H^+ \xrightarrow{\text{cat}_2} H_2 + 2S$$

The oxidized donor reacts with the reduced acceptor:

$$D^+ + A^- \longrightarrow D + A$$

In this scheme the positive or negative charges transferred to water come from the oxidized or reduced storing photosensitizer but alternatively they could have come from the reversible donor and acceptor:

$$4D^+ + 2H_2O \xrightarrow{\text{catalyst}} 4D + 4H^+ + O_2$$

$$2A^- + 2H^+ \xrightarrow{\text{catalyst}} 2A + H_2$$

Here dioxygen is not formed from S^+ but from D^+, and dihydrogen not from S^- but from A^-.

Another conceivable variation of the above scheme is that S^+ and/or S^- (not only S) also act as photosensitizers, and their excited states react further to give species with two charges stored:

$$S^+ \xrightarrow{h\nu} S^{+*} \xrightarrow{A} S^{2+} + A^-$$

$$S^- \xrightarrow{h\nu} S^{-*} \xrightarrow{D} S^{2-} + D^+$$

In any case, the initial mixture is quite complicated and becomes even more complicated because of the transient species formed. However, if the scheme is to work, side reactions should be minimal. Undesired reactions are those between D and A or between S^+ and S^-, the recombination of H_2 with O_2, and the inhibition of the desired reactions by some of the components or by some of the intermediates. The requirements are really quite demanding, and this is perhaps the reason a homogeneous system for cyclic water decomposition, with all the components in the same vessel, has not yet been developed. The chances for avoiding undesired reactions are better if some of the compounds are kept separately, in different compartments of the vessel or in different phases, and this is indeed where progress has been made so far.

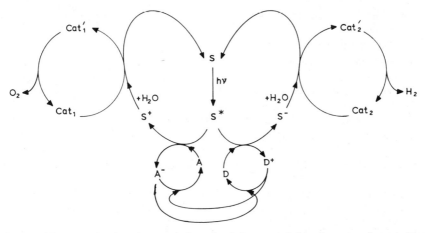

Figure 9.15. Example of a photocatalytic cycle. It is assumed that the same photosensitizer leads to O_2 and H_2, but generally this is not necessarily so. Side reactions are omitted. The cycles for cat_1 and cat_2 are expected to be highly complex, since the first must involve four S^+ and the second, two S^-.

A Return to Catalytic Cycles

In order to put the scheme just described in the format of a catalytic cycle as in Chapter 8, it must be taken into account that now there is not just one catalyst and its metamorphoses, but several catalysts, each with a specific function. Thus, even though the substance that absorbs the radiant energy and then transfers it is called a photosensitizer, it is essentially a catalyst facilitating the overall reaction by going back and forth from the ground state to the excited state. For this reason photosensitizers are also called *photocatalysts*. The reversible donors or acceptors also have the characteristics of a catalyst.

An illustration of a photocatalytic cycle is given in Figure 9.15. This cycle is not simply a network of cyclic paths involving "disguises" of the same catalyst, but a combination of five different cycles, one for S (with two branches), one for cat_1 leading to O_2, one for cat_2 leading to H_2, one for A, and one for D.

GENERAL REFERENCES

A. W. Adamson, J. Namnath, V. J. Shastry, and V. Slawson, Thermodynamic Inefficiency of Conversion of Solar Energy to Work, *J. Chem. Educ.* **61**, 221 (1984).

A. W. Adamson and P. D. Fleischauer, *Concepts of Inorganic Photochemistry,* Wiley-Interscience, New York, 1975.

V. Balzani, L. Moggi, M. F. Manfrin, F. Bolletta, and G. S. Laurence, Quenching and Sensitization Processes of Coordination Compounds, *Coord. Chem. Rev.* **15**, 321 (1975).

V. Balzani and V. Carassiti, *Photochemistry of Coordination Compounds*, Academic Press, New York, 1970.

J. R. Bolton, Solar Fuels, *Science* **202**, 705 (1978).

J. G. Calvert and J. N. Pitts, Jr., *Photochemistry*, Wiley, New York, 1966.

P. C. Ford, Ligand Substituent Effects in Transition Metal Photochemistry. The Tuning of Excited States, *Rev. Chem. Intermed.* **2**, 267 (1979).

L. S. Forster, Environmental Effects on Intra- and Intermolecular Photophysical Processes in Cr^{3+} Complexes, *Adv. Chem. Ser.* **150**, 172 (1976).

G. L. Geoffroy and M. S. Wrighton, *Organometallic Photochemistry*, Academic, New York, 1979.

P. Grutsch and C. Kutal, Mechanistic Inorganic Photochemistry II, *J. Chem. Educ.* **53**, 437 (1976).

P. Grutsch and C. Kutal, Mechanistic Photochemistry II, *J. Chem. Educ.* **53**, 437 (1976).

R. R. Hautala, R. B. King, and C. Kutal (eds.), *Solar Energy: Chemical Conversion and Storage*, Humana Press, Clifton, NJ, 1979.

B. R. Hollebone, C. H. Langford, and N. Serpone, The Mechanisms of Photoreactivity of Coordination Compounds: Limiting Cases of Decay on a Specific Nuclear Coordinate (DOSENCO) or via Random Coordinate Selection (DERCOS), *Coord. Chem. Rev.* **39**, 181 (1981).

M. A. Jamieson, N. Serpone, and M. Z. Hoffman, Advances in the Photochemistry and Photophysics of Chromium(III) Polypyridyl Complexes in Fluid Media, *Coord. Chem. Rev.* **39**, 121 (1981).

Journal of Chemical Education. The October 1983 issue (Vol. 60, No. 10) is dedicated to inorganic photochemistry.

T. J. Kemp, Inorganic Photochemistry, *Prog. Reaction Kinetics* **10**, 301 (1980).

A. D. Kirk, Chromium(III) Photochemistry and Photophysics, *Coord. Chem. Rev.* **39**, 225 (1981).

E. A. Koerner von Gustorf, L. H. G. Leenders, I. Fischler, and R. N. Perutz, Aspects of Organo-Transition-Metal Photochemistry and their Biological Implications, *Adv. Inorg. Chem. Radiochem.* **19**, 65 (1976).

C. Kutal, Mechanistic Inorganic Photochemistry, *J. Chem. Educ.* **52**, 502 (1975).

S. J. Lippard (ed.), *Progress in Inorganic Chemistry*, Vol. 30, Interscience, New York, 1983.

N. Sutin, Light-Induced Electron Transfer Reactions, *J. Photochem.* **10**, 19 (1979).

L. G. Vanquickenborne and A. Ceulemans, Ligand-Field Models and the Photochemistry of Coordination Compounds, *Coord. Chem. Rev.* **48**, 157 (1983).

M. S. Wrighton, Inorganic and Organometallic Photochemistry, *Adv. Chem. Ser.* 168 (1978).

PROBLEMS

1. Describe the spin-allowed transitions in high-spin d^6 and d^7 complexes using the electrostatic model of Section 9.3.

2. The energy difference between the zero vibrational levels of the lowest excited state of $Ru(bipy)^{2+}$ and the ground state is 2.12 eV. The ground-state reduction potentials of the couples $Ru(bipy)_3^{3+}/Ru(bipy)_3^{2+}$ and $Ru(bipy)_3^{2+}/Ru(bipy)_3^{+}$ are 1.26 and -1.28 V, respectively. Estimate the corresponding potentials of the excited state.

3. The following endergonic reactions can in principle be used for energy storage:

Reaction	$\Delta H°$, kJ mol^{-1}	$\Delta G°$, kJ mol^{-1}
$CO_2(g) \longrightarrow CO(g) + \frac{1}{2} O_2(g)$	283	257
$CO_2(g) + H_2O(l) \longrightarrow HCHO(g) + O_2(g)$	563	522
$CO_2(g) + 2H_2O(l) \longrightarrow CH_3OH(l) + \frac{3}{2} O_2(g)$	727	703
$CO_2(g) + 2H_2O(l) \longrightarrow CH_4(g) + 2O_2(g)$	890	818
$N_2(g) + 3H_2O(l) \longrightarrow 2NH_3(g) + \frac{3}{2} O_2(g)$	765	678

How many electrons are transferred in each case? What is the maximum wavelength needed if the energy is supplied by one, two, or four photons?

4. A photosensitizer absorbs light and reacts efficiently with several amine donors but not with acetone or alcohols. On the other hand, all attempts to detect a light-emitting excited state living long enough to undergo bimolecular reactions failed. Offer a plausible explanation.

5. Draw variations of the catalytic cycle of Figure 9.15. Include undesired side reactions.

INDEX

Absorption of light, 337, 356, 359
Accepton, electron, 131
Acceptor:
  binuclear, 271
  pi, 133
  proton, 31
  reversible, 365
  sigma, 133
Acceptor bonds, 133
Acceptor electrons, stabilization, 141
Acceptor LUMO, 137
Acceptor molecule, 132, 348
Acceptor nucleus, proximity, 146
Acceptor orbital, 139
Acceptor potential energy curve, 275
Acceptor sites, 275
Acetylacetonate complexes, 220
Acetyl-coenzyme, 263
Acid:
  isonicotinic, 80
  lactic, 263
  oxaloacetic, 263
  pyruvic, 263
Acid-base catalysis, 292
Acid catalysis, 209, 241
Acid dissociation constant, 24
Acid hydrolysis, 169, 204
Acoustical methods, 99, 104
AC polarography, 116
Action:
  antagonistic, 300
  nonantagonistic, 300
Activated, dioxygen, 161
Activated complex, 55–60, 67, 290
Activated reactants, 305

Activation:
  of carbon monoxide, 319
  of dihydrogen, 310, 312
  enthalpy, 59
  entropy, 59, 175
  of multiple bonds, 329, 311
  of single bonds, 329
  volume, 65, 175–176
Activation energy, 33, 51, 135, 154, 169, 291,
  314, 364
Activation parameters, 208
Activity, antitumor, 196
Addition, 168, 181
  Diels-Alder, 161
  intramolecular oxidative, 251
  multiple bond, 247
  oxidative, 244, 250, 253, 292, 310
  reductive, 292
Adduct, oxidant-reductant, 243
Adduct formation, catalyst-substrate, 303
Adducts, 294
Adiabatic, 271
Adsorption, 8
Alanine, 180
Alchemical, 1
Alkene:
  coordination, 326
  insertion, 326
Alkyl group, coordinated, 314
Allyl-type intermediate, 325
Alumina, 309
Aluminum, dimer, 233
Ammonia, 130, 329
Analytical integration, 11
Anation, 169, 200

Anchored catalysts, 309
Anisotropy, 122
Antagonistic action, 300
Antibonding character, 249
Antibonding LUMO, 189
Antibonding orbital, 275
Antiparallel spins, 153
Antisymmetric, 153
Antitransformation, 217
Antitumor activity, 196
Approximations, 11, 15, 298
Area, surface, 27
Arrangement, geometrical, 129
Arrhenius, 51–52
Associative mechanism, 170, 173, 191, 205
Asymmetric catalysis, 304
Asymmetric donor, 133
Atom-transfer, 162, 262
Attack:
    electrophilic, 293, 320
    nucleophilic, 133, 293, 319–320
Attraction, HOMO–LUMO, 141
Attractive interactions, 141
Average, weighted, 52
Axially coordinated water, 241

Band, intervalence, 281
Bandwidth, 283
Barrier:
    Berry rearrangement, 216
    Franck-Condon, 267
    potential energy, 57, 63
    reduction potential, 160
    thermodynamic, 364
Barrier heights, 63
Base catalysis, 241
Base hydrolysis, 169
Bases, Lewis, 223
Basicity, 31
Beam, crossed molecular, 44
Benzene, 31
Berry rearrangement barrier, 216
Beryllium, 187, 234
Beta-elimination, 255, 325
Bicapped tetrahedron, 221
Big cycles, 295
Bimolecular electron-transfer reaction, 276
Bimolecular reaction, 33, 44
Bimolecular redox reaction, 265
Binuclear acceptor or donor, 271
Bis(2-diphenylphosphinoethyl)-
    phenylphosphine, 329
Bite, 178
Boltzmann constant, 54, 59

Boltzmann distribution, 159
Bond:
    HOMO–LUMO, 139
    three-center, 184
Bond breaking, 8, 74, 231
Bond cleavage, 320
Bond energy, 340
Bond formation, 8, 74, 231
Bonding region, 291
Bond length, 272, 340
Boron, 183
Branched, 29
Bridge, monatomic, 238
Bridged binuclear intermediate, 235
Bridged complex, 170
Broadening region, outer-sphere, 121
Bromate ion, 88
Bromide ion, 88
Bromine, 17
Bromine trifluoride, 31
Bromophenol blue, 66
Bronsted-Bjerrum equation, 33

Carbene intermediates, 327
Carbon-carbon bond, 311
Carbon dioxide, 75
Carbon monoxide, 291, 318–320, 328
Carbon tetrachloride, 31, 129
Carbonylation of methanol, 319, 322–323
Carbonyls, rhodium, 323
Carriers, 28
Catalysis, 289, 291
    acid, 209, 241
    acid-base, 292
    asymmetric, 304
    base, 241
    general acid-base, 292
    heterogeneous, 289, 308
    homogeneous, 308
    by redox couples, 292
    specific acid-base, 292
Catalyst, Wilkinson, 317–318
Catalysts, 305–309, 315, 365
    homogenous, 289
Catalyst-substrate adduct formation, 303
Catalytic cycles, 294, 367
Catalytic decomposition, 6
Catalytic mechanism, 296
Catalytic process, 289
Cell, electrochemical, 114
Cells, photovoltaic, 359
Cellulose, 309
Chain-propagation, 28
Chain reactions, 27–29

Change:
  entropy, 59
  stereochemical, 204
  structural, 244
Character:
  antibonding, 249
  electrophilic, 338
Characterization of intermediates, 301
Charge, effective, 274
Charges, real effective, 234
Charge-transfer transitions, 339, 341
Chelate ring formation, 178
Chemically induced dynamic electron
    polarization, 124
Chemical reaction, unimolecular, 356
Chemical transformation, 295
Chlorate ion, 71
Chlorine, 19
Chlorine dioxide, 26
Chloroalkylphosphinoxides, 185
Chloroform, 47
Chlorous acid, 26
Chromium(II), 61, 265
Chromium(II) in situ, 103
Chromium(III), 103
Chronoamperometry, 116
Chronocoulometry, 116
Chronopotentiometry, 116
CIDEP, 124
Cis-dibromo(tren)cobalt(III), 200
Cis-effect, 195
Classical electrostatic forces, 144
Classification of solvents, 30
Cleavage:
  bond, 320
  heterolytic, 310
  homolytic, 310
Closed shell-repulsion, 5
Cloud, electron, 145
Clusters, 306
Coefficient, transmission, 59, 270
Coefficients, stoichiometric, 296
Collective modes:
  of activation, 306
  of binding, 306
Collision:
  nonreactive, 45
  triple, 160
Collisions, molecular, 44
Collision theory, 53
Combinations, orbital, 151, 274
Competition methods, 68
Competitive reactions, 26, 30
Complementary pairs, 347

Complementary reactions, 261
Complex:
  acetylacetonate, 220
  activated, 55–57, 59–60, 67, 290
  bridged, 170
  dinitrogen, 328
  direct overlap, 170
  inner-sphere bridged activated, 237
  olefinic, 212
  outer-sphere activated, 237
  precursor, 276
  square planar, 144, 191
  square tetragonal, 197
  tetrahedral, 187
  tropolonate, 220
Complex biological systems, 263
Complex formation, 119, 169
Complicated rate laws, 87
Composite rate constants, 52
Compressibility, 65
Concentration, steady state, 28
Concentration-time integral, 93
Concept:
  donor-acceptor, 131
  of force, 143
Concerted chelate ring formation, 182
Concerted dissociative-associative
    mechanism, 211
Condensation, 182, 296
Configuration, 55
  electronic, 129, 152
  inversion, 190, 217
  low spin, 347
  stable, 143–144
Configurational interactions, 351
Configuration d^1, 343
Configuration d^2, 344
Configuration d^3, 345
Configuration d^4, high spin, 346
Configuration d^5, high spin, 347
Configuration d^6, d^7, d^8, d^9, high spin, 347
Conjugate acid-base, 23
Connotations, mystical, 295
Consecutive reactions, 17
Conservation:
  of orbital symmetry, 147
  spin, 154, 156
Constant:
  acid dissociation, 24
  Boltzmann, 54, 59
  composite rate, 52
  coupling, 118
  cross-reaction rate, 280
  diffusion-controlled rate, 79

Constant (*Continued*)
Michaelis-Menten, 298
Planck's, 59
water exchange rate, 207
Contact time, 101
Continuous illumination, 356
Continuous slow passage, 117
Contour diagrams, 172
Conversion:
Fischer-Tropsch, 318
internal, 350
Coordinates, internal, 172
Coordination of alkene, 326
Coordination sphere expansion, 181, 243
Coordinatively unsaturated, 258
Coordinatively unsaturated intermediates, 294
Coplanar geometry, 312
Copper(II), 17, 324
Coulometry, 116
Coupled reactions, 111
Coupling, vibronic, 351
Coupling constants, 118
Criterion, maximum overlap, 132
Critical elementary step, 171
Cross-correlation, Marcus, 279
Crossed molecular beam, 44
Crossing, intersystem, 156, 350–351
Crossing point, 156
Cross-reaction rate constants, 280
Cross section, collision, 53
Curvature, 52
Cycle:
elementary, 294
photocatalytic, 367
Cycles:
big, 295
catalytic, 294, 367
small, 295–296
Cyclic process, 289
Cyclic voltammetry, 116
Cycloaddition, ethylene, 151
Cyclobutane, 149
Cyclohexene, hydrogenation, 318
Cyclometalation, 251
Cylindrical chemical flow reactors, 101

DC polarography, 116
Deactivation, modes, 348
Decomposition reactions, 167
Degeneracy, 343
Degradation, energy, 349
Delphi, 24
Destabilization of the donor electrons, 141

Deviations from equilibrium, 110
Dewar-Chatt-Duncanson model, 211
Diagram:
energy-level, 342
potential energy, 268
Diagrams, contour, 172
Diamagnetic organometallic complexes, 293
Diastereoisomers, 131
Dielectric constant, 31
Diels-Alder addition, 161
Differences in electronegativity, 273
Differential selectivity, 302
Diffuse, molecules, 78
Diffusion, 115
Diffusion-controlled, 72, 79, 349
Dihydrogen, 291
activation, 310, 312
Dihydronicotinamideadeninedinucleotide, 263
Dimerization of ethylene, 148
Dinitrogen, 136
Dinitrogen complex, 328
Dinitrogen tetroxide, 70
Dioxygen, excited, 161
Dioxygen activated, 161
Dioxygen complexes, mononuclear, 163
Dioxygen inert, 158, 161
Dipole force, 145
Dipole moment transitions, 266
Dipole transitions, 341
Direct electron transfer, 237
Direct overlap complex, 170
Disguises, 322
Displacement, 45
electron, 134
intramolecular, 350
Disproportionation, 255, 259
reductive, 256
Dissipation, energy, 350
Dissipation of heat, 349
Dissociative mechanism, 170, 173, 176, 187, 205, 352–353
Distance, equilibrium, 275, 281
Distribution:
Boltzmann, 159
statistical, 58, 132
Donor:
binuclear, 271
electron, 131
nonspherical, 132
pi, 133
proton, 31
reversible, 365
sigma, 133

Donor–acceptor concept, 131
Donor-acceptor model, 270
Donor-acceptor properties, 31
Donor bonding electrons, 244
Donor HOMO, 137
Donor molecule, 348
Donor orbital, 139
Donor overlap, 133
Dynamic electron polarization, 124

Effect, trans, 146, 194
Effective charge, 274
Effective charge force, 145
Effective electron transfer, 275
Effect of leaving ligand, 199
Effects:
    ionic strength, 33
    pressure, 65
    solvent, 30
Electrochemical method, 99, 114, 115
Electrochemical reduction, 242
Electron, promotion, 355
Electron acceptor, 131
Electron affinity, 134
Electron donation, nonbonding, 140
Electron donor, 131, 244
Electronegativity, 134, 272–273, 231
Electroneutrality, tendency, 234
Electron exchange reactions, 244
Electronically excited state, 113
Electronic configuration, 129, 152
    ground-state, 158
Electronic excitation, 337, 348
Electronic transition moment, 266
Electronic transitions, 341
Electron pair, hybrid lone, 145
Electrons:
    solvated, 239
    unpaired, 156
Electron spin resonance, 99
Electron transfer, 7, 115, 230, 244, 268, 277
    inner-sphere, 259
    intramolecular, 259, 276
    optical, 267, 280
    photochemical, 353
    thermal, 280
Electron transfer modes, 268
Electron transfer reactions, 70
Electrophilic, 171
Electrophilic attack, 293, 320
Electrophilic character, 319, 338
Electrostatic interactions, 152
Electrostatic model, 194, 350
Electrostatic repulsion, 5

Elementary conserted reactions, 148
Elementary cycle, 294
Elementary reactions, 9, 24, 34, 51, 171, 289
Elementary redox reactions, 235
Elementary steps, 44
Elimination:
    beta, 255
    gamma, 256
    reductive, 252–253, 255, 315
Emission, photon, 348
Emission of light, 356
Emission spectroscopy, 113
Encounter, 266, 289
Endergonic process, 360
Energetics, 363
Energized substrate, 293
Energy:
    activation, 33, 51, 135, 154, 160, 291, 314, 364
    bond, 340
    forces and potential, 145
    free, 32
    kinetic, 152
    linear free, 36
    transport, 360
Energy barrier, 5
Energy conversion, solar, 359
Energy curves, potential, 270, 274–275, 280–281, 340
Energy degradation, 349
Energy dissipation, 350
Energy-level diagram, 342
Energy matching, 151
Energy pathways, minimum, 173
Energy profiles, potential, 268
Energy storage, solar, 359
Energy surfaces, 56
Energy transfer, 348–349
Enthalpy, 59
Entropy, 59, 175
Enzymatic reactions, 296, 298
Equation:
    Bronsted-Bjerrum, 33
    Eyring, 58–59
    stoichiometric, 69, 290
Equatorially coordinated water, 241
Equilibria, fast, 18
Equilibrium:
    deviations, 110
    thermal, 58
Equilibrium distance, 271, 275, 281
Equilibrium position, 144
Ethyl acetoacetate, 17
Ethylene, 148–151

Europium(II), 80
Events, random, 49
Exchange:
  ligand, 169
  metal ion, 169
  solvent, 139, 169, 176
Exchange-controlled region, 121
Exchange force, 145
Exchange process, 82
Exchange reactions, 69, 81
Excitation:
  electronic, 337, 348
  promotional, 343
  vibrational, 349
Excitation modes, 338
Excited dioxygen, 161
Excited photosensitizer, 366
Excited states, 113, 151, 337, 344, 357
Exothermic, 30
Expansion of coordination sphere, 181
Experimental methods:
  for fast reactions, 99
  for slow reactions, 80
Experimental technique, 79
  choice of, 78
Explosions, 27, 29
External reagents, 362
Eyring equation, 58–59

Factor:
  electronic, 244
  frequency, 51
  geometrical, 130
  preexponential, 51
  steric, 129
Fast-exchange region, 121
Fast reactions, experimental methods, 99
Femtosecond spectroscopy, 99, 104
Fenton's reagent, 6
Final electronic state, 266
Finite differences, method, 83
First coordination sphere, 231
Fischer-Tropsch process, 318–320
Fitting relaxation times, 112
Fixation, nitrogen, 328–329
Flash photolysis, 99, 103
Flooding technique, 87
Flow techniques, 99–100, 102
Fluorescence, 99, 113, 348
Fluoro-bridged species, 187
FMO, 131
Forbidden transitions, 342
Force, 143–146
  Hellmann-Feynman, 188

exchange, 145
Forces:
  in HOMO interactions, 146
  in LUMO interactions, 146
  Pauli, 154
Formal oxidation numbers, 230, 234
Four-electron oxidizing agent, 159
Fraction, mole, 31, 32
Fragmentation, 45
Franck-Condon principle, 266–268, 349
Free energy, 32
Free energy of activation, Gibbs, 59, 64
Free radical, 160, 254
Free radical formation, 258
Free radical mechanism, 315
Frequency factor, 51
Frequency of precession, 116
Frontier molecular orbitals (FMO), 131
Functions, transfer, 32

Gamma-elimination, 256
Gas, synthesis, 321
Gases, kinetic theory, 129
General acid-base catalysis, 292
Geometrical factor, 130
Geometrical structure, 129
Geometry:
  coplanar, 312
  linear, 250
  molecular, 231, 338
  perpendicular, 314
Gibbs free energy, 59, 64
Glycine, 112
Glycolytic path, 263
Ground state, 113, 151, 338
Ground-state electronic configuration, 158
Group transfer, 167, 231
Guggenheim method, 86
Guldberg, 34

Half-lives, 85
Hamiltonian operator, 152
Happenings, molecular, 58
Harmonic frequency, 277
Heat, dissipation, 349
Heights, barrier, 63
Hellmann-Feynman force, 188
Hellmann-Feynman theorem, 144, 188, 194
Hemoglobin, 164
Heraclitus, 24
Heterocyclic rings, 316
Heterogeneous catalysis, 289–290, 308
Heterogenized homogeneous catalysts, 308
Heterolytic cleavage, 8, 310

Higher-energy surface, 155
Highest occupied molecular orbital, *see* HOMO (Highest occupied molecular orbital)
High-temperature limit, 277
Hindrance, stereochemical, 199
Historical note, 229
HOMO (Highest occupied molecular orbital), 131, 270, 331
  donor, 137
Homogeneous catalysis, 289–290, 308
HOMO–HOMO interactions, 223
HOMO–LUMO attraction, 139, 141
HOMO–LUMO interactions, 196, 273
HOMO–LUMO repulsion, 141
HOMO–LUMO symmetry, 337
Homolytic cleavage, 8, 310
Hybrid phase catalysts, 309
Hydrazine, 329
Hydride ion, 133, 315
Hydride ion transfer, 215
Hydridocobalt, 323
Hydrocyanation reaction, 316
Hydroformylation, 291, 319, 322
Hydrogen abstraction, 45
Hydrogenation, 75, 310, 318
  stoichiometric, 81
Hydrogen bonding, 31
Hydrogen fluoride, 130
Hydrogen peroxide, 25, 60–61
Hydrolysis, acid, 169, 204
Hydrolyzed monomers, 62
Hydrosilation reaction, 316
Hypochlorite ion, 24
Hypochlorous acid, 26

Illumination, 356–357
Immobilized catalysts, 309
Impact parameters, 45
Indicators in relaxations studies, 111
Indirect electron transfer, 237, 239
Induction period, 27
Inert, dioxygen, 158, 161
Inhibition, 88
Inhibitors, 291
Initiation, 27
Inner-sphere, 70, 169, 276
Inner-sphere bridged activated complex, 237
Inner-sphere electron transfer, 259
Insertion:
  of alkene, 326
  of carbon monoxide, 320
Insertion reactions, 213
Instability constant, 37

Integral:
  concentration-time, 93
  overlap, 132, 266
Integral selectivity, 302–303
Integration:
  analytical, 11
  method, 84
Interactions:
  attractive, 141
  configurational, 351
  electrostatic, 152
  forces in HOMO, LUMO, 146
  HOMO–HOMO, 223
  HOMO–LUMO, 196
  interelectronic, 343
  repulsive, 141
  two-center, 248
Interchange mechanisms, 170, 176, 184
Intermediate:
  allyl-type, 325
  bridged binuclear, 235
  metal-alkyl, 325
  metal hydride, 325
  unsaturated, 315
Intermediate complex, 299
Intermediates, 15, 26, 57, 62
  carbene, 327
  characterization, 301
  coordinatively unsaturated, 294
  reactive, 364
  stable, 79
Intermolecular, 253–255
Intermolecular energy transfer, 348
Intermolecular reactions, 258
Intermolecular rearrangements, 169
Intermolecular reductive elimination, 315
Internal conversion, 350
Internal coordinates, 172
Internal variables, 172
Internuclear distance, equilibrium, 271
Interpretations, mechanistic, 148
Intersystem crossing, 156, 350–351
Intervalence band, 281
Intervalence compound, 282
Intimate mechanism, 44, 248
Intraconfigurational transitions, 339
Intraligand transitions, 339, 341
Intramolecular displacement, 350
Intramolecular electron transfer, 259, 276
Intramolecular metathesis, 211
Intramolecular methyl transfer, 214
Intramolecular oxidative addition, 251
Intramolecular proton transfer, 222
Intramolecular rearrangements, 169

Intramolecular transfer, 140, 255
Intrinsic mechanism, 6
Inversion of configuration, 190, 217
Inverted donor, 275
Iodide ion, 24–25, 61
Ionic strength effects, 33
Ionization potential, 134
Ion radicals, 257
Iron(III), 18
Iron-dioxygen bond, 164
Isoelectronic carbon monoxide, 328
Isolation technique, 301
Isomerization, 169, 324
Isonicotinic acid, 80
Isotopic composition, 81
Isotopic labeling, 71
Isotopic tracers, 81

Kinetic energy, 152, 269
Kinetics, phenomenological, 48
Kinetic stability, 37
Kinetic theory of gases, 129

Labile, 38
Lability, 71, 206, 239
Labilization, 189
Lactic acid, 263
Laplace transforms, 97
Laporte rule, 342
Large cycle, 332
Lavoiser, 230
L-DOPA, 304
Levels, vibrational, 266, 273
Lewis acids and bases, 223
Lifetime, 113
Lifetime excited state, 357
Ligand:
    diazenido, 328
    effect of leaving, 199
    nature of entering, 174, 192
    nature of leaving, 193
Ligand exchange, 169
Ligands:
    diatomic, 249
    nonleaving, 176, 199
    polyatomic, 239
    shielding, 130
Light:
    absorption, 337, 356, 359
    emission, 356
    scattering, 359
Linear free energy relations, 36–37, 199, 278
Linear geometry, 250
Linear voltammetry, 116

Line broadening, NMR, 12, 112, 177
Lowest unoccupied molecular orbital, see
    LUMO (lowest unoccupied molecular
    orbital)
Low-grade energy, 350
Low-spin configurations, 347
Luminescence, 356–357
LUMO (lowest unoccupied molecular
    orbital), 131, 270, 331
    acceptor, 137
    antibonding, 189

McKay plots, 82
Maleic acid, 81
Marcus cross-correlation, 278–279
Mass action, law, 34
Matching, energy, 151
Matrices, stochastic, 51
Matrix, transition, 50
Maximum overlap criterion, 132, 135
Mechanism:
    associative, 170, 173, 191, 205
    Berry, 216
    catalytic, 296
    chain, 28
    concerted dissociative-associative, 211
    dissociative, 170, 173, 176, 187, 205, 352–
        353
    free radical, 315
    intimate, 44, 248
    intrinsic, 6
    stoichiometric, 44
    topological, 215
Mechanistic interpretations, 148
Metal-alkyl intermediate, 325
Metal hydride intermediate, 325
Metamorphoses, 295, 318, 367
Metathesis, 168
Metathesis mechanisms, 211
Methane, 45, 130
Methanol, carbonylation, 319, 322–323
Methanol exchange, 176
Method:
    of finite differences, 83
    Guggenheim, 86
    of integration, 84
    of isolation, 87
    structure correlation, 189–190
Methods:
    acoustical, 99, 104
    competition, 68
    electrochemical, 99, 114
    flow, 100
    parametric, 68

simulation, 68
Michaelis-Menten constant, 298
Microscopic reversibility, 35
Minimum energy pathways, 132, 173, 190
Mixing chamber, 101
Model:
  donor–acceptor, 270
  electrostatic, 194, 350
Model-simulation, 263–264
Modes of activation, collective, 306
Modes of binding, collective, 306
Modes of deactivation, 348
Modes of electron-transfer, 268
Modes of excitation, 338
Molecular collisions, 44
Molecular geometry, 231, 338
Molecular happenings, 58
Molecularity, 9, 51
Molecular movement, 78
Molecular properties, 63
Mole fraction, 31–32
Moment, electronic transition, 266
Momentum, 147–148
Monatomic bridge, 238
Monomeric hydroxo compounds, 233
Monomers, hydrolyzed and nonhydrolyzed,
    62
Monomolecular reactions, 10, 48
Mononuclear catalysts, 305, 315
Mononuclear dioxygen complexes, 163
Monsanto process, 319
Multibonded terminal oxide, 240
Multiple bond addition, 247
Multiple relaxations, 108
Multiplicity, spin, 156, 158, 343
Mutual polarization, 270
Myoglobin, 164
Mystical connotations, 295

Nature:
  entering ligand, 174, 192
  leaving ligand, 193
Neutralization reactions, 168, 222
Nitrogen fixation, 328–329
NMR line broadening, 112, 177
Nodal surface, 149
Nonadiabatic, 271
Nonantagonistic action, 300
Nonbonding electron donation, 140
Nonbonding orbital, 274
Noncomplementary reactions, 261
Noncomplexing solvents, 31
Nonionized solvents, 31
Nonleaving ligand effects, 176, 199

Nonpolar solvents, 31
Nonprotic solvents, 31
Nonspherical donor, 132
Nonsymmetric mixed valence compound,
    282–283
Nonsymmetric reactions, 273
Normalization condition, 48
Nuclear magnetic resonance, 99, 116
Nuclear rearrangement, 269
Nuclear tunneling, 270
Nucleophilic, 171, 184
Nucleophilic attack, 133, 293, 319–320
Numerical fit, 16
Numerical methods, 11, 14

Olefin:
  activation, 311
  hydroformylation, 319
  hydrogenation, 310
Olefinic complex, 212
Olefinic double bond, 133
Olefin insertion reactions, 213
Oligomerization, 326
One-electron transfer, 262
Opposing reactions, 34, 64
Optical electron transfer, 267, 280
Orbital:
  acceptor, 139
  antibonding, 275
  donor, 139
  nonbonding, 274
Orbital angular momentum, 148
Orbital combinations, 151, 274
Orbital motion, 123
Orbital symmetry conservation rules, 147–
    148
Organochromium(III)complexes, 208
Organotin, pentacoordinated, 189
Orientation, 44, 47
Ostwald's method of isolation, 87
Outer-sphere, 70, 169, 276
Outer-sphere activated complex, 237
Outer-sphere adducts, 294
Overlap:
  HOMO–LUMO, 139
  positive, 132, 134, 147
Overlap integral, 132, 266
Overlap region, 145
Oxidase, enzyme, 164
Oxidation numbers, formal, 230, 234
Oxidative addition, 244, 250, 253, 292, 316
Oxidizing agent, four-electron, 159
Oxo process, 319
Oxovanadium(IV), 112

Palladium(II), 323
Parallel reactions, 15, 17, 177, 302
Paramagnetic ion, 120
Parameters:
  activation, 208
  impact, 45
  Racah, 345
Parametric methods, 68
Pathways, minimum energy, 132, 190
Pauli forces, 154
Pauli principle, 152
Pentacoordinated organotin, 189
Pentacyanocobalt(III), 60
Pentagonal bipyramid, transformation, 221
Perchlorates, 130
Periodates, 130
Peripheral forces, 144
Perpendicular geometry, 314
Phenol red, 111
Phenomenological kinetics, 48
Philosophers stone, 295
Phlogiston theory, 229
Phosphine, 130
Phosphorescence, 113, 348
Photocatalytic cycle, 367
Photochemical electron transfer, 353
Photochemical ethylene cycloaddition, 150
Photochemical process, 151, 352, 358
Photolysis, flash, 99, 103
Photon emission, 348
Photosensitizers, 365–366
Photostereochemistry, 353
Photosubstitution, 353
  stereochemistry, 354
Photosynthesize, 362
Photovoltaic cells, 359
Pi acceptors, 133
Pi donors, 133
Planck's constant, 59
Plato, 1
Polarization, 124
Polarography:
  AC, 116
  DC, 116
Polar solvents, 31
Polyatomic ligands, 239
Polybutadiene, 309
Polymerization, Ziegler-Natta, 326
Polynuclear catalysts, 305–306
Polynuclear homogeneous catalysts, 308
Positive overlap, 132–135, 147
Potassium atoms, 47
Potential, ionization, 134
Potential energy barrier, 57, 63

Potential energy curves, 270, 274, 280–281, 340
Potential energy diagram, 268
Potential energy profile, 192, 268
Potential energy surface, 56, 79
Precession, frequency, 116
Precursor complex, 276
Precursor-to-successor transformation, 269
Preexponential factor, 51
Pressure effects, 65
Pressure jump, 99
Principle:
  Franck-Condon, 266, 267, 268, 349
  Pauli, 152
Probability, 48, 266, 269
Process:
  Fischer-Tropsch, 319
  Monsanto, 319
  Oxo, 319
  photochemical, 352, 358
  Roelen, 319
  Solvay, 295
  Wacker, 213, 323–324
Profile, potential energy, 192, 351
Promoters, 291
Promotional excitation, 343
Promotional transitions, 339
Propagation, 27
Properties:
  donor–acceptor, 31
  magnetic, 154
  molecular, 63
Proposed mechanism, 24
Protic solvents, 30
Proton acceptors, 31
Proton donors, 31
Proton transfer, 30, 140, 168, 222
Pseudo-first-order rate constant, 88
Pulsed illumination, 357
Pulse radiolysis, 99, 103
Pyramidal, trigonal, 69
Pyruvic acid, 263

Quantum yield, 356, 358
Quenching, 356

Racah parameters, 345
Racemization, 169
Radiationless transition, 350
Radiative transitions, 359
Radical, free, 254
Radiolysis, pulse, 99, 103
Rancid butter, 157
Random events, 49

Randomization, 123
Rate constants, composite, 52
Rate-controlling, 17
Rate-determining, 17
Rate equations, 10
Rate of exchange, 177
Rate expression, 18
Rate law, 10
Rate laws, complicated, 87
Reactants, activated, 305
Reaction:
  anation, 200
  bimolecular, 33, 44
  bimolecular electron-transfer, 276
  enzymatic, 296
  exchange, 69
  gasification, 1
  hydrocyanation, 316
  hydrosilation, 316
  photochemical, 151
  reductive elimination, 314
  self-exchange, 274, 281
  solvent exchange, 175
  symmetric, 267
  thermal, 149
Reactions:
  addition combinations, 168
  atom-transfer, 167
  chain, 27, 29
  competitive, 26, 30
  complementary, 261
  consecutive, 17
  coupled, 111
  decomposition, 167
  diffusion-controlled, 72
  electron exchange, 244
  electron-transfer, 2, 70
  elementary, 9, 17, 24, 34, 51, 171, 289
  elementary conserted, 148
  elementary redox, 235
  enzymatic, 296, 298
  exchange, 81
  insertion, 213
  intermolecular, 258, 314
  neutralization, 168
  noncomplementary, 261
  nonelementary, 50, 60
  nonsymmetric, 273
  olefin insertion, 213
  opposing, 34, 64
  oxidation-reduction, 230, 235, 292, 355
  oxidative addition, 316
  parallel, 15, 17
  parallel first-order, 302
  proton transfer, 168, 222
  redox, 230, 235, 292, 355
  substitution, 167, 169
  successive, 15
Reactive intermediates, 364
Reactivity, 67, 193
Reactors, cylindrical chemical flow, 101
Reagents, external, 362
Real effective charges, 234
Rearrangement, 259
  intramolecular, 169
  nuclear, 269
  topological, 216
  trans-to-cis, 315
Reciprocal times, 109
Redox couples, catalysis, 292
Redox reactions, 230, 292, 355
Reductants, sacrificial, 362
Reduction:
  electrochemical, 242
  of multiple bonds, 256
  potential barrier, 160
Reductive addition, 292
Reductive disproportionation, 255–256
Reductive elimination, 252–253, 255, 314
Reference solvent, 32
Region:
  exchange-controlled, 121
  fast-exchange, 121
  overlap, 145
  relaxation-controlled, 121
Relative electronegativities, 231
Relative selectivity, 303
Relaxation:
  multiple, 108
  vibrational, 349–350
Relaxation-controlled region, 121
Relaxation studies, indicators, 111
Relaxation techniques, 118, 222
Relaxation time, spin-lattice, 124
Relaxation times, 106, 118–119
  fitting, 112
Relays, 365
Reorganization, solvent, 66, 270
Repulsion, 313
  closed-shell, 5
  electrostatic, 5
  HOMO–LUMO, 141
Repulsive forces, 145
Repulsive interactions, 141
Reversibility, microscopic, 35
Reversible acceptors, 365
Reversible donors, 365
Rhodium carbonyls, 323

Ring formation, chelate, 178, 182
Rings:
 aromatic, 316
 heterocyclic, 316
Roelen process, 319
Rotating disc electrode voltammetry, 116
Rotation, 354
Rule:
 Laporte, 342
 sixteen or eighteen, 293
Rules:
 orbital symmetry conservation, 148
 selection, 341
 spin conservation, 154, 160
 spin selection, 156
 Wigner-Witmer, 147
 Woodward-Hoffmann, 147

Sacrificial reductants, 362
Scattering of light, 359
Selection rules, 341
Selectivity, 301
 differential, 302
 integral, 302–303
 relative, 303
Self-exchange reaction, 274, 281
Self-ionize, 31
Semisteady state, 12
Seven-coordinated, 69
Shielding by ligands, 130
Shifts, chemical, 118
Sigma acceptors, donors, 133
Silicon tetrachloride, 129
Silver(II), 21
Simulation, model, 68, 263–264, 301
Single bonds, activation, 329
Sites, acceptor, 275
Sixteen or eighteen rule, 293
Slow passage, continuous, 117
Small cycles, 295–296
Small deviations from equilibrium, 110
Solar energy conversion, 359
Solar energy storage, 359
Solar light, 361
Solar protons, 363
Solar spectrum, 361
Solvated electrons, 239
Solvay process, 295
Solvent:
 classification, 30
 noncomplexing, 31
 nonionized, 31
 nonpolar, 31
 nonprotic, 31

 polar, 31
 protic, 30
 reference, 32
 surrounding, 273
Solvent effects, 30
Solvent exchange, 169, 175
Solvent mediation, 239
Solvent molecules, 62
Solvent reorganization, 66, 270
Solvolysis, 169
Species, polynuclear, 306
Specific acid-base catalysis, 292
Spectrophotometric methods, 80
Spectroscopy:
 emission, 113
 femtosecond, 99, 104
 microwave, 100
 picosecond, 99, 104
Spectrum, solar, 361
Spin-allowed excited state, 344
Spin conservation, 154, 156, 160
Spin interaction, 156
Spin-lattice relaxation time, 124
Spin multiplicity, 158, 343
Spins, antiparallel, 153
Spin selection rules, 156
Splitting:
 fine, 122
 photochemical, 361
 zero field, 122
Square planar complexes, shielding, 131
Square planar substitution, 144, 191, 194, 196
Square tetragonal complex, 197
Stability, kinetic, 37
Stabilization of acceptor electrons, 141
Stable configuration, 143–144
Stable intermediates, 79
Stahl, George Ernest, 229
Standard enthalpy, 59
Standard enthalpy difference, 104
Standard thermodynamic state, 59
State, electronically excited, 113
Stations, 79
Statistical activation, 63
Statistical distribution, 58, 132
Steady state, 14, 17, 20, 28, 35
Steady-state approximation, 12, 298
Step:
 critical elementary, 171
 electron-transfer, 268
Steps, elementary, 44
Stereochemical changes, 31, 204
Stereochemical hindrance, 199

Stereochemistry, photosubstitution, 354
Stereoselectivity, 304–305
Steric factor, 129
Stochastic, 47
Stochastic matrices, 51
Stoichiometric coefficients, 9, 296
Stoichiometric equation, 69, 290
Stoichiometric hydrogenation, 81
Stoichiometric mechanism, 44
Stoichiometry, 9, 24
  variable, 26
Stone, philosophers, 295
Stopped flow, 101
Strength of interaction, 272
Structural change, 244
Structural information, 119
Structure, 67
Structure correlation method, 189–190
Studies, mechanistic, 78
Substituted amines, 224
Substitution, 71
  octahedral, 199, 205
  square planar, 131, 194, 196
  tetrahedral, 183, 189
Substitution reactions, 167, 169
Successive reactions, 15
Sulfite ion, 71
Sulfur tetrafluoride, 130
Supported catalysts, 309
Surface:
  nodal, 149
  potential energy, 56, 79
Surfaces, energy, 56, 155
Symmetric reaction, 267
Symmetry, 133, 148
  HOMO–LUMO, 337
  individual orbitals, 148
Synthesis gas, 321
Syn transformation, 217
System of differential equations, 94
Systems, complex biological, 263

Technique:
  flooding, 87
  simple titration, 80
Techniques:
  experimental, 78–79
  flow, 99
  line-broadening, 120
  relaxation, 222
Temperature jump, 99, 104
Terminal oxide, multibonded, 240
Termination, 27–30
Tetragonally distorted octahedral, 232

Tetrahedral phosphorus, 185
Tetrahedral species, 185, 187–188
Tetrahedral substitution, 183, 189
Tetrahedron, bicapped, 221
Tetrahydrofuran, 31
Thalium(I), 21
Theorem, Hellmann-Feynman, 144, 194
Theory:
  phlogiston, 229
  transition state, 57–58, 60, 277
Thermal electron transfer, 280
Thermal equilibrium, 58
Thermal reaction, 149, 269
Thermodynamic barriers, 37, 364
Thermodynamic function, 36
Thexi state, 350
Three-center bond, 184
Three-dimensional process, 290
Time:
  contact, 101
  relaxation, 106
Time dependence, 10
Times:
  reciprocal, 109
  relaxation, 118–119
Time scale, 78, 115
Topological mechanisms, 55, 215
Topological rearrangements, 216
Total angular momentum, 147
Total symmetry, 148
Total wavefunction, 153
Tracers, isotopic, 71, 81
Trans-to-cis rearrangement, 315
Trans effect, 146, 194
Transfer:
  atom, 262
  direct electron, 237
  effective electron, 275
  electron, 115, 230, 244, 260, 262, 268, 277
  energy, 349
  group, 231
  hydride ion, 215
  indirect electron, 237, 239
  intramolecular, 214, 255
  proton, 30
  thermal, 269
Transfer of electron density, 139
Transfer functions, 32
Transformation, 1, 311
  anti, 217
  chemical, 295
  pentagonal bipyramid, 221
  precursor-to-successor, 269
  syn, 217

Transforms, Laplace, 97
Transition, 113
  intraligand, 339
  radiationless, 350
Transition matrix, 50
Transitions:
  charge-transfer, 339, 341
  d-d, 338–339
  dipole moment, 266, 341
  electronic, 341
  forbidden, 342
  intraconfigurational, 339
  intraligand, 341
  promotional, 339
  radiative, 359
Transition state theory, 32–33, 37, 57–58, 60, 277
Translational kinetic energy, 269
Transmission coefficient, 59, 270
Transmutations, chemical, 295
Transport energy, 360
Trigonal pyramidal, 69
Trigonal twist, 218
Triple collision, 160
Tris(2,2′-bipyridine)ruthenium(II), 355
Tropolonate complexes, 220
Tunneling, nuclear, 270
Twist, trigonal, 218
Two-center interactions, 248
Two-electron oxidative addition, 310
Two-term rate laws, 90

Unimolecular chemical reaction, 356
Unpaired electrons, 156
Unsaturated, coordinatively, 258
Unsaturated compounds, 247, 256
Unsaturated intermediate, 315

Unstable free radicals, 160
Uranium(IV), 18, 27–28
Uranium(V), 28, 83
Uranium(VI), 27–28, 177
Uranyl complex, 177

Vacancies, low-energy, 353
Valence compounds, mixed, 260
Vanadium(II), 81, 265
Vanadium(III), 19, 61
Vanishing determinant, 109
Variables, internal, 172
Variable stoichiometry, 26
Vibrational levels, 266, 273
Vibrational relaxation, 349–350
Vibrations, 45
Vibronic coupling, 351
Viscosity, 31
Voltammetry:
  cyclic, 116
  linear, 116
Volume of activation, 65, 175–176

Waage, 34
Wacker process, 213, 323–324
Water, coordinated, 241
Water exchange, 139, 207
Wavefunction, 151, 153
Weighted average, 52
Wigner-Witmer rules, 147
Wilkinson's catalyst, 317–318
Woodward-Hoffmann rules, 147

Zeolites, 309
Zero field splitting, 122
Ziegler-Natta polymerization, 326